Student Solutions Manual for
Wackerly, Mendenhall, and Scheaffer's

Mathematical Statistics
with Applications

Fifth Edition

Charles D. Kincaid
University of Florida

Duxbury Press

An Imprint of Wadsworth Publishing Company

I(T)P® An International Thomson Publishing Company

Belmont • Albany • Bonn • Boston • Cincinnati • Detroit • London • Madrid • Melbourne
Mexico City • New York • Paris • San Francisco • Singapore • Tokyo • Toronto • Washington

CONTENTS

CHAPTER 1 WHAT IS STATISTICS?

1.1 **a.** <u>Population</u>: measurements of weekly water consumption for all single-family dwelling units in the city.
<u>Objective</u>: to estimate μ, the average for all family units.

  b. <u>Population</u>: assign the value 0 if a tire manufactured by the company in the specific year has a safe tread; assign the value 1 if the tire has an unsafe tread. The population consists of a set of 0's and 1's, each corresponding to one of the tires manufactured during the year.
<u>Objective</u>: to estimate p, the proportion of 1's in the population.

  c. <u>Population</u>: a set of 0's and 1's, each corresponding to an adult resident of the state. Assign the value 1 or 0 according as the resident does or does not favor a unicameral legislature.
<u>Objective</u>: to estimate p, the proportion of 1's in the population.

  d. <u>Population</u>: times until recurrence for all people who have had a particular disease.
<u>Objective</u>: to estimate μ, the average time until recurrence.

  e. <u>Population</u>: transistor lifetimes for the hypothetical population of all possible transistors of a certain type.
<u>Objective</u>: to estimate μ, the average life of the transistor.

1.3 Similar to Exercise 1.2. We chose seven intervals of length 2.

Class Boundaries	Tally	Frequency	Relative Frequency
.005– 2.005	~~1111~~ ~~1111~~ 11	12	.48
2.005– 4.005	~~1111~~ 1	6	.24
4.005– 6.005	111	3	.12
6.005– 8.005	1	1	.04
8.005–10.005	11	2	.08
10.005–12.005		0	.00
12.005–14.005	1	1	.04
		25	1.00

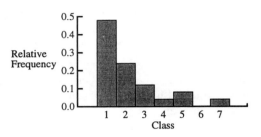

Figure 1.2

1.5   **a.** $\sum\limits_{i=1}^{n} c = c + c + c + \ldots + c$, where the sum involves n elements. Hence $\sum\limits_{i=1}^{n} c = nc$.

b. $\sum\limits_{i=1}^{n} cy_i = cy_1 + cy_2 + cy_3 + \ldots + cy_n = c(y_1 + y_2 + y_3 + \ldots + y_n) = c \sum\limits_{i=1}^{n} y_i$.

c. $\sum\limits_{i=1}^{n} (x_i + y_i) = x_1 + y_1 + x_2 + y_2 + x_3 + y_3 + \ldots + x_n + y_n$

$$= (x_1 + x_2 + x_3 + \ldots + x_n) + (y_1 + y_2 + y_3 + \ldots + y_n)$$

$$= \sum_{i=1}^{n} x_i + \sum_{i=1}^{n} y_i.$$

Consider the numerator of s^2, which is $\sum\limits_{i=1}^{n} (y_i - \bar{y})^2$.

$$\sum_{i=1}^{n} (y_i - \bar{y})^2 = \sum_{i=1}^{n} (y_i^2 - 2y_i + \bar{y})^2 = \sum_{i=1}^{n} y_i^2 - \sum_{i=1}^{n} 2y_i\bar{y} + \sum_{i=1}^{n} \bar{y}^2$$

$\bar{y}$ and $\bar{y}^2$ are constants with respect to the variable of summation (i). Hence

$$\sum_{i=1}^{n} (y_i - \bar{y})^2 = \sum_{i=1}^{n} y_i^2 - 2\bar{y} \sum_{i=1}^{n} y_i + n\bar{y}^2 = \sum_{i=1}^{n} y_i^2 - 2\bar{y}(n\bar{y}) + n\bar{y}^2 = \sum_{i=1}^{n} y_i^2 - n\bar{y}^2$$

with the second equality following from the fact that $\sum\limits_{i=1}^{n} y_i = n\bar{y}$.

Thus, $s^2 = \dfrac{1}{n-1}\left[\sum\limits_{i=1}^{n} y_i^2 - n\bar{y}^2 \right]$, or, since $\bar{y}^2 = \dfrac{1}{n^2}\left[\sum\limits_{i=1}^{n} y_i \right]^2$,

$$s^2 = \frac{1}{n-1}\left[\sum_{i=1}^{n} y_i^2 - \frac{1}{n}\left(\sum_{i=1}^{n} y_i \right)^2 \right]$$

1.7 **a.** The two necessary sums for calculating $\bar{y}$ and s are

$$\sum_{i=1}^{50} y_i = 62{,}622.0 \text{ and } \sum_{i=1}^{50} y_i^2 = 86{,}027{,}202.0.$$

Then $\bar{y} = \dfrac{\sum\limits_{i=1}^{n} y_i}{n} = \dfrac{62{,}622.0}{50} = 1252.44$

and $s^2 = \dfrac{1}{n-1}\left[\sum_{i=1}^{n} y_i^2 - \frac{1}{n}\left(\sum_{i=1}^{n} y_i\right)^2\right]$

$$= \frac{1}{49}\left[86{,}027{,}202.0 - \frac{1}{50}(62{,}622.0)^2\right]$$

$$= 155{,}038.86$$

$$s = \sqrt{s^2} = \sqrt{155{,}038.86} = 393.75$$

b. The table below shows the intervals, the counts, and the expected counts for each interval.

k	$\bar{y} \pm ks$	Interval Boundaries	Frequency	Expected Frequency
1	1252.44 ± 393.75	858.69 to 1646.19	45	34
2	1252.44 ± 787.50	464.94 to 2039.94	48	47.5
3	1252.44 ± 1181.25	71.19 to 2433.69	49	50.0

There are many more values within one standard deviation than expected because the two extremely large values give a large standard deviation.

1.9 Similar to Exercise 1.7.

a. Calculate $\sum\limits_{i=1}^{25} y_i = 141.9$ and $\sum\limits_{i=1}^{25} y_i^2 = 2147.61$. Then $\bar{y} = \frac{1}{n}\sum\limits_{i=1}^{n} y_i = \frac{141.9}{25} = 5.67$

and $s^2 = \dfrac{1}{n-1}\left[\sum_{i=1}^{n} y_i^2 - \frac{1}{n}\left(\sum_{i=1}^{n} y_i\right)^2\right]$

$$= \frac{1}{24}\left[2147.61 - \frac{1}{25}(141.9)^2\right]$$

$$= 55.9244$$

$$s = \sqrt{s^2} = \sqrt{55.9244} = 7.48$$

b.

k	$\bar{y} \pm ks$	Interval Boundaries	Frequency	Expected Frequency
1	5.676 ± 7.48	-1.80 to 13.15	23	17
2	5.676 ± 14.96	-9.28 to 20.63	24	23.75
3	5.676 ± 22.43	-16.75 to 28.11	24	25

The number of measurements falling within each interval does not compare well with the expected number because of the highly skewed nature of the data. The empirical rule works for mound-shaped data. See Exercise 1.10.

1.11 For Exercise 1.2, the approximation is
$$\frac{\text{range}}{4} = \frac{3168 - 565}{4} = 650.75, \text{ while } s = 393.75.$$
Note the poor approximation due to the extreme values.

For Exercise 1.3, the approximation is
$$\frac{\text{range}}{4} = \frac{12.48 - .32}{4} = 3.04, \text{ while } s = 3.17.$$
For Exercise 1.4, the approximation is
$$\frac{\text{range}}{4} = \frac{38.3 - 1.8}{4} = 9.125, \text{ while } s = 7.48.$$

1.13 Similar to Exercise 1.12. The point representing a gain of 20 pounds is shown in Figure 1.5 to be one standard deviation below the mean. Using the same reasoning as in Exercise 1.12, 16% of the measurements fall below 20, and hence $100 - 16 = 84\%$ fall above 20. Hence .84 is the approximate probability that a weight gain exceeds 20, and the manufacturer is probably correct.

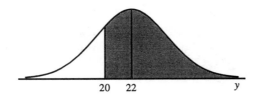

Figure 1.5

1.15 **a.** We assume that the set of 1521 games is the population. Then $\mu = 143$ and $\sigma = 26$. $169 = \mu + \sigma$, and by the empirical rule approximately 32% should be outside the interval $(\mu - \sigma, \mu + \sigma)$. Then half or approximately 16% should be greater than 169.

b. $117 = \mu - \sigma$ and half of approximately 68%, 32%, of the games should be between $\mu - \sigma$ and μ. $195 = \mu + 2\sigma$ and half of approximately 95%, or 47.5%, should be between μ and $\mu + 2\sigma$. Now, $32\% + 47.5\% = 81.5\%$, so that approximately 81.5% of the games should have ended with a total score between 117 and 195 points.

c. No. A score of 225 is greater than $\mu + 3\sigma$. Such a score is very unlikely according to the empirical rule.

1.17 **a.** $s = \dfrac{\text{range}}{4} = \dfrac{716-8}{4} = 177$

b. We chose to use 15 classes of length 50.

Class Boundaries	Tally	Frequency	Relative Frequency
.5– 50.5	~~1111~~ ~~1111~~ 1	11	.125
50.5–100.5	~~1111~~ ~~1111~~ ~~1111~~ 1	16	.182
100.5–150.5	~~1111~~ 111	8	.091
150.5–200.5	~~1111~~ ~~1111~~ ~~1111~~	15	.170
200.5–250.5	~~1111~~ ~~1111~~	10	.114
250.5–300.5	~~1111~~ 1	6	.068
300.5–350.5	~~1111~~	5	.057
350.5–400.5	~~1111~~ 1	6	.068
400.5–450.5	111	3	.034
450.5–500.5	11	2	.023
500.5–550.5	1	1	.011
550.5–600.5	1	2	.023
600.5–650.5	1	1	.011
650.5–700.5	1	1	.011
700.5–750.5	1	1	.011
		88	.999

The last column should sum to 1.000. The value .999 is due to rounding off the relative frequencies. The figure is omitted.

c. Calculate $\Sigma y_i = 18{,}550$ and $\Sigma y_i^2 = 6{,}198{,}356$. Then

$$\bar{y} = \frac{18{,}550}{88} = 210.80, \quad s^2 = \frac{1}{87}(6{,}198{,}356 - 3{,}910{,}255.7) = 26{,}300.00, \text{ and}$$

$$s = 162.17.$$

d.

k	$\bar{y} \pm ks$	Interval Boundaries	Frequency	Expected Frequency
1	210.8 ± 162.17	48.6 to 373.0	63	.72
2	210.8 ± 324.34	−113.5 to 535.1	82	.93
3	210.8 ± 486.52	−275.7 to 697.3	87	.99

1.19 Assuming that the distribution of scores is bell-shaped, the empirical rule provides a means for describing the variability of the data. The results of the empirical rule are shown below.

k	$\bar{y} \pm ks'$	Interval Boundaries	Percentage of Measurements Within the Interval
1	72 ± 8	64 to 80	Approximately 68%
2	72 ± 16	56 to 88	Approximately 95%
3	72 ± 24	48 to 96	Nearly 100%

Hence one would expect 68% of the 340 scores, that is, .68(340) = 231.2 or 231 scores, to fall in the interval 64 to 80. Similarly, 95% of the scores, that is, .95(340) = 323 scores, should fall in the interval 56 to 88.

1.21 It is given that the distribution of bearing diameters has a mean of 3.00 and a standard deviation of .01. We wish to calculate the fraction of this machine's production that will fail to meet specifications, given that the only bearings that <u>will</u> meet specifications are those whose diameters lie in the interval 2.98 to 3.02. Notice that this interval contains the values within two standard deviations of the mean. That is,

$$\bar{y} \pm 2s' = 3.00 \pm 2(.01) = 3.00 \pm .02.$$

Hence, if we assume that the distribution of bearing diameters is approximately bell-shaped, the empirical rule states that the fraction of acceptable bearings will be approximately .95. Consequently, the fraction of this machine's production that will fail to meet specifications is approximately $1 - .95 = .05$.

1.23 Assuming the distribution of yields is mound-shaped, the empirical rule states that approximately 95% of all measurements fall in the interval $\mu + 2\sigma = 60 \pm 20$, or 40 to 80. Only 5% fall outside this interval and, by symmetry, .025 is the fraction below 40. If the yield should fall below 40 (an unlikely event, having probability .025), one would suspect an abnormality in the process.

1.25 Using Tchebysheff's theorem with $k = 2$, at least $1 - \left(\frac{1}{k^2}\right) = 1 - \left(\frac{1}{4}\right) = \frac{3}{4}$ or 75% of all measurements should lie within two standard deviations of the mean. Hence the interval is $\mu \pm 2\sigma = 5.5 \pm 5$, or .5 to 10.5.

1.27 Lead content readings must be nonnegative. Zero is only .33 standard deviations below the mean, so that the population can only extend .33 standard deviations below the mean. This radically skews the distribution, so that it is not normal.

CHAPTER 2 PROBABILITY

2.1 Construct the three sets: $A = \{FF\}$; $B = \{MM\}$; $C = \{MF, FM, MM\}$. Then

$A \cap B = \emptyset$ $A \cap C$: $\emptyset$ $B \cap C$: $\{MM\}$ $C \cap \overline{B}$: $\{MF, FM\}$

$A \cup B$: $\{FF, MM\}$ $A \cup C$: S $B \cup C$: C

2.3 To show $\overline{(A \cup B)} = \overline{A} \cap \overline{B}$, we can draw the following:

$\overline{(A \cup B)}$

$A \cup B \Rightarrow$

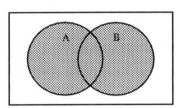

$\overline{A \cup B} \Rightarrow$

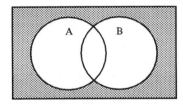

$\overline{A} \cap \overline{B}$

$\overline{A} \Rightarrow$

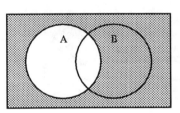

$\overline{B} \Rightarrow$

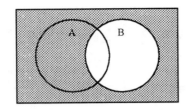

$\overline{A} \cap \overline{B} \Rightarrow$

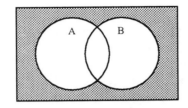

$\overline{A \cap B}$

$A \cap B \Rightarrow$

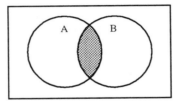

$\overline{A \cap B} \Rightarrow$

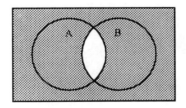

$\overline{A} \cup \overline{B}$

   $\overline{A}, \overline{B}$ shown above

$\overline{A} \cup \overline{B} \Rightarrow$

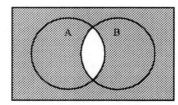

2.5   A typical element in the set S is a pair representing the two applicants selected for the job. Then

A: {two males} = $\{M_1M_2, M_1M_3, M_2M_3\}$

B: {at least one female} = $\{M_1W_1, M_2W_1, M_3W_1, M_1W_2, M_2W_2, M_3W_2, W_1W_2\}$

Hence

$\overline{B}$:   {no females} = {two males} = A

$A \cup B$: {two males or at least one female} = S

$A \cap B$: $\emptyset$ $A \cap \overline{B}$: $A \cap A = A$

2.7   **a.**   We know $P(S) = P(E_1 \cup E_2 \cup E_3 \cup E_4 \cup E_5)$

$\qquad = P(E_1) + P(E_2) + P(E_3) + P(E_4) + P(E_5)$

$\qquad = .15 + .15 + .4 + 2P(E_5) + P(E_5)$

$\qquad = .70 + 3P(E_5) = 1.0$

$3P(E_5) = .30$

$P(E_5) = .10$

Then $\mathrm{P}(E_4) = .20$

b. $P(E_1) = 3P(E_2) = .3$ means that $P(E_2) = .1$. Then

$P(S) = .3 + .1 + P(E_3) + P(E_4) + P(E_5)$

$1 = .4 + 3P$ (assuming $P(E_3) = P(E_4) = P(E_5) = p$)

and $p = .2$

2.9   **a.**   We know that $P(S) = P(E_1 \cup E_2 \cup E_3 \cup E_4) = 1$ and, since the four events are pairwise mutually exclusive, $P(S) = P(E_1) + \ldots + P(E_4) = .01 + ? + .09 + .81$. Thus $? = P(E_2) = 1 - .91 = .09$.

b. P(at least one hit) $= P(E_1 \cup E_2 \cup E_3) = P(E_1) + P(E_2) + P(E_3)$
$\qquad = .01 + .09 + .09 = .19$.

2.11   Let $B =$ assembly has bushing defect and $S =$ shaft defect.

a. $P(B) = .06 + .02 = .08$

b. $P(B \cup S) = .06 + .08 + .02 = .16$

c. $P(B\overline{S} \cup S\overline{B}) = .06 + .08 = .14$

d. $P(\overline{B \cup S}) = 1 - .16 = .84$

2.13 a. Notice that order is important since the pair (V_2, V_3) is different from the pair (V_3, V_2) due to the day on which the order is received. The sample points are

$$(V_1, V_1) \quad (V_1, V_2) \quad (V_1, V_3) \quad (V_2, V_1) \quad (V_2, V_2)$$
$$(V_2, V_3) \quad (V_3, V_1) \quad (V_3, V_2) \quad (V_3, V_3)$$

b. The points are equally probable, with common probability $\frac{1}{9}$.

c. The following sets are of interest:

A: {same vendor gets both} = $\{(V_1, V_1), (V_2, V_2), (V_3, V_3)\}$

B: {V_2 gets at least one} = $\{(V_1, V_2), (V_2, V_1), (V_2, V_3), (V_3, V_2), (V_2, V_2)\}$

Then

$$P(A) = \frac{3}{9} = \frac{1}{3} \qquad\qquad P(B) = \frac{5}{9} \qquad\qquad P(A \cup B) = \frac{7}{9}$$

$$P(A \cap B) = P(V_2, V_2) = \frac{1}{9}$$

2.15 a. $3 \times 3 = 9$

$\{SS, SR, SL, RS, RR, RL, LS, LR, LL\}$

b. $P(SL) + P(RL) + P(LS) + P(LR) + P(LL) = \frac{5}{9}$

c. $P(SS) + P(SR) + P(SL) + P(RS) + P(LS) = \frac{5}{9}$

2.17 a. Two systems are selected from six, two of which are defective. Denote the six systems as $G_1, G_2, G_3, G_4, D_1, D_2$, according to whether they are defective or nondefective. Each sample point will represent a particular pair of systems chosen for testing. The sample space, consisting of 15 pairs, is shown below.

$$G_1G_2 \quad G_1G_3 \quad G_1G_4 \quad G_2G_1 \quad G_2G_4 \quad G_3G_4 \quad G_1G_1 \quad G_1G_2$$
$$G_2G_3 \quad G_2G_2 \quad G_3G_1 \quad G_3G_2 \quad G_4G_1 \quad G_4G_2 \quad G_1G_2$$

Note that the two systems are drawn simultaneously and that order is unimportant in identifying a sample point. Hence the pairs G_1G_2 and G_2G_1 are not considered to represent two different sample points. Then

$$P(\text{at least one defective}) = \frac{9}{15} = \frac{3}{5} \quad \text{and} \quad P(\text{both defective}) = P(D_1D_2) = \frac{1}{15}.$$

b. If four of the six systems are defective, the 15 sample points are

$$G_1G_2 \quad G_1D_1 \quad G_1D_2 \quad G_1D_3 \quad G_1D_4 \quad G_2D_1 \quad G_2D_2 \quad G_2D_3$$
$$G_2D_4 \quad D_1D_2 \quad D_1D_3 \quad D_1D_4 \quad D_2D_3 \quad D_2D_4 \quad D_3D_4$$

Then

$$P(\text{at least one defective}) = \frac{14}{15} \quad \text{and} \quad P(\text{both defective}) = \frac{6}{15} = \frac{2}{5}.$$

2.19 **a.** Define the events: E: family's income exceeds \$35,353

N: family's income does not exceed \$35,353

Then the sample points are

E_1: $(EEEE)$	E_5: $(NEEE)$	E_9: $(ENNE)$	E_{13}: $(NENN)$
E_2: $(EEEN)$	E_6: $(EENN)$	E_{10}: $(NENE)$	E_{14}: $(NNEN)$
E_3: $(EENE)$	E_7: $(ENEN)$	E_{11}: $(NNEE)$	E_{15}: $(NNNE)$
E_4: $(ENEE)$	E_8: $(NEEN)$	E_{12}: $(ENNN)$	E_{16}: $(NNNN)$

b. A: $\{E_1, E_2, \ldots, E_{11}\}$ B: $\{E_6, E_7, \ldots, E_{11}\}$ C: $\{E_2, E_3, E_4, E_5\}$

c. Since $P(E) = P(N)$, each simple event is equally likely and $P(E_i) = \frac{1}{16}$. Then

$$P(A) = \frac{11}{16} \qquad P(B) = \frac{6}{16} = \frac{3}{8} \qquad P(C) = \frac{4}{16} = \frac{1}{4}.$$

2.21 The mn rule is used. The flight from New York to California can be chosen in any one of 6 ways, the flight from California to Hawaii in any one of 7 ways. Thus, the total number of flights will be $(6)(7) = 42$.

2.23 **a.** Use the mn rule. Since the first die can result in one of 6 possible outcomes, and the second can result in one of 6 outcomes, a total of $(6)(6) = 36$ pairs can be formed, each containing one element from the first group and one element from the second.

b. Define the event A, "observe a sum of 7 on the two dice." This event will occur if any one of the following simple events occurs:

$$(1, 6), \quad (2, 5), \quad (3, 4), \quad (4, 3), \quad (5, 2), \quad (6, 1)$$

Then $P(A) = P(\text{observe a sum of } 7) = \frac{6}{36} = \frac{1}{6}.$

2.25 Again, an extension of the mn rule is used, since we are concerned with the number of possible combinations from 7 different groups. Notice that the first digit of the 7-digit number can be chosen in 9 ways; that is, we can choose any of the 9 integers, 1, 2, 3, 4, 5, 6, 7, 8, 9. However, each of the remaining digits can be chosen from any one of the 10 integers, 0, 1, 2, 3, 4, 5, 6, 7, 8, 9. Thus, the total number of arrangements will be

$$9(10)(10)(10)(10)(10)(10) = 9(10)^6$$

2.27 $\frac{9!}{3!\,5!\,1!} = 504$ ways

2.29 $\binom{17}{2\ 5\ 10} = 408{,}408$

2.31 **a.** $\binom{130}{2} = 8385$

    b. $26 \cdot 26 = 676$ two-letter codes
        $26 \cdot 26 \cdot 26 = 17{,}576$ three-letter codes
                  $= 18{,}252$ total major codes available

    c. $8385 + 130 = 8515$ required

    d. Yes.

2.33 There are $\binom{50}{3} = 19{,}600$ ways to choose the three winners. Since the choice is random, each of the 19,600 sample points is equally likely.

    a. There are $\binom{4}{3} = 4$ ways for the organizers to win all of the prizes. Hence, the desired probability is $\dfrac{4}{19{,}600}$.

    b. The organizers can win exactly 2 of the prizes if 1 of the other 46 people wins 1 prize. Using the *mn* rule, there are $\binom{4}{2}\binom{46}{1} = 276$ ways for this to occur. Hence, the desired probability is $\dfrac{276}{19{,}600}$.

    c. $\binom{4}{1}\binom{46}{2} = 4140$. The probability is $\dfrac{4140}{19{,}600}$.

    d. $\binom{46}{3} = 15{,}180$. The desired probability is $\dfrac{15{,}180}{19{,}600}$.

2.35 **a.** Define a sample point as a triplet, the first element being the firm chosen for contract 1, the second element the firm chosen for contract 2, and so on. In choosing 3 of the 5 firms, order is important, and the number of sample points is
$$P_3^5 = \frac{5!}{2!} = 60$$

    b. Assuming that F_3 is awarded a contract, the other two contracts can be awarded in $P_2^4 = \frac{4!}{2!} = 12$ ways. However, since F_3 can be awarded one of three contracts, the total number of ways to award F_3 a contract is $3P_2^4 = 36$ and $P(F_3$ is awarded a contract$) = \frac{36}{60} = .6$.

2.37 In this exercise two cards are drawn from a deck of 52. Define the event A, "the two cards are an ace and a face card." The order of draw is unimportant in considering the event A; hence there are a total of $\binom{52}{2} = \frac{52 \cdot 51}{2} = 1326$ different ways of drawing two cards from the deck. Now in order that the event A occurs, one of the cards must be an ace (of which there are 4) and one must be a face card (of which there are 12). The number of ways of choosing one of the 4 aces and one of the 12 face cards will be, using the rule, $\binom{4}{1}\binom{12}{1} = 4(12) = 48$. Hence the desired probability is $P(A) = \frac{48}{1326}$.

2.39 The situation presented here is analogous to drawing 5 items from a jar (the 5 members voting in favor of the plaintiff). If the jar contains 5 red and 3 white items (5 women and 3 men), what is the probability that all 5 items are red? That is, if there is no sex bias, 5 of the 8 members are randomly chosen to be those voting for the plaintiff. What is the probability that all 5 are women? There are

$$\binom{8}{5} = C_5^8 = \frac{8!}{5!\,3!} = \frac{8(7)(6)}{3(2)(1)} = 56$$

sample points in the experiment, only one of which results in choosing 5 women. Hence

$$P(\text{five women}) = \frac{1}{56}.$$

2.41 $5!\left(\frac{2}{6}\right)\left(\frac{1}{6}\right)^4 = \frac{5}{162}$

2.43 As shown in Example 2.13, $P(E_i) = \frac{1}{M^n} = \frac{1}{10^7}$.

a. Let A denote the event that all of the orders go to different distributors. This will happen if 7 of the 10 distributors are given orders. Thus, A contains

$$n_a = 10 \times 9 \times 8 \times 7 \times 6 \times 5 \times 4 = 604{,}800$$

sample points. The probability is, then,

$$P(A) = \frac{120}{10^7} = .0605$$

b. Let A be the event that Distributor I gets exactly 2 orders and Distributor II gets exactly 3 orders. The 2 orders assigned to Distributor I can be chosen from the 7 in $\binom{7}{2} = 21$ ways. The 3 orders assigned to Distributor II can then be chosen from the remaining 5 in $\binom{5}{3} = 10$ ways. The final 2 orders can be assigned to any of the other 8 distributors in $8^2 = 64$ ways. Therefore, A contains

$$n_a = \binom{7}{2}\binom{5}{3}8^2 = 13{,}440$$

sample points. Then

$$P(A) = \frac{13{,}440}{10^7} = .001344$$

c. Let A be the event that Distributors I, II, and III get exactly 2, 3, and 1 order(s), respectively. Distributor I can get exactly 2 orders in $\binom{7}{2} = 21$ ways. Distributor

II can get exactly 3 orders in $\binom{5}{3} = 10$ ways. Distributor III can get 1 order in $\binom{2}{1} = 2$ ways. The 1 remaining order can be assigned to any of the 7 remaining distributors in 7 ways. Thus, A contains

$$n_a = \binom{7}{2}\binom{5}{3}\binom{2}{1} \cdot 7 = 2940.$$

Therefore, $P(A) = \dfrac{2940}{10^7} = .00029$

2.45 $\binom{n}{k} + \binom{n}{k-1} = \dfrac{n!}{k!\,(n-k)!} + \dfrac{n!}{(k-1)!\,(n-k+1)!}$

$$= \dfrac{n!\,(n-k+1)}{k!\,(n-k+1)!} + \dfrac{n!\,k}{k!\,(n-k+1)!}$$

$$= \dfrac{n!\,(n+1)}{k!\,(n-k+1)!} = \dfrac{(n+1)!}{k!\,(n+1-k)!}$$

$$= \binom{n+1}{k}$$

2.47 a. $P(A|B) = \dfrac{P(A \cap B)}{P(B)} = \dfrac{.1}{.3} = \dfrac{1}{3}$

b. $P(B|A) = \dfrac{P(A \cap B)}{P(A)} = \dfrac{.1}{.5} = \dfrac{1}{5}$

c. $P(A|A \cup B) = \dfrac{P(A)}{P(A \cup B)} = \dfrac{.5}{.5 + .3 - .1} = \dfrac{5}{7}$

Notice that we are using the additive law of probability, given in Section 2.8.

d. $P(A|A \cap B) = \dfrac{P(A \cap B)}{P(A \cap B)} = 1$

e. $P(A \cap B|A \cup B) = \dfrac{P(A \cap B)}{P(A \cup B)} = \dfrac{.1}{.5 + .3 - .1} = \dfrac{1}{7}$

2.49 a. The three tests are independent. Thus, the probability in question is $(.05)^3 = .000125$.

b. The probability of at least one mistake equals 1 minus the probability of no mistakes. The probability of no mistakes is $(.95)^3$. Thus, the probability of at least one mistake is $1 - (.95)^3 = 1 - .857 = .143$.

2.51 Many of the probabilities can be found directly from the table; others require calculation.

a. $P(A) = .40$

b. $P(B) = .37$

c. $P(AB) = .10$

d. $P(A \cup B) = .10 + .30 + .27 = .67$

e. $P(\overline{A}) = P(9 \text{ years or less of education}) = .60$

f. $P(\overline{A \cup B}) = 1 - P(A \cup B) = 1 - .67 = .33$

g. $P(\overline{AB}) = 1 - P(AB) = 1 - .10 = .90$

h. $P(A|B) = \dfrac{P(AB)}{P(B)} = \dfrac{.10}{.37} = .27$

i. $P(B|A) = \dfrac{P(AB)}{P(A)} = \dfrac{.10}{.40} = .25$

2.53 We assume that A and B are mutually exclusive, and that $P(A) > 0$ and $P(B) > 0$. In order to show that A and B are not independent, we use the technique of proof by contradiction. Assume that A and B are independent. Then by definition, $P(A \cap B) = P(A)P(B)$. But since A and B are mutually exclusive, we know that $P(A \cap B) = 0$, which implies $P(A)P(B) = 0$. It is given that $P(A) > 0$ and $P(B) > 0$, which implies $P(A)P(B) > 0$, and we have a contradiction. Hence A and B must not be independent.

2.55 Given that $P(A) < P(A|B)$

$$P(A) < \frac{P(AB}{P(B)} \qquad \text{by definition of conditional probability}$$

$$P(A) < \frac{P(A)P(B|A)}{P(B)} \qquad \text{multiplicative law of probability}$$

$$P(A)P(B) < P(A)P(B|A) \qquad \text{given } P(B) > 0$$

$$P(B) < P(B|A) \qquad \text{given } P(A) > 0$$

2.57 In order to show that A and $\overline{B}$ are independent, it is necessary to show that $P(A \cap \overline{B}) = P(A)P(\overline{B})$. The following relationships are known:

(1) $B \cup \overline{B} = S$, where B and $\overline{B}$ are complementary and hence mutually exclusive events.

(2) $A \cap S = A$.

(3) $P(\overline{B}) = 1 - P(B)$.

Hence we can write

$$P(A) = P(A \cap S) = P\big[A \cap (B \cup \overline{B})\big].$$

Now, using the distributive law given at the end of Section 2.3,

$$P[A \cap (B \cup \overline{B})] = P[(A \cap B) \cup (A \cap \overline{B})].$$

Using the additive law (from Section 2.8) and noting that the events $A \cap B$ and $A \cap \overline{B}$ are mutually exclusive, we have

$$P[(A \cap B) \cup (A \cap \overline{B})] = P(A \cap B) + P(A \cap \overline{B}).$$

Since A and B are independent, $P(A \cap B) = P(A)P(B)$, and the entire equality implies that

$$P(A) = P(A)P(B) + P(A \cap \overline{B}).$$

So,

$$P(A \cap \overline{B}) = P(A) - P(A)P(B) = P(A)[1 - P(B)] = P(A)P(\overline{B}).$$

Hence A and $\overline{B}$ are independent.

In order to show that $\overline{A}$ and $\overline{B}$ are independent, the same reasoning is used.

$$P(\overline{B}) = P(\overline{B} \cap A) + P(\overline{B} \cap \overline{A}) = P(\overline{B})P(A) + P(\overline{B} \cap \overline{A})$$

since A and $\overline{B}$ are independent. Hence

$$P(\overline{A} \cap \overline{B}) = P(\overline{B})[1 - P(A)] = P(\overline{B})P(\overline{A})$$

and the result is proven.

2.59 **a.** $P(A \cup B) = P(A) + P(B) - P(A \cap B)$. Hence

$$P(A \cap B) = P(A) + P(B) - P(A \cup B) = .2 + .3 - .4 = .1.$$

b. From Exercise 2.3, $P(\overline{A} \cup \overline{B}) = P(\overline{A \cap B}) = 1 - P(A \cap B) = 1 - .1 = .9.$

c. $P(\overline{A} \cap \overline{B}) = P(\overline{A \cup B}) = 1 - P(A \cup B) = 1 - .4 = .6.$

d. $P(\overline{A}|B) = \dfrac{P(\overline{A} \cap B)}{P(B)}$. To find $P(\overline{A} \cap B)$, note that $P(B) = P(A \cap B) + P(\overline{A} \cap B)$,

so that $P(\overline{A} \cap B) = P(B) - P(A \cap B) = .3 - .1 = .2.$

Thus, $P(\overline{A}|B) = \dfrac{.2}{.3} = \dfrac{2}{3}.$

2.61 **a.** $P(\text{current flows}) = 1 - P(\text{current is not flowing})$
$= 1 - P(\text{all three relays are open}) = 1 - (.1)^3 = .999$

b. $A = $ current flows; $B = $ relay closed properly

$$P(B|A) = \frac{P(BA)}{P(A)} = \frac{P(B)}{P(A)}; \text{ since } B \subset A$$

$$= \frac{.9}{.999} = .9009$$

2.63 Given $P(\overline{A \cup B}) = a$; $P(B) = b$; A and B are independent.

Thus, $P(A \cup B) = 1 - a$; $P(AB) = P(A)P(B) = P(A)[b]$.

Consider $P(A \cup B) = P(A) + P(B) - P(AB)$

$$1 - a = P(A) + b - P(A)[b]$$
$$1 - a = b + P(A)[1 - b]$$
$$1 - a - b = P(A)[1 - b]$$
$$\frac{1 - b - a}{1 - b} = P(A)$$

2.65 Let A = defective item gets past first inspector and B = defective item gets past second inspector.

$$P(A) = .1 \qquad P(B|A) = .5$$

The event that the defective item gets past both inspectors is $A \cap B$.

$$P(A \cap B) = P(B|A)P(A) = .05$$

2.67 **a.** We assume that the Connecticut and Pennsylvania lotteries are independent. Thus, $P(666 \text{ in Connecticut}|666 \text{ in Pennsylvania}) = P(666 \text{ in Connecticut})$

$$= \frac{1}{10^3} = .001$$

b. $P(666 \text{ in Connecticut} \cap 666 \text{ in Pennsylvania})$

$$= P(666 \text{ in Connecticut})P(666 \text{ in Pennsylvania})$$
$$= \frac{.001}{\left(\frac{1}{8}\right)} = .000125$$

2.69 $P(\text{landing safely on both jumps}) \geq 1 - .05 - .05 \geq .90$

2.71 Define the following events:

A: buyer sees magazine ad
B: buyer sees corresponding ad on television
C: buyer purchases the product

The following probabilities are known:

$$P(A) = .02 \qquad P(B) = .20 \qquad P(A \cap B) = .01.$$

Now $P(A \cup B) = P(A) + P(B) - P(A \cap B) = .02 + .20 - .01 = .21$. Further,

$$P(\overline{A} \cap \overline{B}) = 1 - P(A \cup B) = 1 - .21 = .79$$

where the event $\overline{A} \cap \overline{B}$ is the event that the buyer does not see the ad either on television or in a magazine. Finally, it is given that $P(C|A \cup B) = \frac{1}{3}$ and $P(C|\overline{A} \cap \overline{B}) = \frac{1}{10}$. It is necessary to find $P(C)$.

$$\begin{aligned}
P(C) &= P(\text{buyer purchases the product}) \\
&= P(\text{buyer sees ad and buys}) + P(\text{buyer doesn't see ad and buys}) \\
&= P[C \cap (A \cup B)] + P[C \cap (\overline{A} \cap \overline{B})] \\
&= P(C|A \cup B)P(A \cup B) + P(C|\overline{A} \cap \overline{B})P(\overline{A} \cap \overline{B}) \\
&= \left(\frac{1}{3}\right)(.21) + \left(\frac{1}{10}\right)(.79) = .07 + .079 = .149.
\end{aligned}$$

2.73 Consider an independent sequence of F's and $\overline{F}$'s corresponding to the sets "detection" or "nondetection" of aircraft. The event of interest is $\overline{F}\,\overline{F}\,\overline{F}F$ and has probability

$$P(\overline{F}\,\overline{F}\,\overline{F}F) = \left[P(\overline{F})\right]^{3}P(F) = (.98)^{3}(.02).$$

2.75 Define W to be the event that the team wins. Then

$$P(\text{four wins}) = P(WWWW) = P(W)P(W)P(W)P(W) = (.75)^{4}.$$

2.77 Define the following events:

  R: driver is rejected; that is, he chooses inspection team 2
  $\overline{R}$: driver passes inspection; that is, he chooses inspection team 1

Then $P(R) = P(\overline{R}) = \frac{1}{2}$.

Using a quadruplet, each component of which represents the fate of one of the four drivers, we are interested in the events

  A: three of four drivers are rejected   and B: all four drivers pass

The first event is a union of four mutually exclusive subevents. That is,

$$\begin{aligned}
P(A) &= P(RRR\overline{R}) + P(RR\overline{R}R) + P(R\overline{R}RR) + P(\overline{R}RRR) \\
&= 4[P(R)]^{3}P(\overline{R}) = 4\left(\frac{1}{2}\right)^{4} = \frac{4}{16} = \frac{1}{4}.
\end{aligned}$$

The probability that all four will pass is

$$P(B) = P(\overline{R}\,\overline{R}\,\overline{R}\,\overline{R}) = \left[P(\overline{R})\right]^{4} = \frac{1}{16}.$$

2.79 a. Define the events

  A: Obtain a sum of 3
  B: Do not obtain a sum of 3 or 7.

In Example 2.5 it was shown that there are 36 sample points corresponding to the numbers on the upper faces of two dice. Of these pairs, two sum to 3 and six sum to 7, leaving 28 that do not sum to either 3 or 7. Hence,

$$P(A) = \tfrac{2}{36} \qquad \text{and} \qquad P(B) = \tfrac{28}{36}.$$

Now, obtaining a sum of 3 before obtaining a sum of 7 can happen on the first toss as

$A,$

on the second toss as

$BA,$

on the third toss as

$BBA,$

and, in general, on the ith toss as

$$\underbrace{BBB\cdots B}_{i-1}A.$$

Note that because the tosses are independent the corresponding probabilities are

1st toss: $\quad P(A) = \tfrac{2}{36}$

2nd toss: $\quad P(B)P(A) = \left(\tfrac{28}{36}\right)\left(\tfrac{2}{36}\right)$

3rd toss: $\quad P(B)^2 P(A) = \left(\tfrac{28}{36}\right)^2\left(\tfrac{2}{36}\right)$

$\quad\vdots$

i^{th} toss: $\quad P(B)^i P(A) = \left(\tfrac{28}{36}\right)^i\left(\tfrac{2}{36}\right).$

Then, the overall probability of obtaining a 3 before a 7 is

$$\sum_{i=0}^{\infty} \left(\tfrac{28}{36}\right)^i\left(\tfrac{2}{36}\right) = \left(\tfrac{2}{36}\right)\left(\frac{1}{1-\tfrac{28}{36}}\right) = \tfrac{1}{4}.$$

The first equality is from the sum of a geometric series as shown in Appendix A1.11.

b. Similar to part (a) except replace A and B with

C: obtain a sum of 4
D: do not obtain a sum of 4 or 7

where $P(C) = \tfrac{3}{36}$ and $P(D) = \tfrac{27}{36}.$

Then the overall probability of obtaining a 4 before a 7 is

$$\sum_{i=0}^{\infty} \left(\frac{27}{36}\right)^i \left(\frac{3}{36}\right) = \left(\frac{3}{36}\right)\left(\frac{1}{1-\frac{27}{36}}\right) = \frac{1}{3}.$$

2.81   a. $\frac{1}{n}$

     b. $\left(\frac{n-1}{n}\right) \times \left(\frac{1}{n-1}\right) = \frac{1}{n}$ second try

        $\left(\frac{n-1}{n}\right) \times \left(\frac{n-2}{n-1}\right) \times \left(\frac{1}{n-2}\right) = \frac{1}{n}$ third try

     c. $P[\text{gain access}] = P(\text{first try}) + P(\text{second try}) + P(\text{third try})$
$$= \frac{1}{7} + \frac{1}{7} + \frac{1}{7} = \frac{3}{7}$$

2.83   Define these events:

     D: person has the disease   and H: test says person has the disease

Then $P(H|D) = .9$; $P(\overline{H}|\overline{D}) = .9$; $P(D) = .01$; $P(\overline{D}) = .99$. Using Bayes's Rule,

$$P(D|H) = \frac{P(H|D)P(D)}{P(H|D)P(D) + P(H|\overline{D})P(\overline{D})} = \frac{(.9)(.01)}{(.9)(.01) + (.1)(.99)} = \frac{.009}{.108} = \frac{1}{12}$$

2.85   Define these events:

     P: positive response M: respondent was male F: respondent was female

Then $P(P|F) = .7$; $P(P|M) = .4$; $P(M) = \frac{1}{4}$. Using Bayes's Rule,

$$P(M|\overline{P}) = \frac{P(\overline{P}|M)P(M)}{P(\overline{P}|M)P(M) + P(\overline{P}|F)P(F)} = \frac{(.6)\left(\frac{1}{4}\right)}{(.6)\left(\frac{1}{4}\right) + (.3)\left(\frac{3}{4}\right)} = \frac{.6}{1.5} = .4$$

2.87   Define these events:

     D: item is defective C: item goes through complete inspection

We know $P(D) = .1$, $P(C|D) = .6$, and $P(C|\overline{D}) = .2$. Thus,

$$P(D|C) = \frac{P(C|D)P(D)}{P(C|D)P(D) + P(C|\overline{D})P(\overline{D})} = \frac{(.6)(.1)}{(.6)(.1) + (.2)(.9)} = \frac{.06}{.24} = .25$$

2.89   Define these events:

     G: student guesses C: student correctly answers question

We know $P(G) = .2$, $P(C|\overline{G}) = 1$, and $P(C|G) = .25$. Thus,

$$P(\overline{G}|C) = \frac{P(C|\overline{G})P(\overline{G})}{P(C|\overline{G})P(\overline{G}) + P(C|G)P(G)} = \frac{(1)(.8)}{(1)(.8) + (.25)(.2)} = \frac{.8}{.85} = .9412$$

2.91 Let M = major airline

P = private plane

C = commercial plane

B = travel for business

$P(M) = .6; \; P(P) = .3; \; P(C) = .1$

$P(B|M) = .5; \; P(B|P) = .6; \; P(B|C) = .9$

a. $P(B) = P(MB) + P(PB) + P(CB) = P(M)P(B|M) + P(P)P(B|P)$
$$+ P(C)P(B|C)$$
$$= (.6)(.5) + (.3)(.6) + (.1)(.9) = .57$$

b. $P(PB) = P(P)P(B|P) = (.3)(.6) = .18$

c. $P(P|B) = \dfrac{P(PB)}{P(B)} = \dfrac{.18}{.57} = .3158$

d. $P(B|C) = .9$

2.93 Let A = both balls white

A_i = both balls selected from bowl i are white

B_i = i^{th} bowl is selected

a. $P(A) = \sum P(B_i \cap A_i) = \left(\frac{1}{5}\right)\left[\sum P(A_i|B_i) \right]$ where $i = 1, \ldots, 5$

$$= \left(\frac{1}{5}\right)\left[0 + \left(\frac{2}{5}\right)\left(\frac{1}{4}\right) + \left(\frac{3}{5}\right)\left(\frac{2}{4}\right) + \left(\frac{4}{5}\right)\left(\frac{3}{4}\right) + 1 \right] = \frac{2}{5}$$

b. $P(B_3|A) = \dfrac{P(B_3 \cap A)}{P(A)} = \dfrac{\left(\frac{3}{50}\right)}{\left(\frac{2}{5}\right)} = \dfrac{15}{100} = \dfrac{3}{20}$

2.95 In Exercise 2.72 we calculated $P(Y = 0) = (.02)^3$ and $P(Y = 3) = (.98)^3$. The event that exactly one plane is detected consists of the following three mutually exclusive events:

$F\overline{F}F$ $\overline{F}FF$ $FF\overline{F}$

Hence $P(Y = 1) = 3(.98)(.02)^2$. Similarly, $P(Y = 2) = 3(.98)^2(.02)$.

2.97 The events $Y = 2$, $Y = 3$, and $Y = 4$ were found in Exercise 2.80 to have probabilities $\frac{1}{15}$, $\frac{2}{15}$, and $\frac{3}{15}$, respectively. The event $Y = 5$ can occur in four ways:

$DGGGD$ $GDGGD$ $GGDGD$ $GGGDD$

Each of these has probability

$$\frac{2(4)(3)(2)}{6(5)(4)(3)} \times \frac{1}{2} = \frac{1}{15}$$

Hence $P(Y = 5) = 4\left(\frac{1}{15}\right) = \frac{4}{15}$.

Another way to see this is to note that if $Y = 5$, then the last two observations are, in order, DG. One of the first four positives must be filled by a D (see Exercise 2.80 "alternate solution"). Thus, $P(Y = 5) = \dfrac{\dbinom{4}{1}}{\dbinom{6}{2}} = \dfrac{4}{15}$.

The event $Y = 6$ can occur in five ways:

$$DGGGGD \qquad GDGGGD \qquad GGDGGD \qquad GGGDGD \qquad GGGGDD$$

Each of these has probability

$$\frac{2(4)(3)(2)(1)}{6(5)(4)(3)(2)} = \frac{1}{15}$$

Hence $P(Y = 6) = 5\left(\frac{1}{15}\right) = \frac{5}{15}$.

Note that $P(Y = 6) = \dfrac{\dbinom{5}{1}}{\dbinom{6}{2}}$.

2.99 The law of total probability gives

$$P(B) = P(B \cap A) + P(B \cap \overline{A})$$

$$\frac{P(B)}{P(B)} = \frac{P(B \cap A)}{P(B)} + \frac{P(B \cap \overline{A})}{P(B)}$$

$$1 = P(A|B) + P(\overline{A}|B)$$

$$P(A|B) = 1 - P(\overline{A}|B)$$

2.101 The 18 tests may be performed in any sequence, and hence in order to determine which sequence is most efficient, the efficiency expert must study all possible orderings of 18 tests. Permutations are used, and the total number of sequences is

$$P_{18}^{18} = \frac{18!}{0!} = 18!$$

2.103 There are $\binom{13}{2}$ ways of getting a set of three of a kind and a set of two of a kind. There are $\binom{4}{3}$ ways to obtain the cards that are three of a kind and $\binom{4}{2}$ ways to obtain the cards that are two of a kind. Thus,

$$P(\text{full house}) = \frac{\binom{13}{2}\binom{4}{3}\binom{4}{2}}{\binom{52}{5}}.$$

2.105 The following table displays the probabilities given in the exercise (underlined). Others are obtained by subtraction.

	B	$\overline{B}$	Total
A	.1	.2	.3
$\overline{A}$	.3	.4	.7
Total	.4	.6	

a. $P(\overline{A \cup B}) = P(\overline{A} \cap \overline{B}) = .4$

b. $P(A \cup B) = P(A) + P(B) - P(A \cap B) = .3 + .4 - .1 = .6$

c. $P(A \cap B|B) = \dfrac{P(A \cap B)}{P(B)} = \dfrac{.1}{.4} = .25$

2.107 a. $P(A) = .25 + .10 + .05 + .10 = .50$

b. $P(A \cap B) = .10 + .05 = .15$

c. $P(A \cap B \cap \overline{C}) = .10$

d. Using a result proven in Exercise 2.64,

$$P(\overline{A} \cup \overline{B}|C) = \frac{P[(\overline{A} \cup \overline{B}) \cap C]}{P(C)} = \frac{P(\overline{A} \cap C) + P(\overline{B} \cap C) - P(\overline{A} \cap \overline{B} \cap C)}{P(C)}$$
$$= \frac{.25 + .25 - .15}{.4} = .875$$

2.109 This exercise is similar to Exercise 2.108.

a. $P(\text{white}) = \dfrac{177{,}749}{203{,}212} = .87$

b. $P(\text{central city}) = \dfrac{63{,}922}{203{,}212} = .31$

c. $P(\text{urban fringe}|\text{white}) = \dfrac{51{,}405}{177{,}749} = .29$

d. $P(\text{white}|\text{urban fringe}) = \dfrac{51{,}405}{54{,}525} = .94$

e. $P(\text{outside urban}|\text{nonwhite}) = \dfrac{3057}{25{,}463} = .12$

f. $P(\text{nonwhite and central city } \underline{\text{or}} \text{ white and outside urban}) = \dfrac{14{,}375 + 27{,}281}{203{,}212}$

= .21.

2.111   Refer to Exercise 2.110. It is now necessary to enumerate all possible arrangements of 2 defectives and 10 nondefectives that result in 3 runs. Then $P(R \leq 3) = P(R = 2) + P(R = 3)$, since the minimum value of R is $R = 2$.

The 3 runs of D's and Ns may occur in one of two ways; 1 run of D's and 2 of N's, or 1 run of N's and 2 of D's. The latter will occur only if the following arrangement occurs: $DNNNNNNNNNND$. If the 2 D's occur on consecutive trials (but not in positions 1 and 2 or 11 and 12), the arrangement will result in 1 of D's and 2 of N's. There are 9 such arrangements:

$NDDNNNNNNNNN$ $NNNDDNNNNNNN$ $NNNNNDDNNNNN$

$NNNNNNNDDNNN$ $NNNNNNNNNDDN$ $NNDDNNNNNNNN$

$NNNNDDNNNNNN$ $NNNNNNDDNNNN$ $NNNNNNNNDDNN$

Note that the two D's must be placed together in one of the $(10-1) = 9$ "slots" between the N's. Hence

$$P(R \leq 3) = P(R = 2) + P(R = 3) = \tfrac{2}{66} + \left(\tfrac{1}{66} + \tfrac{9}{66}\right) = \tfrac{12}{66} = \tfrac{2}{11}.$$

2.113   Let RO_i = relay i is open; RC_i = relay i is closed, where $i = 1, 2, 3, 4$. Consider design A:

$$P(\text{current flows}) = 1 - P(\text{current doesn't flow})$$

$$= 1 - P\big[(RO_1 \cap RO_2) \cup (RO_3 \cap RO_4)\big]$$

$$= 1 - \big[P(RO_1 \cap RO_2) + P(RO_3 \cap RO_4) - P(RO_1 \cap RO_2 \cap RO_3 \cap RO_4)\big]$$

$$= 1 - \big[(.1)^2 + (.1)^2 - (.1)^4\big] = 1 - .0199 = .9801$$

Consider design B:

$$P(\text{current flows}) = P\big[(RC_1 \cap RC_3) \cup (RC_2 \cap RC_4)\big]$$

$$= P(RC_1 \cap RC_3) + P(RC_2 \cap RC_4) - P(RC_1 \cap RC_2 \cap RC_3 \cap RC_4$$

$$= (.9)^2 + (.9)^2 - (.9)^4 = .9639$$

Thus, design A yields a higher probability that the current will flow when the relays are activated.

2.115 Y can take on the values 1, 2, 3, 4, and 5. There are always 70 ways that 4 tires can be chosen at random.

$Y = 1$ if the customer chooses the tire ranked #1. There are then $\binom{7}{3} = 35$ ways to choose the other 3 tires with inferior rankings. $P(Y = 1) = \frac{35}{70} = \frac{1}{2}$.

$Y = 2$; given that the customer chooses the tire ranked #2, there are $\binom{6}{3} = 20$ ways to choose the other 3 tires with inferior rankings. $P(Y = 2) = \frac{20}{70}$.

$Y = 4$; given that the customer chooses the tire ranked #4, there are $\binom{4}{3} = 4$ ways to choose the 3 tires with inferior rankings. $P(Y = 4) = \frac{4}{70}$.

$Y = 5$; given that the customer chooses the tire ranked #5, there is $\binom{3}{3} = 1$ way to choose the other 3 tires with inferior rankings. $P(Y = 5) = \frac{1}{70}$.

y	1	2	3	4	5
$P(y)$	$\frac{35}{70}$	$\frac{20}{70}$	$\frac{10}{70}$	$\frac{4}{70}$	$\frac{1}{70}$

Note: the discussion of $Y = 3$ is in Problem 2.114.

2.117 The probability that Skylab will hit someone is unconditionally $\frac{1}{150}$, regardless of where the person lives. If one wants to know the probability condition on living in a certain area, it is not possible to determine. You can say that in an area containing 4 billion inhabitants, the expected number of casualties is (4 billion) times $\frac{1}{150}$.

2.119 Denote the four possible simple events as HH, HT, TH, TT. Then the events A, B, C, $A \cap B$, $B \cap C$, $A \cap B \cap C$ can be written as sets of simple events:

$$A = \{HH, HT\} \qquad B = \{HH, TH\} \qquad C = \{HH, TT\}$$
$$A \cap B = \{HH\} \qquad B \cap C = \{HH\} \qquad A \cap C = \{HH\} \qquad A \cap B \cap C = \{HH\}$$

Notice that

$$P(A) = P(B) = P(C) = \frac{1}{2} \qquad P(A \cap B) = P(A \cap C) = P(B \cap C) = \frac{1}{4}$$
$$P(A \cap B \cap C) = \frac{1}{4}$$

Since $P(A \cap B \cap C) \neq P(A)P(B)P(C)$, the three events are not independent.

2.121 The game that is being played can be represented by a series of W's and L's, W representing a win for gambler Jones (and hence an increase of $1 for his bank) and L representing a loss (and hence a decrease of $1). The tosses are independent and $P(W) = P(L) = \frac{1}{2}$.

a. For this part of the exercise, consider that the coin has been tossed 6 times, producing a total of 6 outcomes, each of which may be either a W or an L.

Each sequence of outcomes is equally likely, and has probability $\left(\frac{1}{2}\right)^6$.

It is necessary only to count the number of such sequences for which the two gamblers break even. If the gamblers are to break even after 6 trials, then each must win 3 times and lose 3 times. If this is the case, the total gain for either gambler will be $3 - $3 = $0, and he will be left with a total of $6 after 6 trials.

Notice that it is irrelevant to consider the order in which the wins and losses occur, since if the gambler is only allowed to lose or win 3 times, he cannot exhaust his bank of $6 before trial 6 and hence end the game.

Consequently, the total number of outcomes consisting of 3 wins and 3 losses can be found by counting the number of ways to choose 3 trials (out of 6) in which to place the 3 wins. That is,

$$n_a = \binom{6}{3} = 20 \qquad \text{and the desired probability is } \binom{6}{3} \times \left(\frac{1}{2}\right)^6.$$

b. For this part of the exercise, consider that the coin has been tossed 10 times. There are 2^{10} possible sequences of W's and L's, each with probability $\left(\frac{1}{2}\right)^{10}$.

It is necessary to enumerate n_a, the number of sequences that result in gambler Jones winning the game at exactly the tenth toss.

In order for the event of interest to occur, gambler Jones must have $11 at trial 9 and must win on trial 10. That is, in 9 trials (during which $9 will chang hands), Jones must win 7 times and lose only twice in order that his total gain be $7 - $2 = $5 and that he be left with a total of $11 after 9 trials.
If the two losses were placed randomly among the 9 trials, which can be done in $\binom{9}{2} = 36$ ways, it is possible that the game might end before the appointed 10 trials. These arrangements must be eliminated from the 36 possibilities. Notice that gambler Jones can only win on an even-numbered trial, since he must gain $6. That is, the difference between the number of wins and losses is 6, so that the sum of the number of wins and losses must be an even number.

The number of arrangements of 2 L's and 7 W's for which Jones would win on trial 6 are

$$W\,W\,W\,W\,W\,W\,W\,L\,L \qquad W\,W\,W\,W\,W\,W\,L\,L\,W \qquad W\,W\,W\,W\,W\,W\,L\,W\,L$$

The number of arrangements for which Jones would win on trial 8 is found by using the same argument as for the event "win on trial 10" above. The arrangements are

$$LWWWWWWWL \qquad WLWWWWWWL \qquad WWLWWWWWL$$
$$WWWLWWWWL \qquad WWWWLWWWL \qquad WWWWWLWWL$$

These 9 arrangements are eliminated from consideration, so that $n_a = \binom{9}{2} - 9$ $= 27$, and the probability of interest is $27 \left(\frac{1}{2} \right)^{10}$.

2.123 Consider using the device of representing the n balls as 0's and creating N boxes by arbitrarily placing bars between the 0's. The space between 2 adjacent bars is a box and the number of 0's between any 2 adjacent bars is the number of balls in the box. Such a device forces us to start and end the sequence with a bar, and in order to create a set of N boxes, a total of $N + 1$ bars is needed. Eliminating the 2 that are forced to be at the beginning and end, we have a total of $N - 1$ bars to be arranged:

$$||00|||000|00|\ldots|0||$$

We have not yet placed any restrictions upon the arrangement of the $N - 1$ bars and n 0's, since more than 1 ball is allowed per box and boxes are allowed to remain empty. Thus picking $N - 1$ of the $N - 1 + n$ positions in which to place the bars, we have a total of $\binom{N + n - 1}{N - 1}$ arrangements.

It is not necessary to determine how many of these result in no box being empty. If no 2 bars are to be placed adjacent to one another, the $N - 1$ bars must be placed in the $n - 1$ spaces between the 0's. Some of the $n - 1$ spaces may be left unfilled, since $n > N$, but none may contain more than 1 bar:

$$|000|00|\ldots|00|0|\ldots0|$$

The total number of ways to do this is $\binom{n - 1}{N - 1}$. Thus the probability that no box remains empty is given by

$$\frac{\binom{n - 1}{N - 1}}{\binom{N + n - 1}{N - 1}}$$

CHAPTER 3 DISCRETE RANDOM VARIABLES AND THEIR PROBABILITY DISTRIBUTIONS

3.1 The table shown below shows the probabilities for the four possible intersections concerning events A and B. Underlined figures are given in the exercise; others are found by subtraction.

	A	$\overline{A}$	Total
B	.10	.40	.50
$\overline{B}$	.30	.20	.50
Total	.40	.60	1.00

Then

$$P(Y = 0) = P(\text{not } A \cap \text{not } B) = .20$$
$$P(Y = 1) = P(\text{not } A \cap B) + P(A \cap \text{not } B) = .3 + .4 = .7$$
$$P(Y = 2) = P(A \cap B) = .10$$

3.3 Similar to Exercise 2.97. The event $Y = 2$ occurs if the first and second components tested are both defective.

$$p(2) = P(DD) = \tfrac{2}{4}\left(\tfrac{1}{3}\right) = \tfrac{1}{6}$$

$$p(3) = P(DGD) + P(GDD) = 2\left(\tfrac{2}{4}\right)\left(\tfrac{2}{3}\right)\left(\tfrac{1}{2}\right) = \tfrac{2}{6}$$

$$p(4) = P(GGDD) + P(DGGD) + P(GDGD) = 3\left(\tfrac{2}{4}\right)\left(\tfrac{1}{3}\right)\left(\tfrac{2}{2}\right) = \tfrac{1}{2}$$

Since there are only four components, $Y = 2$, 3, and 4 are the only possible values for the random variable Y.

3.5 Assume that the correct ordering of the animal words is ABC. If the child is guessing there are 6 possible equally likely permutations he could choose, with $Y = $ number of matches associated with each permutation.

A	B	C	$Y = $ No. of Matches
A	B	C	3
A	C	B	1
B	A	C	1
B	C	A	0
C	A	B	0
C	B	A	1

Then

$$p(0) = \frac{2}{6} = \frac{1}{3}$$

$$p(1) = \frac{3}{6} = \frac{1}{2}$$

$$p(2) = 0$$

$$p(3) = \frac{1}{6}$$

3.7 The random variable Y takes on the values 0, 1, 2, and 3. We can assume that the three entries are independent.

a. Let E denote an error on a single entry; let N denote that there is no error. There are $2^3 = 8$ sample points:

$$EEE \quad EEN \quad ENE \quad NEE \quad ENN \quad NEN \quad NNE \quad NNN$$

Thus,

$$P(Y = 3) = P(EEE) = (.05)^3 = .000125$$

$$P(Y = 2) = P(EEN) + P(ENE) + P(NEE)$$

$$= 3(.05)^2(.95) = .007125$$

$$P(Y = 1) = P(ENN) + P(NEN) + P(NNE)$$

$$= 3(.05)(.95)^2 = .135375$$

$$P(Y = 0) = P(NNN) = (.95)^3 = .857375$$

b. The probability histogram is

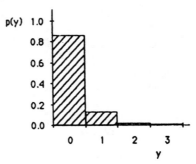

c. $P(Y > 1) = P(Y = 2) + P(Y = 3) = .00725$

3.9 $p(x) = \binom{3}{x}\left(\frac{1}{3}\right)^x\left(\frac{2}{3}\right)^{3-x}$; $x = 0, 1, 2, 3$

x	0	1	2	3
$p(x)$	$\frac{8}{27}$	$\frac{12}{27}$	$\frac{6}{27}$	$\frac{1}{27}$

$p(y) = \binom{3}{y}\left(\frac{1}{15}\right)^y\left(\frac{14}{15}\right)^{3-y}$; $y = 0, 1, 2, 3$

y	0	1	2	3
$p(y)$	$\frac{2744}{3375}$	$\frac{196}{3375}$	$\frac{14}{3375}$	$\frac{1}{3375}$

Considering the distribution of $x + y$, let $x + y = \#$ of people with type O blood, $x + y = 0, 1, 2, 3$.

$p =$ Probability that a person has type O blood

$\quad =$ Probability a person has O^+ blood $+$ Probability a person has O^- blood

$\quad = \frac{1}{3} + \frac{1}{15} = \frac{6}{15}$.

Let $S =$ person has O blood, $F =$ person doesn't have O blood.
The sample space $S = \{SSS, SSF, SFS, SFF, FSS, FSF, FFS, FFF\}$

Thus, the probability distribution for $x + y$ is

$x + y$	$p(x + y)$
0	$\left(\frac{9}{15}\right)^3 = \frac{729}{3375}$
1	$3\left(\frac{6}{15}\right)\left(\frac{9}{15}\right)^2 = \frac{1458}{3375}$
2	$3\left(\frac{6}{15}\right)^2\left(\frac{9}{15}\right) = \frac{972}{33752}$
3	$\left(\frac{6}{15}\right)^3 = \frac{216}{3375}$

3.11 $E(Y) = (-1)\left(\frac{1}{2}\right) + (1)\left(\frac{1}{4}\right) + (2)\left(\frac{1}{4}\right) = \frac{1}{4}$

$E(Y^2) = (-1)^2\left(\frac{1}{2}\right) + (1)^2\left(\frac{1}{4}\right) + (2)^2\left(\frac{1}{4}\right) = \frac{7}{4}$

Cost of play $= C$

Net winnings $= C - Y$

$\quad E(C - Y) = C - E(Y) = C - \frac{1}{4} = 0 \Rightarrow C = \frac{1}{4}$

3.13 The expected value of N is

$$E(N) = E(8\pi r^2) = 8\pi E(r^2)$$

Now,

$$E(r^2) = \Sigma r^2 p(r) = 21^2(.05) + 22^2(.20) + 23^2(.30) + 24^2(.25) + 25^2(.15) + 26^2(.05)$$

$$= 549.1$$

Thus, $E(N) = 8\pi(549.1) = 13,800.388$

3.15 Define G to be the gain to a person in drawing one card. G can take on only three values, \$15, \$5, or \$$-4$, with probabilities as shown in the accompanying table.

G	$p(G)$
15	$\frac{2}{13}$
5	$\frac{2}{13}$
-4	$\frac{9}{13}$

Then $E(G) = \Sigma G p(G) = 15\left(\frac{2}{13}\right) + 5\left(\frac{2}{13}\right) - 4\left(\frac{9}{13}\right) = \frac{4}{13} = .31$

The expected gain is \$.31.

3.17 Let $X_1 = $ # of contracts assigned to firm 1; $X_2 = $ # of contracts assigned to firm 2

There are 9 possible ways the contracts can be assigned to the three firms, each with probability of $\frac{1}{9}$.

Considering the probability distributions for X_1 and X_2, we have

x_1	$p(x_1)$	x_2	$p(x_2)$
0	$\frac{4}{9}$	0	$\frac{4}{9}$
1	$\frac{4}{9}$	1	$\frac{4}{9}$
2	$\frac{1}{9}$	2	$\frac{1}{9}$

The $E(X_1) = \Sigma x_1 p(x_1) = 0\left(\frac{4}{9}\right) + 1\left(\frac{4}{9}\right) + 2\left(\frac{1}{9}\right) = \frac{6}{9} = \frac{2}{3}$

Let $Y = $ profit for firm 1 $= 90,000 X_1$.

The $E[\text{profit for firm 1}] = E(Y) = E[90,000 X_1] = 90,000 E(X_1)$

$= 90,000\left(\frac{2}{3}\right) = 60,000.$

Let $W = $ profit for both firm 1 and 2 $= 90,000(X_1 + X_2)$.

Since X_2 is distributed like X_1, we know that $E(X_2) = \frac{2}{3}$.

Thus, the $E(W) = E[90{,}000(X_1 + X_2)] = 90{,}000E(X_1 + X_2)$

$$= 90{,}000[E(X_1) + E(X_2)]$$

$$= 90{,}000\left[\left(\tfrac{2}{3}\right) + \left(\tfrac{2}{3}\right)\right] = 120{,}000$$

3.19 The object of this exercise is to determine the value of the premium an insurance company should charge so that, over a long period of time, the expected loss or the expected gain the insurance company sustains will be 0; that is, the company will break even. As in Exercise 3.18, we determine the probability distribution for Y, defined to be the insurance company's loss. We then equate $E(Y)$, the expected loss, to 0 in order to find the value of the premium required to break even. The following information is given:

$$P(\text{total loss}) = .001 \qquad\qquad P(50\% \text{ loss}) = .01$$

Let x be the value of the premium the insurance company charges. There are then three possible situations that must be considered. If a customer sustains a total loss, the loss to the insurance company will be $y = 85{,}000 - x$, since they will have gained x dollars from the premium. If a customer sustains a 50% loss, the value of y will be $y = 42{,}500 - x$. Finally, if the customer does not sustain a loss, which occurs with probability $1 - .01 - .001 = .989$, the insurance company will lose $0 - x$ dollars. Thus, the probability distribution for Y is as shown in the table.

y	$p(y)$
$85{,}000 - x$	.001
$42{,}500 - x$	.01
$0 - x$	.989

$E(Y) = \sum\limits_{y} yp(y)$

$$= (85{,}000 - x)(.001) + (42{,}500 - x)(.01) + (0 - .989)$$

$$= 85 - .001x + 425 - .01x - .989x.$$

Setting $E(Y)$ equal to 0 yields $510 - x = 0$, or $x = 510$. Thus, the desired premium is \$510.

3.21 a. We let $g_1(Y) = aY$ and $g_2(Y) = b$.

Then, by Theorem 3.5, $E(aY + b) = E[g_1(Y) + g_2(Y)]$

$$= E[g_1(Y)] + E[g_2(Y)] = E[aY] + E[b].$$

We now use Theorems 3.4 and 3.3 to get $E(aY + b) = aE(Y) + E(b) = a\mu + b$.

b. By Definition 3.5,

$$V(aY + b) = E[aY + b - (a\mu + b)]^2$$
$$= E[aY - a\mu + b - b]^2 = E[a(Y - \mu)]^2$$
$$= E[a^2(Y - \mu)^2].$$

Using Theorem 3.4, this equals $a^2 E(Y - \mu)^2$ which equals $a^2 V(Y) = a^2 \sigma^2$, by Definition 3.5.

3.23 $B = SS \cup FS$

$$P(B) = P(SS) + P(FS) \qquad\qquad SS \text{ and } SF \text{ are mutually exclusive}$$
$$= \frac{2000}{5000} \times \frac{1999}{4999} + \frac{3000}{5000} \times \frac{2000}{4999}$$
$$= \frac{2000}{5000}\left(\frac{1999}{4999} + \frac{3000}{4999}\right) = \frac{2000}{5000} = 0.4.$$

$$P(B|\text{first trial success}) = \frac{1999}{4999} = 0.3999$$

which is <u>not</u> markedly different from 0.4.

3.25 Define Y to be the number of components failing in less than 1000 hours. Then

$$p = P(\text{component fails in less than 1000 hours}) = .2.$$

There are $n = 4$ independent components.

a. $P(Y = 2) = p(2) = \binom{4}{2}p^2 q^2 = \binom{4}{2}(.2)^2(.8)^2 = .1536.$

b. Since the subsystem will operate if 2, 3, or 4 of the components last longer than 1000 hours, it will operate if 2, 1, or none of the components fail in less than 1000 hours. Calculate

$$p(0) = \binom{4}{0}(.2)^0(.8)^4 = .4096 \qquad \text{and} \qquad p(1) = \binom{4}{1}(.2)^1(.8)^3 = .4096.$$

Then

$$P(\text{subsystem operates}) = p(0) + p(1) + p(2) = .4096 + .4096 + .1536 = .9728.$$

3.27 Let Y be the number answered correctly. Then $p = P(\text{correct answer}) = \frac{1}{5}$ and $n = 15$.

$$P(Y \geq 10) = 1 - P(Y \leq 9) = 1 - 1.000 = .000 \qquad \text{(to three decimal places)}$$

using Table 1, Appendix III.

3.29 Let Y be the number of qualifying subscribers. Then Y has a binomial distribution with $p = .7$ and $n = 5$.

a. $P(Y = 5) = \binom{5}{5}(.7)^5 = .1681$

b. $P(Y \geq 4) = P(Y = 4) + P(Y = 5)$
$$= \binom{5}{4}(.7)^4(.3) + \binom{5}{5}(.7)^5$$
$$= .3601 + .1681 = .5282$$

3.31   For $n = 3$ and $p = .8$

$P(Y = 0) = (.2)^3 = .008$ $P(Y = 2) = \binom{3}{2}(.8)^2(.2) = .384$

$P(Y = 1) = \binom{3}{1}(.8)(.2)^2 = .096$ $P(Y = 3) = (.8)^3 = .512$

The alarm will function if $Y = 1$, 2, or 3. Hence

$P(\text{alarm functions}) = p(1) + p(2) + p(3) = .096 + .384 + .512 = .992.$

3.33   Refer to Table 1, Appendix III, indexing $n = 20$ and $p = \frac{1}{2}$. The values of $p(y)$ for $y = 0, 1, 2, \ldots, 20$ are obtained by calculating the differences between tabled values for $a = r$ and $a = r - 1$ for $r = 1, 2, \ldots, 20$. Note that $p(0)$ is obtained directly from the table. The binomial probability distribution and its probability histogram are shown below.

y	$p(y)$	y	$p(y)$
0	.000	11	.160
1	.000	12	.120
2	.000	13	.074
3	.001	14	.037
4	.005	15	.015
5	.015	16	.005
6	.037	17	.001
7	.074	18	.000
8	.120	19	.000
9	.160	20	.000

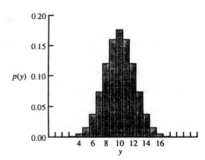

Figure 3.1

3.35   Let Y be the number of housewives preferring brand A. Under the assumption that there is no difference between brands, $p = P(\text{prefer brand } A) = .5$ and $n = 15$.

a.   Using Table 1, Appendix III,

$$P(Y \geq 10) = 1 - P(Y \leq 9) = 1 - .849 = .151$$

b. $P(10 \text{ or more prefer } A \text{ or } B) = P(Y \leq 5 \text{ or } Y \geq 10) = P(Y \leq 5) + [1 - P(Y \leq 9)]$
$= .151 + (1 - .849) = .302$, since 10 or more preferring B is equivalent to 5 or less preferring A.

3.37   Refer to Exercise 3.36. The number of successful explorations is Y and the number of unsuccessful explorations is $(10 - Y)$. Hence the cost is

$$C = 20,000 + 30,000Y + 15,000(10 - Y)$$

and

$$E(C) = 20,000 + 30,000(1) + 15,000(10 - 1) = \$185,000$$

3.39 Let Y be the number of defective motors out of 10. Then Y is binomial with $p = .08$ and $n = 10$. By Theorem 3.7, $E(Y) = (.08)(10) = .8$.

The seller gains \$1000 on the sale of 10 motors and loses \$200 for each defective. The seller's expected net gain is

$$\$1000 - \$200\,E(Y) = \$1000 - \$200(.8) = \$1000 - \$160 = \$840.$$

3.41 Let $Y = \#$ who have Rh^+ blood. Y is binomial with $n = 5$ and $p = .8$.

   a. $P(\text{at least one does not have Rh}^+) = P(Y \le 4) = .672$

   b. $P(Y \le 4) = .672$

   c. Y is now binomial with unknown n and $p = .8$. $P(Y \ge 5) = 1 - (Y \le 4) > .9$.

   Therefore, $P(Y \le 4) < .1$. Using Table 1, we find that 'n' is between $n = 5$ and $n = 10$. Specifically, $n = 8$.

3.43 **a.** For any $y = 1, 2, 3, \ldots, n - 1$

$$p(y) = \binom{n}{y} p_y q^{n-y}$$

and

$$p(y - 1) = \binom{n}{y-1} p^{y-1} q^{n-y+1},$$

Then

$$\frac{p(y)}{p(y-1)} = \frac{\binom{n}{y} p^y q^{n-y}}{\binom{n}{y-1} p^{y-1} q^{n-y+1}}$$

$$= \frac{n!}{y!\,(n-y)!} \frac{(y-1)!\,(n-y+1)!}{n!} \frac{p}{q}$$

$$= \frac{(n-y+1)p}{yq}.$$

   b. We want $\dfrac{(n-y+1)p}{yq} > 1$

   or $\qquad (n+1)p - yp > yq$

   or $\qquad\quad (n+1)p > yq + yp$

   or $\qquad\quad (n+1)p > y.$

   We want $\dfrac{(n-y+1)p}{yq} < 1$

   or $\qquad (n+1)p - yp < yq$

or $\qquad (n+1)p < yq + yp$

or $\qquad (n+1)p < y.$

Also, we want

$$\frac{(n-y+1)p}{yq} = 1$$

or $\qquad (n+1)p - yp = yq$

or $\qquad (n+1)p = yq + yp$

or $\qquad (n+1)p = y$

which is only possible if $(n+1)p$ is an integral, since y only takes on integer values.

c. From the results above,

if $y \le (n+1)p$ then $p(y) \ge p(y-1) > p(y-2) > \ldots$
and if $y \ge (n+1)p$, then $p(y) \ge p(y+1) > p(y+2) > \ldots$.

Therefore, $p(y)$ is maximized when y is as close to $(n+1)p$ as possible.

3.45 Let Y be the number of the interview on which the first applicant having advanced training is found. Then Y has a geometric distribution. Thus,

$$P(Y = 5) = (.7)^{5-1}(.3) = .072$$

3.47 Since $Y =$ the number of calls until the first person is found who is satisfied with the state of the nation, a success occurs when a person is found who is satisfied with the state of the nation. Assuming $P(s) = 0.27$, Y has the geometric distribution with $P = .27$.

3.49 a. $P(Y > a) = 1 - P(Y \le a) = 1 - \sum\limits_{y=1}^{a} q^{y-1}p = 1 - p\left(\sum\limits_{y=1}^{a} q^{y-1} \right)$

The sum of the first a terms of the geometric series given above can be found to be $\dfrac{1-q^a}{1-q}$ by writing

$$\sum\limits_{y=1}^{a} q^{y-1} = S = 1 + q + q^2 + \ldots + q^{a-1}$$

$$qS = q + q^2 + \ldots + q^{a-1} + q^a$$

Subtracting the two equations, we have $(1-q)S = 1 - q^a$. Hence, the series sum is $S = \dfrac{1-q^a}{1-q}$. Thus,

$$P(Y > a) = 1 - p\left(\frac{1 - q^a}{1 - q}\right) = 1 - (1 - q^a) = q^a$$

b. Using the result of part (a),

$$P(Y > a + b | Y > a) = \frac{P(Y > a + \text{b}, \, Y > a)}{P(Y > a)} = \frac{q^{a+b}}{q^a} = q^b = P(Y > b)$$

3.51 Define Y to be the number of the first account containing substantial errors. Then Y has a geometric distribution with $p = .9$.

a. $P9Y = 3) = (.1)^2(.9) = .009$

b. $P(Y \geq 3) = 1 - P(Y \leq 2) = 1 - P(Y = 1) - P(Y = 2) = 1 - .9 = (.1)(.9) = .01$

3.53 Let Y be the number of one-second intervals until the first arrival, so that $p = P(\text{arrival}) = .1$.

a. $P(Y = 3) = q^{3-1}p = q^2 p = (.9)^2(.1) = .081$.

b. $P(Y \geq 3) = 1 - P(Y \leq 2) = 1 - \left(q^{1-1}p + q^{2-1}p\right) = 1 - (.1) - (.9)(.1) = .81$.

3.55 Define Y to be the number of people questioned before a "yes" answer is given. Then

$$\begin{aligned} p = P(\text{yes}) &= P(\text{smoker} \cap \text{yes}) + P(\text{nonsmoker} \cap \text{yes}) \\ &= P(\text{yes}|\text{smoke})P(\text{smoker}) + 0 \\ &= .3(.2) = .06. \end{aligned}$$

Thus,

$$p(y) = pq^{y-1} = .06(.94)^{y-1}, \qquad y = 1, 2, 3, \ldots$$

3.57 The number of tosses until the first head appears is a geometric random variable with $p = \frac{1}{2}$. Using Theorem 3.8, $E(Y) = \frac{1}{p} = \frac{1}{\left(\frac{1}{2}\right)} = 2$.

3.59 Note first that $\dfrac{d^2}{dq^2} q^y = y(y - 1)q^{y-2}$. Hence

$$\frac{d^2}{dq^2} \sum_{y=2}^{\infty} q^y = \sum_{y=2}^{\infty} y(y - 1)q^{y-2}$$

(The interchange of derivative and sum can be justified.) Then

$$E(Y(Y - 1)) = \sum_{y=1}^{\infty} y(y - 1)pq^{y-1} = pq \sum_{y=2}^{\infty} y(y - 1)q^{y-2} = pq \frac{d^2}{dq^2} \sum_{y=2}^{\infty} q^y$$

$$= pq \frac{d^2}{dq^2}\left\{\frac{1}{1 - q} - 1 - q\right\} = pq \frac{d}{dq}\left\{\frac{1}{(1 - q)^2} - 1\right\} = \frac{2pq}{(1 - q)^3} = \frac{2q}{p^2}$$

The variance of Y is then

$$V(Y) = E(Y(Y-1)) + E(Y) - [E(Y)]^2 = \frac{2q}{p^2} + \frac{1}{p} - \frac{1}{p^2} = \frac{2(1-p)+p-1}{p^2} = \frac{q}{p^2}$$

3.61 The total cost of conducting the tests to locate 3 positives is $20X$ dollars. Refer to Theorem 3.9. The expected value of the total cost is

$$E(20X) = 20E(X) = 20\,\frac{r}{p} = \frac{20(3)}{.4} = 150 \text{ dollars}$$

and the variance of the total cost is

$$V(20X) = 400V(X) = 400\,\frac{r(1-p)}{p^2} = \frac{400(3)(.6)}{(.4)^2} = 4500.$$

3.63 Let $Y = \#$ of trials until the rth nondefective engine is found. Y is negative binomial with $r = 3$ and $p = .9$.

a. $P(Y = 5) = \binom{4}{2}(.9)^3(.1)^2 = .04374$

b. $P(Y \le 5) = P(y = 3) + P(Y = 4) + P(Y = 5$
$$= \binom{2}{2}(.9)^3(.1)^2 + \binom{3}{2}(.9)^3(.1) + \binom{4}{2}(.9)^3(.1)^2$$
$$= .729 + .2187 + .04374 = .99144$$

3.65 Recall that Y is geometric with $p = .9$.

$$P(Y \ge 4 | Y > 2) = \frac{P(Y \ge 4)}{P(Y > 2)} = \frac{[1 - P(Y \le 3)]}{[1 - P(Y \le 2)]}$$

$$= \frac{1 - [P(Y = 1) + P(Y = 2) + P(Y = 3)]}{1 - [P(Y = 1) + P(Y = 2)]}$$

$$= 1 - \frac{P(Y = 3)}{1 - [P(Y = 1) - P(Y = 2)]}$$

$$= 1 - \frac{p(1-p)^2}{1 - p - p(1-p)}$$

$$= 1 - \frac{p(1-p)^2}{(1-p)(1-p)}$$

$$= 1 - p = 1 - .9 = .1$$

3.67 a. Let $Y = \#$ of wells drilled until the first strike of oil. Y is geometric with $p = .2$.

$$P(Y = 3) = (.2)(.8)^2 = .128.$$

b. Let $Y = \#$ of wells drilled until the third strike of oil. Y is negative binomial with $r = 3$, $p = .2$.

$$P(Y = 7) = \binom{6}{2}(.2)^3(.8)^4 = .049$$

c. (1)   One of two possible outcomes
    (2)   The probability of success, p, remains constant from oil well to oil well
    (3)   The trials are independent

d. $Y = \#$ of wells drilled until three producing wells are found. Y is negative
   binomial with $r = 3$ and $p = .2$.

$$\mu = \frac{r}{p} = \frac{3}{.2} = 15; \quad \sigma^2 = \frac{r(1 - p)}{p^2} = \frac{2.4}{.04} = 60$$

3.69 There are y trials before the rth success, if the rth success occurs on the $(y + 1)^{\text{st}}$
trial. Let $X = $ the number of the trial on which the rth success occurs, and X is
negative binomial.

Then $X = Y + 1$ and

$$P(Y = y) = P(x = y + 1) = \binom{y + 1 - 1}{r - 1} p^r q^{y+1-r}$$

$$= \binom{y}{r - 1} p^r q^{y+1-r} \qquad y = r - 1, \, r, \, r + 1, \, \ldots \, .$$

3.71 Use the hypergeometric probability distribution.

$$P(5 \text{ nondefectives}) = \frac{\binom{6}{5}\binom{4}{0}}{\binom{10}{5}} = \frac{6}{252} = \frac{1}{42}$$

3.73 Think of a bowl with 6 balls, 2 red (for programs 1 and 2) and 4 black (for
programs 3–6). Let $N = 6$, $n = 2$, and $r = 4$. Then Y follows the hypergeometric
distribution with

$$P(y) = \frac{\binom{6}{y}\binom{2}{2 - y}}{\binom{6}{2}}, \qquad y = 0, \, 1, \, 2.$$

3.75 The probability of an event as rare or rarer than the one observed can be calculated
by using the hypergeometric distribution.

$$P(\text{one or fewer black members}) = \frac{\binom{8}{1}\binom{12}{5}}{\binom{20}{6}} + \frac{\binom{8}{0}\binom{12}{6}}{\binom{20}{6}} = \frac{8(792)}{38,760} + \frac{924}{38,760} = .187$$

This is not a very unlikely event, since it has probability close to $\frac{1}{5}$. It could very
well have happened by chance. There is little reason to doubt the randomness of the
selection.

3.77 The random variable Y follows the hypergeometric distribution with

$$p(y) = \frac{\binom{2}{y}\binom{4}{3-y}}{\binom{6}{3}} \qquad y = 0, 1, 2$$

The probability distribution for Y and the probability histogram are shown below.

y	$p(y)$
0	$\frac{1}{5}$
1	$\frac{3}{5}$
2	$\frac{1}{5}$

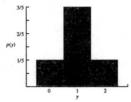

Figure 3.2

3.79 $N = 20$, $n = 5$, $r = 2$. Let $Y =$ the number of improperly drilled gearboxes in the sample of 5. Then Y follows a hypergeometric distribution.

a. $P(Y = 0) = \dfrac{\binom{2}{0}\binom{18}{5}}{\binom{20}{5}} = .553$

b. The total time, T, that it takes to install the boxes (in minutes) is

$$T = 10Y + (5 - Y)$$
$$= 9Y + 5.$$

First,

$$E(Y) = \frac{nr}{N} = 5\left(\frac{2}{20}\right) = .5$$
$$V(Y) = n\left(\frac{r}{N}\right)\left(\frac{N-r}{N}\right)\left(\frac{N-n}{N-1}\right)$$
$$= 5(.1)(1-.1)\left(\frac{20-5}{20-1}\right)$$
$$= .355.$$

It follows that

$$E(T) = 9E(Y) + 5 = 9.5$$

and

$$V(T) = 9^2 V(Y) = 81(.355) = 28.755.$$

Thus, installation time should average 9.5 minutes, with a standard deviation of $\sqrt{28.755} = 5.362$ minutes.

3.81 **a.** $P(Y = 4) = p(4) = \frac{2^4}{4!}e^{-2} = .090$

or, using Table 3, Appendix III,

$P(Y = 4) = P(Y \le 4) - P(Y \le 3) = .947 - .857 = .090$

b. $P(Y \ge 4) = 1 - P(Y \le 3) = 1 - \sum_{y=0}^{3} \frac{2^y}{y!}e^{-2} = 1 - .857 = .143$

Note that $P(Y \le 3) = .857$ can also be obtained from Table 3, Appendix III.

c. $P(Y < 4) = P(Y \le 3) = .857$

d. Recall from part (b) that $P(Y \ge 4) = .143$

and note that $P(Y \ge 2) = 1 - P(Y \le 1) = 1 - .406 = .594$

Hence,

$$P(Y \ge 4 | Y \ge 2) = \frac{P(Y \ge 4 \text{ and } Y \ge 2)}{P(Y \ge 2)}$$

$$= \frac{P(Y \ge 4)}{P(Y \ge 2)} = \frac{.143}{.594} = .241$$

3.83 Let S = total service time in a given hour = $10Y$. Then

$E(S) = 10E(Y) = 10(\lambda) = 10(8) = 80.$

$V(S) = (10)^2 V(Y) = 100\lambda = 800.$

2.5 hours = 150 minutes = $70 + 80 = \mu + 2.475\sigma.$

$P(S > 150) \le \frac{1}{(2.475)^2} = .163.$

So, we infer that it is not unlikely that S will exceed 150.

3.85 Let Y be the number of typing errors per page. Then Y has a Poisson distribution with $\lambda = 4$.

$$P(Y \le 4) = \sum_{y=0}^{4} \frac{4^y e^{-4}}{y!} = e^{-4} + 4e^{-4} + \frac{16}{2}e^{-4} + \frac{4^3}{6}e^{-4} + \frac{4^4}{24}e^{-4} = e^{-4}(34.333)$$

$$= .6288$$

3.87 Let Y be the number of knots in the wood. Then Y has a Poisson distribution with $\lambda = 1.5$.

$$P(Y \le 1) = \frac{(1.5)^0 e^{-1.5}}{0!} + \frac{(1.5)^1 e^{-1.5}}{1!} = 2.5e^{-1.5} = .5578$$

3.89 Define a random variable X, the number of times $Y > 3$ in 10 trials. This random variable will have a binomial distribution if we define a success to be the event "$Y > 3$" on a single trial and

$$p = P(Y > 3) = 1 - \frac{8}{3}e^{-1}$$

from Exercise 3.88. There are $n = 10$ trials and the probability of interest is

$$P(X \geq 1) = 1 - P(X = 0) = 1 - p^0 q^{10} = 1 - \left(\frac{8}{3}e^{-1}\right)^{10} = .1745$$

3.91 This exercise is similar to previous exercises. The random variable is Y, the number of sales in 100 contacts, and possesses a binomial distribution with $p = P(\text{sale}) = .03$. The probability of interest is

$$P(Y \geq 1) = 1 - P(Y = 0) = 1 - \binom{100}{0}(.03)^0(.97)^{100} = 1 - (.97)^{100} = .9524$$

If one chooses to use the Poisson approximation with $\lambda = np = 3$, then one obtains

$$P(Y \geq 1) \approx 1 - \frac{3^0 e^{-3}}{0!} = 1 - e^{-3} = .9502$$

3.93 Use the Poisson approximation to the binomial with $\lambda = np = 30(.2) = .6$. Then

$$p(Y \leq 3) = \sum_{y=0}^{3} \frac{6^y e^{-6}}{y!} = e^{-6} + 6e^{-6} + \frac{36}{2}e^{-6} + \frac{216}{6}e^{-6} = 61e^{-6} = .1512$$

3.95 For a Poisson random variable with $\lambda = 2$,

$$E(Y) = \lambda = 2 \qquad \text{and} \qquad E(Y^2) = V(Y) + [E(Y)]^2 = \lambda + \lambda^2 = 2 + 4 = 6$$

Then $E(X) = 50 - 2E(Y) - E(Y^2) = 50 - 2(2) - 6 = 40$.

3.97 Let $Y = \#$ of breakdowns per day. Y is Poisson with $\lambda = 2$.

$$\mu_y = E(Y) = 2; \text{ Var } (Y) = E(Y^2) - \mu_y^2 = 2, \text{ thus } E(Y^2) = 2 + 4 = 6$$

Now, $E(R) = E(1600 - 50Y^2) = E(1600) - 50E(Y^2) = 1600 - 50(6) = \$1300$.

3.99 $m(t) = E(e^{ty}) = \sum_{y=0}^{n} \binom{n}{y}(pe^t)^y q^{n-y} = (pe^t + q)^n$.

3.101 $m(t) = E(e^{ty}) = \sum_{y=1}^{\infty} pe^{ty}q^{y-1} = pe^t \sum_{y=1}^{\infty} e^{t(y-1)}q^{y-1} = pe^t \sum_{z=0}^{\infty} (qe^t)^z = \frac{pe^t}{1 - pe^t}$

if $qe^t < 1$ or, equivalently, $t < -\ln q$.

3.103 The distributions can be recognized by recalling the moment-generating functions of some common random variables given in this section.

a. Y has a binomial distribution with $n = 5$, $p = \frac{1}{3}$.

b. The form is closest to the geometric except for the "2" in the denominator. In order to comply with the form of this moment-generating function, we can multiply and divide by $\frac{1}{2}$.

$$m(t) = \frac{e^t}{2 - e^t} = \frac{\left(\frac{1}{2}\right)e^t}{\left(\frac{1}{2}\right)(2 - e^t)} = \frac{\left(\frac{1}{2}\right)e^t}{1 - \left(\frac{1}{2}\right)e^t}$$

which is the m.g.f. for a geometric random variable with $p = \frac{1}{2}$.

c. Refer to Example 3.21. Y is a Poisson random variable with $\lambda = 2$.

3.105 a. Differentiate $m(t)$ to find the necessary moments.

$$E(Y) = \frac{d}{dt} m(t) \Big|_{t=0} = \frac{1}{6} e^t + \frac{4}{6} e^{2t} + \frac{9}{6} e^{3t} \Big|_{t=0} = \frac{14}{6} = \frac{7}{3}$$

b. $E(Y^2) = \frac{d^2}{dt^2} m(t) \Big|_{t=0} = \frac{1}{6} + \frac{8}{6} + \frac{27}{6} = 6, \quad V(Y) = 6 - \left(\frac{7}{3}\right)^2 = \frac{5}{9}.$

c. Since $m(t) = E(e^{ty})$, Y must take only the values $Y = 1$, 2, and 3, with probabilities $\frac{1}{6}, \frac{2}{6},$ and $\frac{3}{6}$, respectively.

3.107 $m'_W(t) = be^{bt} m_Y(at) + e^{bt} m'_Y(at) a$

$m''_W(t) = b^2 e^{bt} m_Y(at) + abe^{bt} m'_Y(at) + abm'_Y(at) + a^2 e^{bt} m''_Y(at)$

Note that $m_Y(0) = E(e^{0Y}) = E(1) = 1$.

$$m'_W(t) \Big|_{t=0} = bm_Y(0) + am'_Y(0)$$
$$= at(Y) + b$$
$$= E(W).$$

$$m''_W(t) \Big|_{t=0} = b^2 + 2abE(Y) + a^2 E(Y^2)$$
$$= E(W^2).$$

$$V(W) = E(W^2) - (E(W))^2$$
$$= a^2 E(Y^2) + 2abE(Y) + b^2 - \left(a^2 [E(Y)]^2 + 2abE(Y) + b^2\right)$$
$$= a^2 \left[E(Y^2) - (E(Y))^2\right]$$
$$= a^2 V(Y).$$

3.109 Since $m(t) = e^{5(e^t - 1)}$, $r(t) = 5(e^t - 1)$. Then

$$E(Y) = r^{(1)}(0) = 5e^t \Big|_{t=0} = 5 \qquad V(Y) = r^{(2)}(0) = 5e^t \Big|_{t=0} = 5$$

3.111 $P(t) = \sum_{y=0}^{\infty} \frac{\lambda^y e^{-\lambda} t^y}{y!} = \frac{e^{-\lambda}}{e^{-\lambda t}} \sum_{y=0}^{\infty} \frac{(\lambda t)^y e^{-\lambda t}}{y!} = e^{-\lambda + \lambda t} = e^{\lambda(t-1)}.$

Differentiating with respect to t,

$$E(Y) = \frac{d}{dt}P(t)\Big|_{t=1} = \lambda e^{\lambda(t-1)}\Big|_{t=1} = \lambda, \quad E(Y(Y-1)) = \frac{d^2}{dt^2}P(t)\Big|_{t=1}$$

$$= \lambda^2 e^{\lambda(t-1)}\Big|_{t=1} = \lambda^2$$

Thus $V(Y) = E(Y(Y-1)) + E(Y) - [E(Y)]^2 = \lambda^2 + \lambda - \lambda^2 = \lambda$.

3.113 a. The point $Y = 6$ lies $\frac{11-6}{3} = \frac{5}{3}$ standard deviations below the mean.

Similarly, the point $Y = 16$ lies $\frac{16-11}{3} = \frac{5}{3}$ standard deviations above the

mean. According to Tchebysheff's theorem with $k = \frac{5}{3}$, at least

$$1 - \left(\frac{1}{k^2}\right) = 1 - \frac{9}{25} = .64$$ of the measurements will be in the interval 6 to 16.

b. The second statement of Tchebysheff's theorem states that $P(|Y - \mu| > k\sigma)$

$\leq \frac{1}{k^2}$. To find C, let $.09 = \frac{1}{k^2}$. Then $k^2 = \frac{1}{.09} = \frac{100}{9}$ and $k = \frac{10}{3}$. Since $\sigma = 3$,

$k\sigma = 31\left(\frac{10}{3}\right) = 10 = C$.

3.115 a. $E(Y) = (-1)p(-1) + (0)p(0) + (1)p(1)$

$$= -1\left(\frac{1}{18}\right) + 0\left(\frac{16}{18}\right) + 1\left(\frac{1}{18}\right)$$

$$= 0.$$

$E(Y^2) = (-1)^2 p(-1) + (0)^2 p(0) + (1)^2 p(1)$

$$= (-1)^2\left(\frac{1}{18}\right) + (0)^2\left(\frac{16}{18}\right) + (1)^2\left(\frac{1}{18}\right)$$

$$= \frac{1}{9}$$

$V(Y) = E(Y^2) - (E(Y))^2$

$$= \frac{1}{9} - 0 = \frac{1}{9}$$

b. $\sigma = \sqrt{V(Y)} = \sqrt{\frac{1}{9}} = \frac{1}{3}$.

By Tchebysheff's theorem,

$$P(|y - \mu| \geq 3\sigma) \leq \frac{1}{3^2} = \frac{1}{9}.$$

According to the probability distribution of Y,

$P(|y - \mu| \geq 3\sigma) = P(|y| \geq 1)$

$$= p(-1) + p(1)$$

$$= \frac{1}{18} + \frac{1}{18} = \frac{1}{9}$$

so that the bound is attained when $k = 3$.

c.   Let x have the probability distribution

$$p(-1) = \tfrac{1}{8} \qquad p(0) = \tfrac{6}{8} \qquad \text{and} \qquad p(1) = \tfrac{1}{8}$$

so that $E[x] = 0$ and $V(x) = E[x^2] = \tfrac{1}{4}$.

It follows that

$$P\Big(|X - \mu_x| \ge 2\sigma_x\Big) = P\big(|x| \ge 1\big) = p(-1) + p(1) = \tfrac{1}{4},$$

as desired.

d.   Letting all the probability mass be on values $-1, 0, 1$, $E(W) = 0$ if

$$p(-1) = p \qquad p(0) = 1 - 2p \qquad \text{and} \qquad p(1) = p$$

for some probability p. We want $k\sigma_W = 1$ so that $\sigma_W = \tfrac{1}{k}$ and $\sigma_W^2 = \tfrac{1}{k^2}$.

With $E(W) = 0$, the $V(W) = E(W^2) = 2p$. Setting $2p = \tfrac{1}{k^2}$ gives $p = \tfrac{1}{2k^2}$.

Therefore for any specified $k > 1$, $P\Big(|W - \mu_W| \ge k\sigma_W\Big) = \tfrac{1}{k^2}$ if

$$p(-1) = \tfrac{1}{2k^2} \qquad p(0) = 1 - \tfrac{1}{k^2} \qquad \text{and} \qquad p(1) = \tfrac{1}{2k^2}.$$

Alternatively, we can show the same result using complements. We want, with $k\sigma_W = 1$,

$$P(W = 0) = P\Big(|W - \mu_W| < k\sigma_W\Big) = 1 - \tfrac{1}{k^2}$$

Then, in order for $E(W) = 0$, we must have the same distribution as above.

3.117 Using Tchebysheff's theorem in its first form, we find that the lower bound, $\tfrac{5}{9}$, must equal $1 - \left(\tfrac{1}{k^2}\right)$. That is,

$$k^2 = \tfrac{9}{4} \qquad \text{and} \qquad k = \tfrac{3}{2}$$

The interval of interest is $\mu \pm \left(\tfrac{3}{2}\right)\sigma = 100 \pm \left(\tfrac{3}{2}\right)(10) = 100 \pm 15$, or 85 to 115.

3.119 The random variable Y is defined to be the number of heads observed when a coin is flipped three times. Then $p = P(\text{head}) = \tfrac{1}{2}$.

a.   The binomial probabilities are as follows:

$$P(Y = 0) = p(0) = \binom{3}{0}\left(\tfrac{1}{2}\right)^0 \left(\tfrac{1}{2}\right)^3 = \tfrac{1}{8} \qquad P(Y = 2) = p(2) = \binom{3}{2}\left(\tfrac{1}{2}\right)^2 \left(\tfrac{1}{2}\right)^1 = \tfrac{3}{8}$$

$$P(Y=1) = p(1) = \binom{3}{1}\left(\frac{1}{2}\right)^1\left(\frac{1}{2}\right)^2 = \frac{3}{8} \qquad P(Y=3) = p(3) = \binom{3}{3}\left(\frac{1}{2}\right)^3\left(\frac{1}{2}\right)^0 = \frac{1}{8}$$

b.   The associated probability distribution is shown in Figure 3.4.

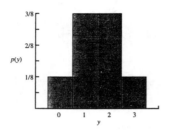

Figure 3.4

c. $\mu = E(Y) = np = 3\left(\frac{1}{2}\right) = 1.5$

$$\sigma = \sqrt{V(Y)} = \sqrt{npq} = \sqrt{3\left(\frac{1}{2}\right)\left(\frac{1}{2}\right)} = .866$$

d.   The desired intervals are

$$\mu \pm \sigma = 1.5 + .866 \qquad \text{or} \qquad .634 \text{ to } 2.366$$
$$\text{and } \mu \pm 2\sigma = 1.5 + 1.732 \qquad \text{or} \qquad -.232 \text{ to } 3.232$$

The values of the random variable Y that fall within the first interval are the values 1 and 2. Thus the fraction of measurements within this interval will be $\frac{3}{8} + \frac{3}{8} = \frac{3}{4}$. The second interval encloses all four values of Y, and thus the fraction of measurements within two standard deviations of the mean will be 1, or 100% of the measurements. These results are consistent with both Tchebysheff's theorem and the empirical rule.

3.121   Let Y be the number of fatalities. Then Y is binomial with $p = .0006$ and $n = 40,000$.

a. $E(Y) = np = 40,000(.0006) = 24$

b. $V(Y) = npq = 24(.9994) = 23.9856 = \sigma^2$

$$\sigma = \sqrt{23.9856} = 4.898$$

c.   No. The value 40 is $\frac{40-24}{4.898} = 3.26$ standard deviations above the mean.

3.123   The mean of C is $E(C) = \$50 + \$3\,E(Y) = \$50 + \$3(10) = \$80$. The variance is $V(C) = V(50 + 3Y) = 9V(Y) = 9(10) = 90$, so that $\sigma = \sqrt{90} = 9.487$.

Using Tchebysheff's theorem with $k = 2$, we have $P\big(|Y - 80| < 2(9.487)\big) \geq .75$ so that the required interval is $(80 - 2(9.487),\ 80 + 2(9.487))$ or $(61.03, 98.97)$.

3.125 Using Tchebysheff's theorem, consider $P(Y \geq \mu + k\sigma) \leq \frac{1}{k^2}$.

$$P(Y \geq 350) \leq \frac{1}{k^2}$$

We evaluate $150 + k(67.081) = 350$, which gives $k = 2.98$. Thus,

$$P(Y \geq 350) \leq \frac{1}{(2.98)^2} = .1126.$$

No, this is not highly unlikely.

3.127 This exercise asks for the probability of accepting a lot of items when the following sampling plan is used: draw a sample of five items and accept the lot if no defectives are observed. Thus,

$$P(\text{acceptance}) = P(\text{observe no defectives}) = \binom{5}{0} p^0 q^5$$

where p is the probability of observing a defective. By substituting the five specific values for p in the above formula, the various probabilities of acceptance are obtained.

$p = $ Fraction Defective	P(Acceptance)
.0	1.0000
.1	.5905
.3	.1681
.5	.0312
1.0	0.0000

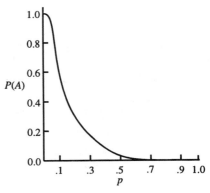

Figure 3.6

Notice that when the fraction defective is 0 (that is, there are no defectives in the lot), the lot will always be accepted and P(acceptance) $= 1$. The operating characteristic curve for this plan is shown in Figure 3.6.

3.129 Proceed by using the binomial tables in Appendix III, indexing $n = 5$, $a = 1$ in the first case and $n = 25$, $a = 5$ in the second.

a. If the fraction defective in the lot ranges from $p = 0$ to $p = .10$, the seller would want the probability of accepting in this interval to be as high as possible. Hence he would choose the second plan.

b. If the buyer wishes to be protected against accepting lots with fraction defective greater than .3, he would want the probability of acceptance when p is greater than .3 to be as small as possible. Thus he would also choose the second plan.

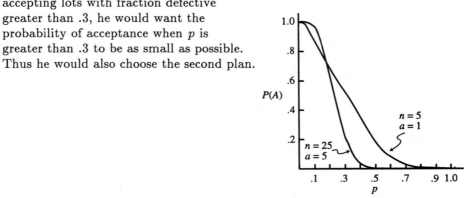

Figure 3.8

3.131 Let Y = the number of rolls the player throws until the player stops. Then Y is a distributed geometric with $p = \frac{5}{6}$.

a. $P(Y = 3) = q^2 p = \left(\frac{1}{6}\right)^2 \left(\frac{5}{6}\right) = .023$

b. $E(Y) = \frac{1}{p} = \frac{6}{5} = 1.2$

c. Let X = amount paid to player. Then

$$X = 2^{Y-1} \qquad \text{and} \qquad E(X) = E\left(2^{Y-1}\right)$$

$$E(X) = \sum_{y=1}^{\infty} 2^{y-1} q^{y-1} p$$

$$= p \sum_{y=0}^{\infty} (2q)^y$$

$$= p \frac{1}{1 - 2q}$$

as the sum of a geometric series, since $2q < 1$

$$= \frac{5}{6}\left(\frac{1}{1 - \frac{2}{6}}\right) = \frac{5}{4}$$

$$= \$1.25.$$

3.133 The random variable is Y, the number of failures in 10,000 starts, and it possesses a binomial distribution with $p = P(\text{failure}) = .00001$. The probability of interest is

$$P(Y \geq 1) = 1 - P(Y = 0) = 1 - \binom{10,000}{0}(.00001)^0(.99999)^{10,000}$$

$$= 1 - (.99999)^{10,000} = .095$$

3.135 The population variance for Exercise 3.77 is calculated as usual and is found to be

$$V(Y) = \Sigma y^2 p(y) - 1^2 = 0\left(\tfrac{1}{5}\right) + 1\left(\tfrac{3}{5}\right) + 4\left(\tfrac{1}{5}\right) - 1 = \tfrac{2}{5}$$

The sample variance is calculated as

$$s^2 = \frac{\sum\limits_{j=1}^{n} y_j^2 - \dfrac{\left(\sum\limits_{j=1}^{n} y_j\right)^2}{n}}{n-1}$$

Refer to Exercise 3.78 and note the number of times that each value of Y occurred. Thus $y = 0$ must be squared 21 times, $Y = 1$ squared 56 times, and so on. The value of s^2 then is

$$s^2 = \frac{[0^2(21) + 1^2(56) + 2^2(23)] - \dfrac{[0(21) + 1(56) + 2(23)]^2}{100}}{99} = \frac{148 - 104.04}{99}$$

$$= \frac{43.96}{99} = .444$$

which is a good estimate for the population variance σ^2. Again, s^2 could be calculated by using the values of f_y, as follows:

$$s^2 = \frac{\sum\limits_{y=0}^{2} y^2 f_y - \dfrac{\left(\sum\limits_{y=0}^{2} y f_y\right)^2}{n}}{n-1}$$

3.137 Let Y_1 = number of defectives out of the 5 from line I,

and Y_2 = number of defectives out of the 5 from line II.

Then $Y_1 \sim$ binomial $(5, p)$ and $Y_2 \sim$ binomial $(5, p)$, where p is the common probability of a defective regulator. Also,

$Y_1 + Y_2 \sim$ binomial $(10, p)$.

$$P(Y_1 = 2 | Y_1 + Y_2 = 4) = \frac{P(Y_1 = 2 \cap Y_1 + Y_2 = 4)}{P(Y_1 + Y_2 = 4)} = \frac{P(Y_1 = 2)P(Y_2 = 2)}{P(Y_1 + Y_2 = 4)}$$

$$= \frac{\binom{5}{2}p^2 q^3 \binom{5}{2}p^2 q^3}{\binom{10}{4}p^4 q^6} = \frac{\binom{5}{2}\binom{5}{2}}{\binom{10}{4}} = .476$$

3.139 **a.** Define X to be the number of imperfections in one square yard of weave. Then

P(one-square-yard sample will contain at least one imperfection)

$$= P(X \geq 1) = 1 - P(X = 0) = 1 - \frac{4^0}{0!}e^{-4} = 1 - .018 = .982$$

b. Now let X_1, X_2, and X_3 be the number of imperfections in the first, second, and third one square yards of weave, respectively. Then X_1, X_2, and X_3 are independent, each having a Poisson distribution with $\lambda = 4$. So

P(three-square-yard sample will contain at least one imperfection)

$$= P(X_1 + X_2 + X_3 \geq 1)$$
$$= 1 - P(X_1 + X_2 + X_3 = 0)$$
$$= 1 - P(X_1 = 0)P(X_2 = 0)P(X_3 = 0)$$
$$= 1 - (.018)^3$$

3.141 **a.** Let X be the number of bacteria colonies in a one-cubic-centimeter sample. Then X has a Poisson distribution with $\lambda = 2$. Now let Y be the number of the four one-cubic-centimeter samples that have one or more bacterial colonies. Note that Y has a binomial distribution with $n = 4$ and

$p = P$(a sample contains one or more bacteria colonies)

$$= P(X \geq 1) = 1 - P(X = 0) = 1 - .135 = .865$$

Thus,

P(at least one sample will contain one or more bacteria colonies)

$$= P(Y \geq 1) = 1 - P(Y = 0) = 1 - \binom{4}{0}(.865)^0(.135)^4 = .9997$$

b. We need to find the number of samples, n, such that approximately

$$P(Y \geq 1) = 1 - P(Y = 0) = 1 - \binom{n}{0}(.865)^0(.135)^n = .95$$

or $(.135)^n = .05$

which implies that $\ln(.135)^n = n \ln(.135) = \ln(.05)$

so $n = \dfrac{\ln(.05)}{\ln(.135)} = 1.496$

So, being conservative, we take $n = 2$.

3.143 Let Y be the number of defective machines. Then Y follows the hypergeometric distribution with

$$P(y) = \frac{\binom{4}{y}\binom{6}{5-y}}{\binom{10}{5}}$$

The mean of the repair cost is

$$E(50Y) = 50E(Y) = 50n\left(\frac{K}{N}\right) = 50(5)\left(\frac{4}{10}\right) = 100$$

and the variance of the repair cost is

$$V(50Y) = 2500V(Y) = 2500n\left(\frac{K}{N}\right)\left(1 - \frac{K}{N}\right)\left(\frac{N-n}{N-1}\right)$$

$$= 2500(5)\left(\frac{4}{10}\right)\left(1 - \frac{4}{10}\right)\left(\frac{10-5}{10-1}\right) = 1666.67$$

According to Tchebysheff's theorem, with probability of at least .75, $50Y$ will be within two standard deviations of the mean. Using this, one can construct the interval

$$\mu \pm 2\sigma = 100 \pm 2\sqrt{1666.67}, \text{ or } 18.35 \text{ to } 181.65$$

3.145 Note that $Y(t)$ has a negative binomial distribution with parameter $r = k$, $p = e^{-\lambda t}$.

a. $E[Y(t)] = \frac{r}{p} = ke^{\lambda t}$, $\qquad V[Y(t)] = \frac{rq}{p^2} = \frac{k\left(1 - e^{-\lambda t}\right)}{e^{-2\lambda t}} = k\left(e^{2\lambda t} - e^{\lambda t}\right)$

b. We are given that $k = 2$, $\lambda = 0.1$, and $t = 5$. Therefore, $E[Y(t)] = ke^{\lambda t}$

$$= 2e^{(5)(0.1)} = 2e^{1/2} = 3.2974, \; V[Y(t)] = 2\left[e - e^{.5}\right] = 2.139$$

3.147 P(toss a die ten times before observing four 6's and a 6 occurs on the ninth and tenth tosses)

$$= P(\text{that two 6's occur in the first eight tosses of the die}) \times \left(\frac{1}{6}\right)^2$$

$$= \left[P(Y = 2) \text{ where } Y \text{ is binomial with } n = 8 \text{ and } p = \frac{1}{6}\right] \times \left(\frac{1}{6}\right)^2.$$

Now, $P(Y = 2) = \binom{8}{2}\left(\frac{1}{6}\right)^2\left(\frac{5}{6}\right)^6 = .26$.

Our desired probability is given by $.26 \times \left(\frac{1}{6}\right)^2 = .00722$.

3.149 Use the Poisson distribution with $\lambda = 5$.

a. $p(2) = \frac{5^2 e^{-5}}{2!} = 12.5e^{-5} = .084$

$$P(Y \le 2) = p(0) + p(1) + p(2) = \frac{5^0 e^{-5}}{0!} + \frac{5^1 e^{-5}}{1!} + \frac{5^2 e^{-5}}{2!} = 18.5e^{-5} = .125$$

b. $P(Y > 10) = 1 - P(Y \le 10) = 1 - .986 = .014$ (using Table 3, Appendix III). Yes, it is unusual that Y will exceed 10.

3.151 We are interested in Y, the number of contacts necessary to obtain the third sale, which is an example of the negative binomial random variable described in Section 3.6. In this case, $r = 3$, $p = .3$, and we have

$$P(Y < 5) = p(3) + p(4) = \binom{2}{2}(.3)^3(.7)^{3-3} + \binom{3}{2}(.3)^3(.7)^1 = (.3)^3 + 3(.3)^3(.7)$$
$$= .0837$$

3.153 There are three possible schemes and with each scheme we can create a probability distribution for $X =$ net profit. Then the merchant's expected net profit will be $E(X)$.

(1) Suppose first of all that she stocks 2 items. Since she knows that she will have a demand for at least 2 items, she will sell both with probability 1. That is, her profit will inevitably be $.40. The profit table and calculation of $E(X)$ are trivial.

x	$.40
$p(x)$	1

$$E(X) = \sum_x xp(x) = .40(1) = \$.40$$

(2) Suppose that the merchant stocks 3 items. She has spent $3 and will realize either $3.60 or $2.40 depending on the number of sales. The profit table with associated probabilities is shown below. Notice that she will sell 3 items ($X = .60$) if either 3 or 4 items are demanded. Hence $P(X = .60) = .4 + .5 = .9$.

x	$p(x)$
.60	.9
−.60	.1

$$E(X) = \sum_x xp(x) = .60(.9) + (-.60)(.1) = .54 - .06 = \$.48.$$

(3) Finally, if she stocks 4 items, she has spent $4.00 and will realize either $2.40, $3.60, or $4.80, depending on demand. The net profits with their associated probabilities are as follows:

x	$p(x)$
.80	.5
−.40	.4
−1.60	.1

$$E(X) = \sum_x xp(x) = .40 - .16 - .16 = \$.08.$$

In order to maximize her expected net profit, she should stock 3 items.

3.155 a. The probability of interest is

$$p(10) = \frac{\binom{40}{10}\binom{60}{10}}{\binom{100}{20}}$$

$$= \frac{40(39)(38)(37)(36)(35)(34)(33)(32)(31)(60)(59)(58)(57)(56)(55)(54)(53)(52)}{10(9)(8)(7)(6)(5)(4)(3)(2)(1)(100)(99)(98)}$$

$$\times \frac{(51)(20)(19)(18)(17)(16)(15)(14)(13)(12)(11)(10)(9)(8)(7)(6)(5)(4)(3)(2)}{(97)(96)(95)(94)(93)(92)(91)(90)(89)(88)(87)(86)(85)(84)(83)(82)(81)}$$

The student can simplify this expression first by direct cancellation and then by using an electronic calculator. However, care must be taken to perform operations in such a way that the capacity of the machine is not exceeded. The resulting numerical value will be $p(10) = .119$.

b. Using the binomial approximation with $p = \frac{r}{N} = \frac{40}{100} = .4$ and $n = 20$ yields

$$p(10) = \binom{20}{10}(.4)^{10}(.6)^{10} = P(Y \le 10) - P(Y \le 9) = .872 - .755 = .117$$

This result was obtained by using the binomial tables in the back of the text. Note the accuracy of the approximation and also the comparative ease of calculation.

3.157 a. If method (1) is used, N tests are required, regardless of the number of people having the disease. However, for method (2) there are two possible values for n', the number of tests required for a group of k people. If all k people are healthy, then only one test is required. The probability that $n' = 1$ is

$$P(n' = 1) = P(k \text{ people healthy}) = (.95)^k$$

If at least one of the k people has the disease, then the test is positive and k more tests (making a total of $k + 1$) are required. Note that

$$P(n' = k + 1) = P(\text{at least one diseased person}) = 1 - P(\text{all healthy})$$
$$= 1 - (.95)^k$$

Hence

$$E(n') = \sum_{n'} n' p(n') = 1(.95)^k + (k + 1)(1 - .95^k)$$
$$= .95^k + k + 1 - k(.95)^k - .95^k$$
$$= 1 + k(1 - .95^k)$$

This expectation holds for each group, so that for n groups the expected number of tests is $n[1 + k(1 - .95^k)]$.

b.   It is necessary to choose k so that the expected number of tests is minimized. Write

$$g(k) = \frac{N}{k}\left[1 + k\left(1 - .95^k\right)\right] = \frac{N}{k} + N\left(1 - .95k^k\right)$$

where $n = \frac{N}{k}$. This quantity must be minimized. Differentiating with respect to k and setting $\frac{d[g(k)]}{dk} = 0$, we obtain

$$h(k) = \frac{d[g(k)]}{dk} = \frac{-N}{k^2} + N\left(-.95^k\right)\ln(.95) = -\frac{N}{k^2} - N\left(.95^k\right)\ln(.95) = 0$$

which implies

$$h(k) = \frac{1}{k^2} + \left(.95^k\right)\ln(.95) = 0$$

This function $h(k)$ is a strictly decreasing function of k. Hence, for various integer values of k, we can find two values of k for which the function $h(k)$ is closest to zero. These are shown in the accompanying table.

k	$h(k)$
2	.2037
3	.0671
4	.0207
5	.00031
6	−.0097

The minimum value is between $k = 5$ and $k = 6$. However, since k can be only integer-valued, we could evaluate the function $g(k)$ at $k = 5$ and $k = 6$ to determine which value of $g(k)$ is smaller. Evaluating, we have

$$g(5) = \frac{N}{5}\left[1 + 5\left(1 - .95^5\right)\right] = N(.4262)$$

$$g(6) = \frac{N}{6}\left[1 + 6\left(1 - .95^6\right)\right] = N(.4316)$$

so that $g(k)$ will be minimized using $k = 5$.

c.   The expected number of tests is $.4262N$, compared to N tests if method (1) is used. The number of tests saved is then $N - .4262N = .5738N$.

3.159 $E(Y) = \sum_{y=0}^{n} \frac{y\binom{r}{y}\binom{N-r}{n-y}}{\binom{N}{n}} = r\sum_{y=1}^{n}\left[\frac{(r-1)!}{(y-1)!\,(r-y)!}\right]\left[\frac{\binom{N-r}{n-y}}{\binom{N}{n}}\right]$

$$= r\sum_{y=1}^{n} \frac{\binom{r-1}{y-1}\binom{N-r}{n-y}}{\binom{N}{n}} \qquad \text{let } x = y - 1$$

$$= r \sum_{x=0}^{n-1} \frac{\binom{r-1}{x}\binom{N-r}{n-1-x}}{\binom{N}{n}}$$

$$= r \sum_{x=0}^{n-1} \frac{\binom{r-1}{x}\binom{N-r}{n-1-x}}{\binom{N}{n}\binom{N-1}{n-1}}$$

$$= \left(\frac{rn}{N}\right) \sum_{x=0}^{n-1} \frac{\binom{r-1}{x}\binom{N-r}{n-1-x}}{\binom{N-1}{n-1}}$$

$$= \frac{nr}{N}, \text{ since the summation sums to 1.}$$

CHAPTER 4 CONTINUOUS RANDOM VARIABLES AND THEIR PROBABILITY DISTRIBUTIONS

4.1 By definition, $F(y) = P(Y \leq y)$ $\qquad y = 1, 2, 3, \ldots$

Then

$$P(Y = y) = P(Y \leq y) - P(Y \leq y - 1)$$
$$= F(y) - F(y - 1) \qquad y = 2, 3, \ldots .$$

Also, $P(Y = 1) = P(Y \leq 1)$
$$= F(1).$$

4.3 **a.** We must find the value of c such that

$$F(\infty) = \int_{-\infty}^{\infty} f(y)\, dy = 1$$

That is,

$$F(\infty) = \int_{0}^{2} cy\, dy = c \left[\frac{y^2}{2} \right]_{0}^{2} = c \left(\frac{4}{2} \right) = 1$$

Hence $c = \frac{2}{4} = \frac{1}{2}$. The density function for Y is

$$f(y) = \begin{cases} \dfrac{y}{2}, & 0 \leq y \leq 2 \\ 0, & \text{elsewhere} \end{cases}$$

b. $F(y) = \displaystyle\int_{-\infty}^{y} f(t)\, dt = \int_{0}^{y} \frac{t}{2}\, dt = \left. \frac{t^2}{4} \right]_{0}^{y} = \frac{y^2}{4} \qquad$ for $0 \leq y \leq 2$

Note that $F(y) = 0$ for $y < 0$ and $F(y) = 1$ for $y > 2$.

c. The graphs of $f(y)$ and $F(y)$ are shown in Figures 4.1 and 4.2.

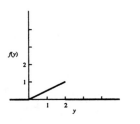

Figure 4.1   Figure 4.2

d. From part (b) we have

$$P(1 \leq Y \leq 2) = P(Y \leq 2) - P(Y \leq 1) + P(Y = 1) = F(2) - F(1) + 0$$

$$= \frac{2^2}{4} - \frac{1^2}{4} = \frac{1}{2}$$

e. Refer to Figure 4.1. It is necessary to calculate the area under the density function $f(y)$ from 1 to 2. Note that the total area under $f(y)$ is 1 and that the area from 0 to 1 is the area of a triangle with base 1 and height $\frac{1}{2}$. Hence the area from 1 to 2 is

$$1 = \left(\frac{1}{2}\right)(1)\left(\frac{1}{2}\right) = 1 - \frac{1}{4} = \frac{3}{4}$$

4.5 **a.** $F(y) = \displaystyle\int_{-\infty}^{y} f(t) \, dt.$ For $y < 0$, $F(y) = P(Y \leq y) = 0$

For $0 \leq y \leq 1$, $F(y) = \displaystyle\int_{0}^{y} t \, dt = \frac{y^2}{2}$

For $1 \leq y \leq 1.5$, $F(y) = P(Y \leq y) = P(0 \leq Y \leq 1) + P(1 < Y \leq y)$

$$= \frac{1}{2} + \int_{1}^{y} dt = \frac{1}{2} + (y - 1) = y - \frac{1}{2}$$

For $y > 1.5$, $F(y) = P(Y \leq y) = P(Y \leq 1.5) + P(1.5 < Y \leq y) = F(1.5) + 0 = 1$
Hence

$$F(y) = \begin{cases} 0, & y < 0 \\ \frac{y^2}{2}, & 0 \le y \le 1 \\ y - \left(\frac{1}{2}\right), & 1 \le y \le 1.5 \\ 1, & 1.5 < y \end{cases}$$

b. $P(0 \le Y \le .5) = F(.5) = \frac{1}{8} = .125$

c. $P(.5 \le Y \le 1.2) = F(1.2) - F(.5) = .7 - .125 = .575$

4.7   **a.**   For $b \ge 0$ and for any value of y, $f(y) \ge 0$. Moreover,

$$\int_{-\infty}^{\infty} f(y)\, dy = \int_{b}^{\infty} \frac{b}{y^2}\, dy = -\frac{b}{y}\Big]_{b}^{\infty} = \frac{b}{b} = 1$$

Since b is defined as a traversal time, it will always be positive. Thus $f(y) \ge 0$ and $f(y)$ has the properties of a density function.

b. $F(y) = \int_{b}^{y} \frac{b}{t^2}\, dt = -\frac{b}{t}\Big]_{b}^{y} = 1 - \frac{b}{y}$ for $y \ge b$; $F(y) = 0$ for $y < 0$.

c. $P(Y > b + c) = 1 - F(b + c) = 1 - \left(1 - \frac{b}{b+c}\right) = \frac{b}{b+c}$

4.9   **a.** $F(\infty) = \int_{-\infty}^{\infty} f(y)\, dy = \int_{0}^{1} (cy^2 + y)\, dy = c\left[\frac{y^3}{3}\right]_{0}^{1} + \left[\frac{y^2}{2}\right]_{0}^{1} = \frac{c}{3} + \frac{1}{2} = 1$

Hence $\frac{c}{3} = \frac{1}{2}$ and $c = \frac{3}{2}$.

b. $F(y) = \int_{-\infty}^{y} f(t)\, dt = \int_{0}^{y} \left(\frac{3}{2}t^2 + t\right) dt = \frac{t^2}{2}\Big]_{0}^{y} + \frac{t^2}{2}\Big]_{0}^{y} = \frac{y^3}{2} + \frac{y^2}{2}$ for $0 \le y \le 1$

and $F(y) = 0$ for $y < 0$, $F(y) = 1$ for $y > 1$.

c. The graphs of $F(y)$ and $f(y)$ are shown in Figures 4.6 and 4.7.

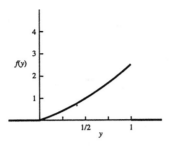

Figure 4.6   Figure 4.7

d. $F(-1) = 0$ since $y < 0$; $F(0) = 0$; $F(1) = \frac{1}{2} + \frac{1}{2} = 1$

e. $P(0 \le Y \le .5) = F(.5) - F(0) = \frac{(.5)^3}{2} + \frac{(.5)^2}{2} - 0 = \frac{1}{16} + \frac{1}{8} = \frac{3}{16}$

f. $P\left(Y > \frac{1}{2}\middle| Y > \frac{1}{4}\right) = \frac{P\left(Y > \frac{1}{2}\right)}{P\left(Y > \frac{1}{4}\right)} = \frac{1 - \left(\frac{3}{16}\right)}{1 - \left(\frac{1}{128} + \frac{1}{32}\right)} = \frac{\frac{13}{16}}{\frac{123}{128}} = \frac{104}{123}$

4.11 a. Differentiating $F(y)$ with respect to y, we have

$$f(y) = \begin{cases} 0, & y \le 0 \\[4pt] \frac{1}{8}, & 0 < y < 1 \\[4pt] \frac{y}{8}, & 2 \le y < 4 \\[4pt] 0, & y \ge 4 \end{cases}$$

Notice that $F(y)$ is not differentiable at $y = 0$, 2, and 4.

b. $P(1 \le Y \le 3) = F(3) = F(1) = \frac{9}{16} - \frac{2}{16} = \frac{7}{16}$

c. $P(Y \ge 1.5) = 1 - F(1.5) = 1 - \frac{1.5}{8} = \frac{13}{16}$

d. $P(Y \ge 1 | Y \le 3) = \frac{P(1 \le Y \le 3)}{P(Y \le 3)} = \frac{\frac{7}{16}}{\frac{9}{16}} = \frac{7}{9}$

4.13 Refer to Exercise 4.9.

$$E(Y) = \int_0^1 \left(\tfrac{3}{2}y^3 + y^2\right) dy = \left[\tfrac{3}{8}y^4 + \tfrac{y^3}{3}\right]_0^1 = \tfrac{3}{8} + \tfrac{1}{3} = \tfrac{9+8}{24} = \tfrac{17}{24} = .708$$

$$E(Y^2) = \int_0^1 \left(\tfrac{3}{2}y^4 + y^3\right) dy = \left[\tfrac{3}{10}y^5 + \tfrac{1}{4}y^4\right]_0^1 = \tfrac{3}{10} + \tfrac{1}{4} = .55$$

so that $V(Y) = E\left(Y^2\right) - \mu^2 = .55 - (.708)^2 = .0487.$

4.15 1. $E(c) = \displaystyle\int_{-\infty}^{\infty} cf(y)\,dy = c \int_{-\infty}^{\infty} f(y)\,dy = c$ since $\displaystyle\int_{-\infty}^{\infty} f(y)\,dy = 1.$

2. $E[cg(y)] = \displaystyle\int_{-\infty}^{\infty} cg(y)f(y)\,dy = c \int_{-\infty}^{\infty} g(y)f(y)\,dy = cE[g(y)].$

3. $E[g_1(y) + g_2(y) + \ldots + g_k(y)]$

$$= \int_{-\infty}^{\infty} [g_1(y) + g_2(y) + \ldots + g_k(y)]f(y)\,dy$$

$$= \int_{-\infty}^{\infty} g_1(y)f(y)\,dy + \int_{-\infty}^{\infty} g_2(y)f(y)\,dy + \ldots + \int_{-\infty}^{\infty} g_k(y)f(y)\,dy$$

$$= E[g_1(y)] + E[g_2(y)] + \ldots + E[g_k(y)].$$

4.17 First calculate

$$E(Y) = \int_0^1 \left(\tfrac{3}{2}y^3 + y^2\right) dy = \left[\tfrac{3}{8}y^4 + \tfrac{y^3}{3}\right]_0^1 = \tfrac{3}{8} + \tfrac{1}{3} = \tfrac{17}{24} = .708$$

Then $E(Y) = 5 - .5E(Y) = 5 - .5\left(\tfrac{17}{24}\right) = \$4.65.$ Now

$$E(Y^2) = \int_0^1 \left(\tfrac{3}{2}y^4 + y^3\right) dy = \left[\tfrac{3}{10}y^5 + \tfrac{y^4}{4}\right]_0^1 = \tfrac{3}{10} + \tfrac{1}{4} = .55$$

Thus, $V(Y) = V(5 - .5Y) = (.5)^2 V(Y) = (.5)^2\left[E\left(Y^2\right) - [E(Y)]^2\right] = (.5)^2[.55 - (.708)^2]$

$= .012$

4.19 $E(Y) = \int\limits_{59}^{61} y\left(\frac{1}{2}\right) dy = \left(\frac{1}{4}\right) y^2 \Big]_{59}^{61} = 60$

$V(Y) = E(Y^2) - [E(Y)]^2 = \left[\int\limits_{59}^{61} y^2 \left(\frac{1}{2}\right) dy \right] - (60)^2 = \left(\frac{1}{6}\right) y^3 \Big]_{59}^{61} - 3600 = \frac{1}{3}$

4.21 $E(Y) = \int\limits_{2}^{6} y\left(\frac{3}{32}\right)(y-2)(6-y)\ dy = \left(\frac{3}{32}\right)\left[\left(-\frac{y^4}{4}\right) + \left(\frac{8y^3}{3}\right) - 6y^2\right]_{2}^{6} = \left(\frac{3}{32}\right)(42.67) = 4$

a. $E(Y) = \left(\frac{3}{64}\right) \int\limits_{0}^{4} y^3(4-y)\ dy = \left(\frac{3}{64}\right)\left[y^4 - \left(\frac{y^5}{5}\right)\right]_{0}^{4} = \left(\frac{3}{64}\right)(256 - 204.8) = 2.4$

$V(Y) = E(Y^2) - [E(Y)]^2 - \left[\left(\frac{3}{64}\right) \int\limits_{0}^{4} y^4(4-y)\ dy\right] - (2.4)^2$

$= \left(\frac{3}{64}\right)\left[\left(\frac{4}{5}\right)y^5 - \left(\frac{y^6}{6}\right)\right]_{0}^{4} - 5.76 = \left(\frac{3}{64}\right)(819.2 - 682.66) - 5.76 = .64$

b. Let Y = weekly cost for CPU time = $200Y$

$E(Y) = E(200Y) = 200E(Y) = 200(2.4) = 480$

$V(Y) = V(200Y) = (200)^2 V(Y) = 40{,}000(0.64) = 25{,}600$

c. $P(Y > 600) = P(Y > 3) = \left(\frac{3}{64}\right) \int\limits_{0}^{4} y^2(4-y)\ dy = \left(\frac{3}{64}\right)\left[\left(\frac{4}{3}\right)y^3 - \left(\frac{1}{4}\right)y^4\right]_{0}^{4}$

$= \left(\frac{3}{64}\right)(5.58) = .26$

We expect the weekly cost to exceed \$600 about 26% of the time.

4.23 a. $E(Y) = \left(\frac{3}{8}\right) \int\limits_{5}^{7} y(7-y)^2\ dy = \left(\frac{3}{8}\right)\left[\left(\frac{49}{2}\right)y^2 - \left(\frac{14}{3}\right)y^3 + \frac{y^4}{4}\right]_{5}^{7} = \left(\frac{3}{8}\right)(14.66) = 5.5$

$V(Y) = E(Y^2) - [E(Y)]^2 = \left[\left(\frac{3}{8}\right) \int\limits_{5}^{7} y^2(7-y)^2\ dy\right] - (5.5)^2$

$= \left(\frac{3}{8}\right)\left[\left(\frac{49}{3}\right)y^3 - \left(\frac{14}{4}\right)y^4 + \left(\frac{1}{5}\right)y^5\right]_{5}^{7} - 30.25 = \left(\frac{3}{8}\right)(81.06) - 30.25 = .1475$

b. Using Tchebysheff's inequality with $k = 2$, we evaluate

$$\left(\mu_x \pm 2\sqrt{V(Y)}\right) = \left(5.5 \pm 2\sqrt{.1475}\right) = (5.5 \pm .768) = (4.732, 6.268) = (5, 6.268)$$

c. $P(Y < 5.5) = \left(\frac{3}{8}\right) \int_{5}^{5.5} (7 - y)^2 \, dy = \left(\frac{3}{8}\right)\left[49y - 7y^2 + \frac{y^3}{3}\right]_{5}^{5.5} = \left(\frac{3}{8}\right)(1.54) = .5775$

Yes, we would expect to see a pH measurement below 5.5 about 58% of the time.

4.25 Since the parachutist is landing at a random point in the interval (A, B), the point of landing is a continuous random variable Y, with a uniform distribution over (A, B). Hence

$$f(y) = \frac{1}{B - A} \qquad A \le y \le B$$

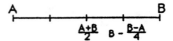

Figure 4.10

a. Refer to Figure 4.10. If he lands closer to A than to B, he has landed in the interval $\left(A, \frac{A + B}{2}\right)$. The probability is

$$\int_{A}^{(B+A)/2} \frac{1}{B - A} \, dy = \frac{\left(\frac{A + B}{2}\right) - A}{B - A} = \frac{1}{2}$$

b. The point at which the distance to A is exactly 3 times the distance to B is the point $B - \left(\frac{1}{4}\right)(B - A) = \frac{3B + A}{4}$. Then

$$P(\text{distance to } A \text{ is more than 3 times distance to } B) = P\left(\frac{3B + A}{4} \le Y \le B\right)$$

$$= \frac{B - \left(\frac{3B + A}{4}\right)}{B - A} = \frac{\left(\frac{B - A}{4}\right)}{B - A} = \frac{1}{4}$$

4.27 $\sigma^2 = E\left(Y^2\right) - \mu^2 = \left[\int_{\theta_1}^{\theta_2} y^2 \left(\frac{1}{\theta_2 - \theta_1} \, dy\right)\right] - \left(\frac{\theta_2 - \theta_1}{2}\right)^2$

$$= \left[\frac{1}{3(\theta_2 - \theta_1)}\right] y^3 \Big|_{\theta_1}^{\theta_2} - \left(\frac{1}{4}\right)(\theta_2 - \theta_1)^2$$

$$= \left[\frac{1}{3(\theta_2 - \theta_1)}\right](\theta_2 - \theta_1)\left(\theta_2^2 + \theta_1\theta_2 + \theta_1^2\right) - \left(\frac{1}{4}\right)\left(\theta_2^2 + 2\theta_1\theta_2 + \theta_1^2\right)$$

$$= \left(\tfrac{4}{12}\right)\left(\theta_2^2 + \theta_1\theta_2 + \theta_1^2\right) - \left(\tfrac{3}{12}\right)\left(\theta_2^2 + 2\theta_1\theta_2 + \theta_1^2\right)$$

$$= \frac{\theta_2^2 - 2\theta_1\theta_2 + \theta_1^2}{12} = \frac{\left(\theta_2 - \theta_1\right)^2}{12}$$

4.29 Let $Y =$ cycle time. Then

$$f(y) = \frac{1}{70 - 50} = \frac{1}{20} \text{ for } 50 \le y \le 70$$

and

$$F(y) = \int_{50}^{y} \frac{1}{20}\, dt = \frac{y - 50}{20} \text{ for } 50 \le y \le 70;\ F(y) = 0 \text{ for } y < 50;$$

$$F(y) = 1 \text{ for } y > 70.$$

Thus

$$P(Y > 65 | Y > 55) = \frac{P(Y > 65)}{P(Y > 55)} = \frac{1 - \left(\frac{65 - 50}{20}\right)}{1 - \left(\frac{55 - 50}{20}\right)} = \frac{20 - 15}{20 - 5} = \frac{1}{3}$$

4.31 Let $Y =$ time the defective board is detected.

a. $P(0 < Y < 1) = \int_0^1 \left(\tfrac{1}{8}\right) dx = \tfrac{1}{8}$

b. $P(7 < y < 8) = \int_7^8 \left(\tfrac{1}{8}\right) dx = \tfrac{1}{8}$

c. $P(4 < Y < 5 | Y > 4) = \dfrac{\int_4^5 \left(\tfrac{1}{8}\right) dx}{\int_4^8 \left(\tfrac{1}{8}\right) dx} = \dfrac{\left(\tfrac{1}{8}\right)}{\left(\tfrac{1}{2}\right)} = \dfrac{1}{4}$

4.33 Let $r =$ radius and $d = 2r$ be the diameter of sphere. d is $U(.01, .05)$.

$$V = \left(\tfrac{4}{3}\right)\pi r^3 = \left(\tfrac{4}{3}\right)\pi \left(\tfrac{d}{2}\right)^3 = \left(\tfrac{\pi}{6}\right) d^3$$

$$\mu = E(Y) = E\left[\left(\tfrac{\pi}{6}\right) d^3\right] = \left(\tfrac{\pi}{6}\right) \int_{.01}^{.01} \left(\tfrac{d^3}{.04}\right) dd = \left(\tfrac{\pi}{6}\right)\left(\tfrac{d^4}{.16}\right)\Big]_{.01}^{.05}$$

$$= \left(\frac{25\pi}{24}\right)\left[(.05)^4 - (.01)^4\right] = \left(6.5 \times 10^{-6}\right)\pi$$

$$V(Y) = V\left[\left(\frac{4\pi}{3}\right)\left(\frac{d}{2}\right)^3\right] = \left(\frac{\pi^2}{36}\right)V(d^3) = \left(\frac{\pi^2}{36}\right)\left[E(d^3)^2 - \left(E(d^3)\right)^2\right]$$

$$= \left(\frac{\pi^2}{36}\right)\left[E(d^6) - \left(E(d^3)\right)^2\right]$$

$$= \left(\frac{\pi^2}{36}\right)\left[\int_{.01}^{.05} \left(\frac{d^6}{.04}\right)dd - \left(E(d^3)\right)^2\right]$$

$$= \left(\frac{\pi^2}{36}\right)\left\{\left[\frac{d^7}{.28}\right]_{.01}^{.05} - \left[\frac{(.05)^4 - (.01)^4}{.16}\right]^2\right\}$$

$$= \left(\frac{\pi^2}{36}\right)\left[2.780 \times 10^{-9}\right]$$

$$= \left(\frac{\pi^2}{36}\right)\left[2.790 \times 10^{-9} - 1.521 \times 10^{-9}\right] = \left[3.525 \times 10^{-4}\right]\pi^2$$

4.35 **a.** The procedure is reversed now, because the area under the curve is known. The objective is to determine the particular values, z_0, that will yield the given probability. In this exercise it is necessary to find a z_0 such that $P(Z > z_0) =$ .5000. By the symmetry of the normal distribution, half of the area falls on each side of the mean. Thus, $P(Z > 0) = .5000$ and the desired value of z_0 is 0.

b. A value of z_0 is desired such that $P(Z < z_0)$

= .8643. Thus the probability, .8643, will

be the entire area under the curve to the left

of the value $z = z_0$. Notice that the probability

is greater than .5, so that z_0 must be in the

right-hand half of the curve (i.e., $z_0 > 0$).

See Figure 4.15. Then

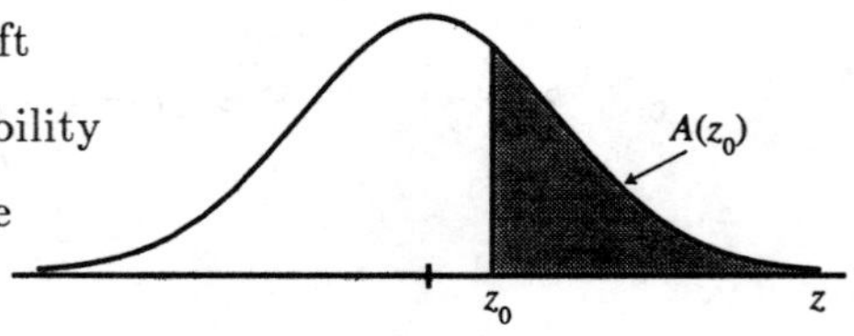

$$P(Z < z_0) = 1 - A(z_0) = .8643$$

$$A(z_0) = .1357$$

From Table 4 we get $z_0 = 1.10$. If the exact probability cannot be found in the

table, we may choose to search for the probability closest to the one desired and

perform an interpolation that will determine the exact value of z_0.

Figure 4.15

c.   It is given that $P(-z_0 < Z < z_0) = .9000$.

See Figure 4.16. That is,

$$A_1 + A_2 = .9000 = 1 - 2A(z_0)$$

$$2A(z_0) = .1$$

$$A(z_0) = .05$$

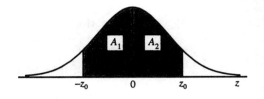

Figure 4.16

The desired value is not tabulated in Table 4 but falls between two tabulated values, .0505 and .0495. Hence z_0 will lie between 1.64 and 1.65, which are the z values associated with the above probabilities. Intuitively, one can see that the value .05 is halfway between the two tabulated values, and thus the desired value of z_0 will be halfway between 1.64 and 1.65, or $z_0 = 1.645$. This method of evaluation is called "linear interpolation."

d.   Similar to part (c). In this case,

$$P(-z_0 < Z < z_0) = 1 \ - 2A(z_0) = .9900$$

so that $A(z_0) = .0050$. Linear interpolation may now be used to determine the value of z_0, which will be between $z_1 = 2.57$ and $z_2 = 2.58$. Hence

$$z = 2.57 + \frac{.0050 - .0049}{.0051 - .0049}(2.58 - 2.57) = 2.57 + \frac{.0001}{.0002}(.01) = 2.57 + .005 = 2.575$$

4.37   Refer to Figure 4.17. It is necessary to find a value for y such that $P(Y > y_0) = .1$. In terms of the corresponding z value,

$$z = \frac{y_0 - \mu}{\sigma} \frac{y_0 - 400}{20}$$

It is necessary to have

$$P(Z > z) = .10 \qquad \text{or} \qquad A(z) = .1$$

Using Table 4, the necessary value for z is $z = 1.28$ and

Figure 4.17

$$z = \frac{y_0 - 400}{20} = 1.28$$

Solving for y_0, $y_0 = 400 + 1.28(20) = \$425.60$, or $\$426$.

4.39   By inspection of the normal curve tail areas, one can see that if one marks off a fixed distance (in this case, $2(.001)$ $= .004$) along the y axis and considers the fraction outside this interval, the minimum area is obtained if the interval is centered about the mean. See Figure 4.18. Hence if this normal distribution

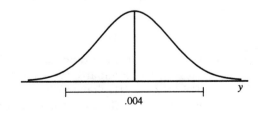

Figure 4.18

is centered on the midpoint of the interval $3.000 \pm .002$, the fraction scrapped will be minimum. The mean diameter should be $\mu = 3$.

4.41 The probability of interest is $P(Y < 1.9)$ with corresponding z value

$$z = \frac{y - \mu}{\sigma} = \frac{1.9 - 2.4}{.8} = -.625$$

(Recall that a negative value of z implies a value to the left of the mean.)   Then

$A_2 = P(Y < 1.9) = P(Z < -.625) = P(Z > .625) = A(.625) = .2660$
(after interpolating).

4.43 Let Y be the measured resistance of a randomly selected wire.

a.   The required probability is

$$P(.12 \le Y \le .14) = P\left(\frac{.12 - .13}{.005} \le \frac{Y - \mu}{\sigma} \le \frac{.14 - .13}{.005}\right)$$

$$= P(-2 \le Z \le 1) = 1 - 2P(Z > 2) = 1 - 2A(z)$$
$$= 1 - 2(.0228) = .9544$$

b.   Define X to be the number of wires that meet the specifications. Then X has a Binomial distribution with $n = 4$ and $p = .9544$. Thus,

$$P(\text{all four will meet the specifications}) = P(X = 4) = \binom{4}{4}(.9544)^4(.0456)^0 = .83$$

4.45 a.   The z values corresponding to $y_1 = 947$ and $y_2 = 958$ are

$$z_1 = \frac{947 - 950}{10} = -.3 \qquad \text{and} \qquad z_2 = \frac{958 - 950}{10} = .8$$

Then

$$P(947 \le Y \le 958) = P(-.3 \le Z \le .8) = 1 - P(Z < -.3) - P(Z < .8)$$
$$= 1 - .3821 - .2119 = .4060$$

b.   It is necessary that $P(Y \le C) = .8531$. Apparently, then, C must be to the right of the mean, $\mu = 950$, and must be associated with a z value such that

$$A(z) = A\left(\frac{C - 950}{10}\right) = .1469$$

This value of z is $z = 1.05$ (from Table 4).   Hence

$$\frac{C - 950}{10} = 1.05 \qquad \text{and} \qquad C = 960.5$$

4.47 It is given that the random variable Y (ounces of fill) is normally distributed with mean μ and standard deviation $\sigma = .3$. The objective is to find a value of μ so that $P(Y > 8) = .01$. That is, an 8-ounce cup will overflow when $Y > 8$, and this should happen only 1% of the time. The z value corresponding to $Y = 8$ is

$$z = \frac{y - \mu}{\sigma} = \frac{8 - \mu}{.3}$$

Thus,

$$P(Y > 8) = P\left(Z > \frac{8 - \mu}{.3}\right) = .01 \qquad \text{or} \qquad A\left(\frac{8 - \mu}{.3}\right) = .01$$

Consider $z_0 = \frac{8 - \mu}{.3}$ and determine the value of z_0 that satisfies the equality shown above. This value is 2.33. Hence the value for μ can be obtained as

$$\frac{8 - \mu}{.3} = 2.33 \qquad \text{or} \qquad \mu = 7.301$$

4.49 $f'(y) = \left(\frac{-y + \mu}{\sigma^3\sqrt{2\pi}}\right)e^{-(y-\mu)^2/2\sigma^2} = \left(\frac{1}{\sigma^3\sqrt{2\pi}}\right)\left[-ye^{-(y-\mu)^2/2\sigma^2} + \mu e^{-(y-\mu)^2/2\sigma^2}\right]$

$$f''(y) = \left(\frac{1}{\sigma^3\sqrt{2\pi}}\right)\left\{-e^{-(y-\mu)^2/2\sigma^2} - ye^{-(y-\mu)^2/2\sigma^2}\left[\left(\frac{1}{2\sigma^2}\right)(-2y + 2\mu)\right]\right.$$

$$\left. + \mu e^{-(y-\mu)^2/2\sigma^2}\left[\left(\frac{1}{2\sigma^2}\right)(-2y + 2\mu)\right]\right\}$$

$$= \left(\frac{1}{\sigma^3\sqrt{2\pi}}\right)e^{-(y-\mu)^2/2\sigma^2}\left[-1 + \frac{(-y + \mu)(-y + \mu)}{\sigma^2}\right] = 0$$

We evaluate

$$-1 + \left(\frac{y^2 - 2y\mu + \mu^2}{\sigma^2}\right) = 0$$

$$\frac{y^2 - 2y\mu + \mu^2}{\sigma^2} = 1$$

$$y = \mu - \sigma \qquad \text{or} \qquad y = \mu + \sigma$$

4.51 Let $Y =$ magnitude of earthquake. Y is exponential with $\beta = 2.4$.

a. $P(Y > 3) = \int_3^\infty \left(\frac{1}{2.4}\right)e^{-y/2.4}\, dy = -e^{-3/2.4}\big]_3^\infty = .2865$

b. $P(2 < Y < 3) = \int_2^3 \left(\frac{1}{2.4}\right)e^{-y/2.4}\, dy = -e^{-y/2.4}\big]_2^3 = .1481$

4.53 **a.** Let Y = demand for water. Y is exponential with $\beta = 100$.

$$P(Y > 200) = \int_{200}^{\infty} \left(\tfrac{1}{100}\right) e^{-y/100}\, dy = e^{-y/100}\Big]_{200}^{\infty} = .1353$$

b. Let C = capacity. $P(Y > c) = \int_{c}^{\infty} \left(\tfrac{1}{100}\right) e^{-y/100}\, dy = -e^{-y/100}\Big]_{c}^{\infty}$

$= e^{-c/100} = .01.$ $c = 360.52$ cfs

4.55 **a.** $P(Y \le 31) = \int_{0}^{31} \left(\tfrac{1}{44}\right) e^{-y/44}\, dy = 1 - e^{-31/44} = .5057$

b. $\sigma^2 = \beta^2 = (44)^2 = 1936$

4.57 **a.** For any $k = 1, 2, 3, \ldots$

$$P(X = k) = P(k - 1 \le Y < k)$$
$$= P(Y < k) - P(Y \le k)$$
$$= \left(1 - e^{-k/\beta}\right) - \left(1 - e^{-(k-1)/\beta}\right)$$
$$= e^{-(k-1)/\beta} - e^{-k/\beta}$$

b. For any $k = 1, 2, 3, \ldots$

$$P(X = k) = e^{-(k-1)/\beta} - e^{-k/\beta}$$
$$= e^{-(k-1)/\beta} - e^{-(k-1)/\beta} e^{-1/\beta}$$
$$= e^{-(k-1)/\beta}\left(1 - e^{-1/\beta}\right)$$
$$= \left(e^{-1/\beta}\right)^{k-1}\left(1 - e^{-1/\beta}\right).$$

Thus, X has a geometric distribution with $p = 1 - e^{-1/\beta}$.

4.59 The density function for Y is $f(y) = \left(\tfrac{1}{4}\right) e^{-y/4}$ for $y \ge 0$. Then

$$P(Y > 4) = \int_{4}^{\infty} \tfrac{1}{4} e^{-y/4}\, dy = -e^{-y/4}\Big]_{4}^{\infty} = 0 - \left(e^{-1}\right) = e^{-1}$$

4.61 $P(Y > 1) = \int\limits_{1}^{\infty} \frac{y^{\alpha-1}e^{-y}}{\Gamma(\alpha)} \, dy = \sum\limits_{y=0}^{\alpha-1} \frac{1^{y}e^{-1}}{y!}$

For $\lambda = 1$ and $\alpha = 2$, this becomes $P(Y > 1) = \sum\limits_{y=0}^{1} \frac{e^{-1}}{y!} = e^{-1} + e^{-1} = 2e^{-1} = .7358$

4.63 $\mu = E(Y) = \alpha\beta = (1.6)(2) = 3.2.$

$\sigma_{\mu}^{2} = \alpha\beta^{2} = (1.6)(4) = 6.4.$

4.65 Using Tchebycheff's theorem, with $k = 2$, consider the interval
$(\mu \pm k\sigma) = (4 \pm 2(\sqrt{8})) = (4 \pm 5.6569) = (-1.6569, 9.6569)$. Since the lower bound for y is 0, give as our interval (0, 9.6569).

4.67 From Theorem 4.8, a gamma-type random variable with $\alpha = 3$ and $\beta = 2$ has

$E(Y) = \alpha\beta = 6$ and $E(Y^2) = \alpha(\alpha + 1)\beta^2 = 3(4)(4) = 48.$ Hence

$E(L) = 30E(Y) + 2E(Y^2) = 30(6) + 2(48) = 276$

Since

$V(L) = E(L^2) - [E(L)]^2 = E(L^2) - 76{,}176 = E(900Y^2 + 120Y^3 + 4Y^4) - 76{,}176$

we need the third and fourth moments about the origin.

$E(Y^3) = \int\limits_{0}^{\infty} \frac{y^5 e^{-y/2}}{\Gamma(3)2^3} \, dy = \frac{\Gamma(6)2^3}{\Gamma(3)} \int\limits_{0}^{\infty} \frac{y^5 e^{-y/2}}{\Gamma(6)2^6} \, dy = 5(4)(3)2^3 = 480$

$E(Y^4) = \int\limits_{0}^{\infty} \frac{y^6 e^{-y/2}}{\Gamma(3)2^3} \, dy = \frac{\Gamma(7)2^4}{\Gamma(3)} \int\limits_{0}^{\infty} \frac{y^6 e^{-y/2}}{\Gamma(7)2^7} \, dy = \frac{6! \, 2^4}{2!} = 5760$

Then $V(L) = 900(48) + 120(480) + 4(5760) - 76{,}176 = 47{,}664.$

4.69 a. $E(Y^a) = \int\limits_{0}^{\infty} \frac{y^a y^{\alpha-1}}{\beta^\alpha \Gamma(\alpha)} e^{-y/\beta} \, dy = \int\limits_{0}^{\infty} \frac{y^{\alpha+a-1}e^{-y/\beta}}{\beta^\alpha \Gamma(\alpha)} \, dy$

$= \frac{\beta^{\alpha+a}\Gamma(\alpha+a)}{\Gamma(\alpha)} \int\limits_{0}^{\infty} \frac{y^{\alpha+a-1}e^{-y/\beta}}{\Gamma(\alpha+a)\beta^{\alpha+a}} \, dy = \frac{\beta^a \Gamma(\alpha+a)1}{\Gamma(\alpha)}(1) \quad \text{if } \alpha+a > 0.$

b. Because $\int\limits_{0}^{\infty} \frac{y^{\alpha+a-1}e^{-y/\beta}}{\Gamma(\alpha+a)\beta^{\alpha+a}} \, dy = 1$ requires $\alpha+a > 0$ and $\beta > 0$.

c.   With $a = 1$

$$E(Y^I) = \frac{\beta^1\Gamma(\alpha+1)}{\Gamma(\alpha)} = \beta\alpha\frac{\Gamma(\alpha)}{\Gamma(\alpha)} = \alpha\beta.$$

d. $E(\sqrt{Y}) = E(Y^{1/2}) = \dfrac{\beta^{1/2}\Gamma\left(\alpha+\frac{1}{2}\right)}{\Gamma(\alpha)}.$

We assume $\alpha > 0$.

e. $E\left(\dfrac{1}{Y}\right) = E(Y^{-1}) = \dfrac{\beta^{-1}\Gamma(\alpha-1)}{\Gamma(\alpha)} = \dfrac{1}{\beta(\alpha-1)}.$

We assume $\alpha > 1$.

$$E\left(\frac{1}{\sqrt{Y}}\right) = E(Y^{-1/2}) = \frac{\beta^{-1/2}\Gamma\left(\alpha-\frac{1}{2}\right)}{\Gamma(\alpha)} = \frac{\Gamma\left(\alpha-\frac{1}{2}\right)}{\sqrt{\beta}\,\Gamma(\alpha)}.$$

We assume $\alpha > \frac{1}{2}$.

$$E\left(\frac{1}{Y^2}\right) = E(Y^{-2}) = \frac{\beta^{-2}\Gamma(\alpha-2)}{\Gamma(\alpha)} = \frac{1}{\beta^2(\alpha-2)}.$$

We assume $\alpha > 2$.

4.71 Similar to Exercise 4.58. The value of k is the constant part of the beta density with $\alpha = 4$ and $\beta = 3$. Hence

$$k = \frac{\Gamma(\alpha+\beta)}{\Gamma(\alpha)\Gamma(\beta)} = \frac{\Gamma(7)}{\Gamma(4)\Gamma(3)} = \frac{6!}{3!\,2!} = 60$$

4.73 Since Y has a beta distribution with $\alpha = 3$, $\beta = 2$,

$$E(Y) = \frac{\alpha}{\alpha+\beta} = \frac{3}{5} \quad\text{and}\quad V(Y) = \frac{\alpha\beta}{(\alpha+\beta)^2(\alpha+\beta+1)} = \frac{6}{25(6)} = \frac{1}{25}$$

4.75 For $\alpha = \beta = 1$, $f(y) = \dfrac{\Gamma(2)}{\Gamma(1)\Gamma(1)}y^{1-1}(1-y)^{1-1} = 1$ for $0 \le y \le 1$, which is the uniform distribution.

4.77 $E(C) = 10 + 20E(X) + 4E(X^2)$

$$= 10 + 20\left(\frac{\alpha}{\alpha+\beta}\right) + 4\left[\frac{\alpha(\alpha+1)}{(\alpha+\beta)(\alpha+\beta+1)}\right] \qquad \text{(see Exercise 4.78)}$$

$$= 10 + 20\left(\frac{1}{3}\right) + 4\left(\frac{2}{12}\right) = \frac{52}{3}$$

To compute $V(C)$ we will need $E(X^3)$ and $E(X^4)$.

$$E(X^3) = \int\limits_0^1 (2x^3 - 2^4)\, dx = \left[\tfrac{1}{2}x^4 - \tfrac{2}{5}x^5\right]_0^1 = \tfrac{1}{2} - \tfrac{2}{5} = \tfrac{1}{10}$$

$$E(X^4) = \int\limits_0^1 (2x^4 - 2^5)\, dx = \left[\tfrac{2}{5}x^5 - \tfrac{1}{3}x^6\right]_0^1 = \tfrac{2}{5} - \tfrac{1}{3} = \tfrac{1}{15}$$

$$V(C) = V(20X + 4X^2)$$

$$= E\left[(20X + 4X^2)^2\right] - \left[E(20X + 4X^2)\right]^2$$

$$= 400E(X^2) + 160E(X^3) = 16E(X^4) - \left(\tfrac{22}{3}\right)^2$$

$$= 400\left(\tfrac{1}{6}\right) + 160\left(\tfrac{1}{10}\right) + 16\left(\tfrac{1}{15}\right) - \left(\tfrac{22}{3}\right)^2 = 29.96$$

4.79 Let $Y =$ measurement error.

a. $$P(Y < .5) = \int\limits_0^{.5} \left[\frac{\Gamma(1+2)}{\Gamma(1)\Gamma(2)}\right] y^{1-1}(1-y)^{2-1}\, dy$$

$$= \int\limits_0^{.5} 2(1-y)\, dy = 2y - y^2\big]^{.5} = 1 - .25 = .75$$

b. $\mu = E(Y) = \dfrac{\alpha}{\alpha + \beta} = \dfrac{1}{3}.$

$$V(Y) = \frac{\alpha\beta}{(\alpha + \beta)^2(\alpha + \beta + 1)} = \frac{2}{3^2(4)} = \frac{1}{18};\ \sigma = \frac{1}{\sqrt{18}} = .2357$$

4.81 a. The variable factor is that of a beta density with $\alpha = 3$ and $\beta = 5$. Hence

$$c = \frac{\Gamma(\alpha + \beta)}{\Gamma(\alpha)\Gamma(\beta)} = \frac{\Gamma(8)}{\Gamma(3)\Gamma(5)} = \frac{7!}{2!\,4!} = 105$$

b. For a beta random variable, $E(Y) = \dfrac{\alpha}{\alpha + \beta} = \dfrac{3}{8}.$

4.83 Let $m_Y(t)$ be the moment-generating function of Y. Then

$$m_u(t) = E\left(e^{tu}\right) = E\left(e^{bt+atY}\right) = e^{bt} E\left(e^{atY}\right)$$

$$= e^{bt} m_Y(at).$$

Differentiating $m_u(t)$, we obtain

$$E(u) = m'_u(t)\big|_{t=0} = ae^{bt}m'_Y(at) + bm_Y(at)e^{bt}\big|_{t=0}$$

$$= am'_Y(0) + bE(e^{aY0}) = am'_Y(0) + bE(1)$$

$$= b + aE(Y).$$

$$E(u^2) = m''_u(t)\big|_{t=0} = a^2e^{bt}m''_Y(at) + m'_Y(at)bae^{bt} + b^2m_Y(at)e^{bt} + e^{bt}bam'_Y(at)\big|_{t=0}$$

$$= a^2m''_Y(0) + 2m'_Y(0)ab + m_Y(0)b^2$$

$$= a^2E(Y^2) + 2abE(Y) + b^2.$$

Hence,

$$V(Y) = a^2E(Y^2) + 2abE(Y) + b^2 - [a + bE(Y)]^2$$

$$= b^2\Big[E(Y^2) - E(Y)^2\Big] = b^2V(Y).$$

4.85 Differentiating $m(t)$ with respect to t, where $m(t) = \dfrac{1}{(1-\beta t)^\alpha} = (1-\beta t)^{-\alpha}$

$$E(Y) = m'(0) = \frac{\alpha\beta}{(1-\beta t)^{2\alpha+1}}\Bigg|_{t=0} = \alpha\beta$$

$$E(Y^2) = m''(0) = \frac{d}{dt}\alpha\beta(1-\beta t)^{-\alpha-1}\Big|_{t=0} = \frac{\alpha\beta^2(\alpha+1)}{(1-\beta t)^{\alpha+2}}\Bigg|_{t=0}$$

and $V(Y) = \alpha\beta^2(\alpha+1) - \alpha^2\beta^2 = \alpha\beta^2$.

4.87 a. $E(e^{3Y/2})\displaystyle\int_\infty^0 e^{(3y/2)+y}\,dy = \frac{2}{5}e^{5y/2}\Big]_{-\infty}^0 = \frac{2}{5}$

b. $m(t) = E(e^{tY}) = \displaystyle\int_{-\infty}^0 e^{(t+1)y}\,dy = \frac{e^{(t+1)y}}{t+1}\Bigg]_{-\infty}^0 = \frac{1}{t+1}$

c. Differentiating with respect to t, we obtain

$$m'(t)\big|_{t=0} = E(Y) = \frac{-1}{(t+1)^2}\Bigg|_{t=0} = -1 \qquad E(Y^2) = m''(0) = \frac{2}{(t+1)^3}\Bigg|_{t=0} = 2$$

$$V(Y) = 2 - (-1)^2 = 1$$

4.89 Similar to Exercise 4.88. It is necessary to have $P(|Y - \mu| \leq 1) \geq .75$. Hence, $1 - \dfrac{1}{k^2} = .75$ and $k = 2$. According to Tchebysheff's inequality, then, $1 = k\sigma$ and $\sigma = \dfrac{1}{k} = \dfrac{1}{2}$.

4.91 From Table A2.2, Appendix II, we find that when Y is uniform over (θ_1, θ_2)

$$E(Y) = \frac{\theta_1 + \theta_2}{2} \qquad V(Y) = \frac{(\theta_2 - \theta_1)^2}{12}$$

Thus,

$$2\sigma = 2\sqrt{V(Y)} = \frac{2(\theta_2 - \theta_1)}{\sqrt{12}} = \frac{\theta_2 - \theta_1}{\sqrt{3}}$$

The probability of interest is $P(|Y - \mu| \leq 2\sigma) = P(\mu - 2\sigma \leq Y \leq \mu + 2\sigma)$. Now

$$\theta_2 - \frac{\theta_1 + \theta_2}{2} = \frac{\theta_2 - \theta_1}{2} < \frac{\theta_2 - \theta_1}{\sqrt{3}} \qquad \text{and hence} \qquad \theta_2 < \frac{\theta_1 + \theta_2}{2} + \frac{\theta_2 - \theta_1}{\sqrt{3}} = \mu + 2\sigma$$

Similarly,

$$\frac{\theta_1 + \theta_1}{2} - \theta_1 = \frac{\theta_2 - \theta_1}{2} < \frac{\theta_2 - \theta_1}{\sqrt{3}} \qquad \text{so that} \qquad \theta_1 > \frac{\theta_1 + \theta_2}{2} - \frac{\theta_2 - \theta_1}{\sqrt{3}} = \mu - 2\sigma$$

But θ_1 and θ_2 are the upper and lower limits on Y. Hence

$$P(\mu - 2\sigma \leq Y \leq \mu + 2\sigma) = P\left(\frac{\theta_1 + \theta_2}{2} - \frac{\theta_2 - \theta_1}{\sqrt{3}} \leq Y \leq \frac{\theta_1 + \theta_2}{2} + \frac{\theta_2 - \theta_1}{\sqrt{3}} \right)$$

$$= P(\theta_1 \leq Y \leq \theta_2) = 1$$

Tchebysheff's theorem is satisfied, but the approximation suggested by the empirical rule is inaccurate. This is because the probability distribution for the uniform random variable is far from mound-shaped.

4.93 From Exercise 4.54, $E(C) = 1100$ and $V(C) = 2,920,000 = \sigma^2$ so that $\sigma = 1708.8$. The value 2000 is $\dfrac{2000 - 1100}{1708.8} = .53$ standard deviations above the mean. Thus we would expect C to exceed 2000 fairly often.

4.95 We need $P(|C - \mu| < k\sigma) \geq .75$. Hence, $1 - \dfrac{1}{k^2} = .75$ and $k = 2$. From Exercise 4.77, $\mu = \dfrac{52}{3}$ and $\sigma = \sqrt{29.96} = 5.474$. The necessary interval is

$$|C - 17.33| < 10.95 \qquad \text{or} \qquad (6.38, 28.28)$$

4.97 The random variable Y has a uniform distribution over the interval $(1, 4)$. That is, $f(y) = \frac{1}{3}$ for $1 \leq y \leq 4$. The cost of delay is given as

$$c(y) = \begin{cases} 100, & \text{for } 1 \leq y \leq 2 \\ 100 + 20(y - 2), & \text{for } 2 \leq y \leq 4 \end{cases}$$

Then

$$E(c(y)) = \int_{-\infty}^{\infty} c(y)f(y)\, dy = \int_{1}^{2} \frac{100}{3}\, dy + \int_{2}^{4} \frac{60 + 20y}{3}\, dy = \frac{100}{3} + \left[20y + \frac{20}{6}y^2\right]_{2}^{4}$$

$$= \frac{100}{3} + (80 - 40) + \frac{320 - 80}{6} = \frac{100}{3} + (80 - 40) + 40 = \$113.33$$

4.99 a.
$$f(x) = \begin{cases} 0, & x < 0 \\ \int_{0}^{x} \left(\frac{1}{100}\right)e^{-y/100}\, dy = 1 - e^{-x/100}, & 0 \leq x < 200 \\ 1, & x \geq 200 \end{cases}$$

b. $E(X) = \int_{0}^{200} \left(\frac{1}{100}\right)xe^{-x/100}\, dx + (.1353)[E(200)]$ where $.1353 = P(x \geq 200)$.

Using integration by parts, we find that $\int_{0}^{200} \left(\frac{1}{100}\right)xe^{-x/100}\, dx = 59.4$.

Thus, $E(X) = 59.4 + (.1353)(200) = 59.4 + 27.06 = 86.47$.

4.101 For this exercise $\mu = 70$ and $\sigma = 12$. The objective is to determine a particular value, y_0, for the random variable Y so that $P(Y < y_0) = .90$ (i.e., 90% of the students will finish the examination before the set time limit). Referring to Figure 4.22, we obtain $A_1 = P(Y \geq y_0) = .10$. Corresponding to the value $Y = y_0$ is the z value

$$z = \frac{y - \mu}{\sigma} = \frac{y_0 - 70}{12}$$

Hence

$$A\left[\frac{y_0 - 70}{12}\right] = .1 \quad \text{or} \quad \frac{y_0 - 70}{12} = 1.28.$$

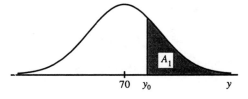

Thus $y_0 = 85.36$

Figure 4.22

4.103 Refer to Exercise 4.38, and let X be the number of defective bearings. Then X will have a binomial distribution with $n = 5$, $p = P(\text{defective}) = .073$, and

$$P(X > 1) = 1 - P(X = 0) = 1 - \binom{5}{0}p^0 q^5 = 1 - (.927)^5 = .3155$$

4.105 **a.** The variable factor of $f(y)$ is that of a gamma density with $\alpha = 2$ and $\beta = \frac{1}{2}$. Hence

$$c = \frac{1}{\Gamma(\alpha)\beta^\alpha} = \frac{1}{\Gamma(2)\left(\frac{1}{2}\right)^2} = \frac{4}{1!} = 4$$

b. Since Y has a gamma distribution with $\alpha = 2$, $\beta = \frac{1}{2}$,

$$E(Y) = \alpha\beta = 1 \qquad V(Y) = \alpha\beta^2 = 2\left(\frac{1}{4}\right) = \frac{1}{2}$$

c. From Exercise 4.13, the moment-generating function of a gamma random variable is

$$m(t) = \frac{1}{(1 - \beta t)^\alpha} \qquad \text{and in this case} \qquad m(t) = \frac{1}{\left[1 - \left(\frac{t}{2}\right)\right]^2} = \left(1 - \frac{t}{2}\right)^{-2}$$

4.107 For the beta random variable given in Section 4.7,

$$\mu'_k = E(Y^k) = \int_{-\infty}^{\infty} y^k f(y) \, dy = \int_0^1 y^k \frac{\Gamma(\alpha + \beta)}{\Gamma(\alpha)\Gamma(\beta)} y^{\alpha - 1}(1 - y)^{\beta - 1} \, dy$$

$$= \frac{\Gamma(\alpha + \beta)}{\Gamma(\alpha)\Gamma(\beta)} \int_0^1 y^{\alpha + k - 1}(1 - y)^{\beta - 1} \, dy$$

The quantity $y^{\alpha + k - 1}(1 - y)^{\beta - 1}$ is the variable factor of a beta density function with parameters $\alpha + k$ and β. Hence

$$\mu'_k = \frac{\Gamma(\alpha + \beta)}{\Gamma(\alpha)\Gamma(\beta)} \times \frac{\Gamma(\alpha + k)\Gamma(\beta)}{\Gamma(\alpha + k + \beta)} \int_0^1 \frac{y^{\alpha + k - 1}(1 - y)^{\beta - 1}}{B(\alpha + k, \, \beta)} \, dy = \frac{\Gamma(\alpha + \beta)\Gamma(\alpha + k)}{\Gamma(\alpha)\Gamma(\alpha + k + \beta)}$$

since the integral of a complete density function is 1.

4.109 Let Y be the time between the arrival of two calls, measured in hours. We require $P\left(Y > \frac{1}{4}\right)$. Since $\lambda t = 10$ and $t = 1$ (hour), $\frac{1}{\lambda} = \frac{1}{10}$, and

$$f(y) = \frac{1}{.1} e^{-y/.1} = 10 e^{-10y}$$

and

$$P\left(Y > \tfrac{1}{4}\right) = \int_{1/4}^{\infty} 10e^{-10y} \, dy = -e^{-10y}\big]_{1/4}^{\infty} = e^{-2.5} = .082$$

4.111 Let R be the distance to the nearest neighbor. Then

$$P(R > r) = P(\text{no plants in a circle of radius } r) = P(\text{no plants in an area of } \pi r^2)$$

Since the number of plants in an area of 1 unit has a Poisson distribution with mean λ, the number of plants in an area of πr^2 units has a Poisson distribution with mean $\lambda \pi r^2$. Hence

$$P(R > r) = \frac{\left(\lambda \pi r^2\right)^0 e^{-\lambda \pi r^2}}{0!} = e^{-\lambda \pi r^2}$$

Then

$$F(r) = P(R \le r) = 1 - e^{-\lambda \pi r^2} \quad \text{and} \quad f(r) = F'(r) = 2\lambda \pi r e^{-\lambda \pi r^2} \quad r > 0$$

4.113 From Exercise 4.3, we have the following:

$$F(y) = \begin{cases} 0, & y < 0 \\ \dfrac{y^2}{4}, & 0 \le y \le 2 \\ 1, & y > 2 \end{cases}$$

The median value of Y will be that value of y such that

$$F(y) = .5 \quad \text{or} \quad \frac{y^2}{4} = .5 \quad \text{or} \quad y^2 = 2$$

Since the range of y for which $0 < F(y) < 1$ is $0 \le y \le 2$, the median must be $y = +\sqrt{2} = 1.414$.

4.115 Let X be the grocer's profit. In general, her profit (in cents) on an order of $100k$ pounds of food will be $X = 1000Y - 600k$, as long as $Y < k$, that is, as long as the demand does not exceed the amount ordered. If $Y > k$, however, the grocer's profit will be $X = 1000k - 600k = 400k$. Define Y' as Y for $0 \le Y \le k$ and $Y' = k$ for $Y \ge k$. Then we can write $g(Y') = X = 1000Y' - 600k$ for all values of Y'. Y' has a mixed distribution with one discrete point, namely, $Y' = k$. Note that

$$c_1 = P(Y' = k) = P(Y \ge k) = \int_k^1 3y^2 \, dy = 1 - k^3$$

so that $c_2 = 1 - c_1 = k^3$. We can write Y' as a mixture of Y_1' and Y_2', where

$$F_1(y) = \begin{cases} 0, & y < k \\ 1, & y \geq k \end{cases}$$

$$F_2(y) = P\left(Y'_2 \leq y | 0 \leq Y' < k\right) = \frac{\int\limits_0^y f(t)\, dt}{k^3} = \frac{\int\limits_0^y 3t^2\, dt}{k^3} = \frac{y^3}{k^3} \qquad 0 < y < k$$

Note that

$$f_2(y) = \begin{cases} \dfrac{3y^2}{k^3}, & 0 \leq y \leq k \\ 0, & \text{elsewhere} \end{cases}$$

Now using Definition 4.14,

$$E(X) = E\left[g(Y')\right] = c_1 E\left[g(Y'_1)\right] + c_2 E\left[g(Y'_2)\right]$$

$$= (1 - k^3)\, 400k + k^3 \int\limits_0^k (1000y - 600k)\, \frac{3y^2}{k^3} = 400k - 400k^4 + 1000 \left[\frac{3y^4}{4}\right]_0^k$$

$$= 400k - 400k^4 + 750k^4 - 600k^4 = 400k - 250k^4$$

To maximize expected profit, then, set $\frac{d}{dk}[E(X)] = 400 - 1000k^3$ equal to 0. The desired value of k is $\left(\frac{400}{1000}\right)^{1/3} = (.4)^{1/3}$. (Note that the second derivative is negative, so that we do have a maximum.)

4.117 From Example 4.16 we know that the moment-generating function of $U = Y - \mu$ is

$$m_U = e^{t^2 \sigma^2 / 2}$$

Then

$$m_Z(t) = m_{U/\sigma}(t) = m_U\left(\frac{t}{\sigma}\right) = e^{(t/\sigma)^2 (\sigma^2/2)} = e^{t^2/2}$$

$$E(Y) = m'_Z(t)\Big|_{t=0} = te^{t^2/2}\Big|_{t=0} = 0$$

$$E(Y^2) = m''_Z(t)\Big|_{t=0} = t^2 e^{t^2/2} + e^{t^2/2}\Big|_{t=0} = 1$$

$$V(Y) = 1 - 0 = 1$$

4.119 **a.** $E(Y) = e^{\mu+\sigma^2/2} = e^{3+16/2} = e^{11}(10^{-2}g) = 598.74g$

$V(Y) = e^{2\mu+\sigma^2}\left(e^{\sigma^2} - 1\right) = e^{6+16}\left(e^{16} - 1\right) = e^{22}\left(e^{16} - 1\right)$

$= \left(e^{38} - e^{22}\right)\left(10^{-4}g^2\right) = 3.1856 \times 10^{12}$

b.   With $k = 2$, consider the interval $\left(E[Y] \pm 2\sqrt{V[Y]}\right)$

$(598.74 \pm 3,569,637.4) = (-3,569,038.7, \, 3,570,236.1)$

Since weights are positive, give as the final interval $(0, \, 3,570,236.1)$.

c. $P(Y < 598.74) = P(Y < 6.3948) = P\left[Z < \dfrac{6.3948 - 3}{4}\right] = P(Z < .8487) = .8023.$

4.121 **a.**   Note that $f_i(y) \geq 0$ and $\displaystyle\int_{-\infty}^{\infty} f_i(y) \, dy = 1 \qquad i = 1, 2$

Since $0 \leq a \leq 1$, it immediately follows that

$f(y) = af_1(y) + (1-a)f_2(y) \geq 0.$ Also,

$\displaystyle\int_{-\infty}^{\infty} f(y) \, dy = \int_{-\infty}^{\infty} [af_1(y0 + (1-a)f_2(y) \, dy]$

$\displaystyle = e \int_{-\infty}^{\infty} f_1(y) \, dy + (1-a) \int_{-\infty}^{\infty} f_2(y) \, dy$

$= a + 1 - a = 1.$

b.   **(i)** $E(Y) = \displaystyle\int_{-\infty}^{\infty} yf(y) \, dy = \int_{-\infty}^{\infty} y[af_1(y) + (1-a)f_2(y)] \, dy$

$\displaystyle = a \int_{-\infty}^{\infty} yf_1(y) \, dy + (1-a) \int_{-\infty}^{\infty} yf_2(y) \, dy$

$= a\mu_1 + (1-a)\mu_2.$

(ii) $\sigma^2 = E(Y^2) - [E(Y)]^2$

Proceeding as in part (b)(i), $E(Y^2) = \displaystyle\int_{-\infty}^{\infty} y^2 f(y) \, dy$

$\displaystyle = a \int_{-\infty}^{\infty} y^2 f_1(y) \, dy + (1-a) \int_{-\infty}^{\infty} y^2 f_2(y) \, dy$

Using the hint, $E(Y^2) = a\left[\mu_1^2 + \sigma_1^2\right] + (1-a)\left[\mu_2^2 + \sigma_2^2\right]$

Thus,

$$\text{Var}(Y) = a\left[\mu_1^2 + \sigma_1^2\right] + (1-a)\left[\mu_2^2 + \sigma_2^2\right] - \left[a\mu_1 + (1-a)\mu_2\right]^2$$

$$= a\sigma_1^2 + (1-a)\sigma_2^2 + \left(a - a^2\right)\mu_1^2 - 2a(1-a)\mu_1\mu_2 + \left[1 - a - (1-a)^2\right]\mu_2^2$$

$$= a\sigma_1^2 + (1-a)\sigma_2^2 + a(1-a)[\mu_1 - \mu_2]^2.$$

4.123 The density for Y, the life length of a resistor in thousands of hours, is

$$f(y) = \frac{2ye^{-y^2/10}}{10}, \qquad 0 \le y < \infty$$

a. $\displaystyle P(Y > 5) = 1 - P(Y \le 5) = 1 - \int_0^5 \frac{2ye^{-y^2/10}}{10}\, dy = 1 - \left[-e^{-y^2/10}\right]_0^5$

$$= 1 - \left[-e^{-2.5} + 1\right] = e^{-2.5}$$

b. Let X be the number of resistors that burn out prior to 5000 hours. Then X is a binomial random variable with $n = 3$ and $p = 1 - e^{-2.5}$. Thus,

$$P(X = 1) = \binom{3}{1}\left(1 - e^{-2.5}\right)\left(e^{-2.5}\right)^2 = .0186$$

4.125 a. $\displaystyle P(Y \le y | Y \ge c) = \frac{P(c \le Y \le y)}{P(Y \ge c)} = \frac{F(y) - F(c)}{1 - F(c)}$

b. Refer to the properties of distribution functions given in Section 4.2. Define $P(Y \le y | Y \ge c) = G(y)$. Then, since we are given $Y \ge c$,

$$G(-\infty) = G(c) = \frac{F(c) - F(c)}{1 - F(c)} = 0 \qquad G(\infty) = \frac{F(\infty) - F(c)}{1 - F(c)} = \frac{1 - F(c)}{1 - F(c)} = 1$$

since F is a distribution function. Finally, for $b \ge a$,

$$G(b) - G(a) = \frac{F(b) - F(a)}{1 - F(c)} \ge 0$$

since F is a distribution function. Hence G must also be a distribution function.

c. It is given that $f(y) = \dfrac{2ye^{-y^2/3}}{3}$. From Exercise 4.124, $1 - F(y) = e^{-y^2/3}$, so that

$$F(y) = 1 - e^{-y^2/3}.$$

Using the result of part (a),

$$P(Y \le 4 | Y \ge 2) = \frac{1 - e^{-4^2/3} - \left(1 - e^{-2^2/3}\right)}{e^{-2^2/3}} = \frac{e^{-4/3} - e^{-16/3}}{e^{-4/3}} = 1 - e^{-4}$$

4.127 It is given that $f(y) = \left(\frac{1}{100}\right)e^{-y/100}$. Then $1 - F(y) = e^{-y/100}$. Using the given result, we have

$$E(Y | Y \ge 50) = \frac{1}{e^{-1/2}} \int\limits_{50}^{\infty} \frac{ye^{-y/100}}{100} \, dy$$

Integrating by parts with $u = y$, $du = dy$, $dv = \left(\frac{e^{-y/100}}{100}\right) dy$, and

$$v = \int \left(\frac{e^{-y/100}}{100}\right) dy = -e^{-y/100},$$ we obtain

$$E(Y | Y \ge 50) = \frac{1}{e^{-1/2}} \left(\left[-ye^{-y/100}\right]_{50}^{\infty} - \int\limits_{50}^{\infty} -e^{-y/100} \, dy \right)$$

$$= \frac{1}{e^{-1/2}} \left[50e^{-1/2} + 100e^{-1/2}\right] = 150$$

Note that the quantity ye^{-y} when evaluated at $y = \infty$ is of the indeterminate form $(0)(\infty)$, and L'Hôpital's rule is used to evaluate it.

4.129 Suppose we have two variables v and w. Integrating by parts, we may write $\int w \, dv$

$= wv - \int v \, dw$, where w and v are suitably chosen to simplify the integration. The integral that must be evaluated is, in this case,

$$\Gamma(u) = \int\limits_{0}^{\infty} y^{u-1} e^{-y} \, dy$$

Let $w = y^{u-1}$ and $dv = e^{-y} \, dy$. Then

$$dw = (u-1)y^{u-2} \, dy \qquad \text{and} \qquad v = \int e^{-y} \, dy = -e^{-y}$$

Integrating by parts, we have

$$\Gamma(u) = -e^{-y}y^{u-1}\Big]_0^\infty + \int_0^\infty (u-1)y^{u-2}e^{-y}\,dy = 0 + (u-1)\int_0^\infty y^{u-2}e^{-y}\,dy$$

$$= (u-1)\Gamma(u-1)$$

Note that the quantity $e^{-y}y^{u-1}$ when evaluated at $y = \infty$ is of the indeterminate form $(0)(\infty)$, and L'Hôpital's rule is used to evaluate it.

4.131 **a.** Let $y = \sin^2\theta$, so that $dy = 2\sin\theta\cos\theta\,d\theta$. Now

$$B(\alpha,\,\beta) = \int_0^1 y^{\alpha-1}(1-y)^{\beta-1}\,dy$$

$$= 2\int_0^{\pi/2} \sin^{2\alpha-2}\theta\big(1-\sin^2\theta\big)^{\beta-1}\sin\theta\cos\theta\,d\theta$$

$$= 2\int_0^{\pi/2} \sin^{2\alpha-2}\theta\,\cos^{2\beta-2}\theta\,\sin\theta\cos\theta\,d\theta$$

$$= 2\int_0^{\pi/2} \sin^{2\alpha-1}\theta\,\cos^{2\beta-1}\theta\,d\theta$$

b. Following the instructions given in the text, we consider

$$\Gamma(\alpha)\Gamma(\beta) = \int_0^\infty y^{\alpha-1}e^{-y}\,dy\int_0^\infty z^{\beta-1}e^{-z}\,dz = \int_0^\infty\int_0^\infty y^{\alpha-1}z^{\beta-1}e^{-(y+z)}\,dy\,dz$$

and transform to polar coordinates using the transformation $y = r^2\cos^2\theta$ and $z = r^2\sin^2\theta$. Using this second transformation, the Jacobian is

$$|J| = \begin{vmatrix} \dfrac{dy}{dr} & \dfrac{dy}{d\theta} \\[2mm] \dfrac{dz}{dr} & \dfrac{dz}{d\theta} \end{vmatrix} = \begin{vmatrix} 2r\cos^2\theta & -2r^2\cos\theta\sin\theta \\ 2r\sin^2\theta & 2r^2\cos\theta\sin\theta \end{vmatrix}$$

$$= 4r^3\cos^3\theta\sin\theta + 4r^3\sin^3\theta\cos\theta = 4r^3\cos\theta\sin\theta$$

Then

$$\Gamma(\alpha)\Gamma(\beta) = \int_0^\infty\int_0^\infty y^{\alpha-1}z^{\beta-1}e^{-(y+z)}\,dy\,dz$$

$$= \int_0^{\pi/2}\int_0^\infty r^{2\alpha-2}\cos^{2\alpha-2}\theta\,r^{2\beta-2}\sin^{2\beta-2}\theta\,e^{-r^2}4r^3\cos\theta\sin\theta\,dr\,d\theta$$

$$= 2 \int_0^{\pi/2} \cos^{2\alpha-1}\theta \, \sin^{2\beta-1}\theta \, d\theta \int_0^\infty 2r^{2\alpha+2\beta-1}e^{-r^2} \, dr$$

$$= B(\alpha, \, \beta) \int_0^\infty r^{2(\alpha+\beta-1)}e^{-r^2}(2r) \, dr$$

Using the transformation $x = r^2$ so that $dx = 2r \, dr$, we have

$$\Gamma(\alpha)\Gamma(\beta) = B(\alpha, \, \beta) \int_0^\infty x^{\alpha+\beta-1}e^{-x} \, dx = B(\alpha, \, \beta)\Gamma(\alpha+\beta)$$

and the result is proven.

CHAPTER 5 MULTIVARIATE PROBABILITY DISTRIBUTIONS

5.1 Denote a sample space in terms of the firm that received the first and second contracts:

S	(y_1, y_2)
AA	$(2, 0)$
AB	$(1, 1)$
AC	$(1, 0)$
BA	$(1, 1)$
BB	$(0, 2)$
BC	$(0, 1)$
CA	$(1, 0)$
CB	$(0, 1)$
CC	$(0, 0)$

The probability the first contract will be assigned to any of the three firms is $\frac{1}{3}$. Likewise for the second contract. The probability for each of the outcomes in the sample space S is $\frac{1}{9}$. Setting up a table for the joint function for Y_1 and Y_2,

			y_1	
		0	1	2
	0	$\frac{1}{9}$	$\frac{2}{9}$	$\frac{1}{9}$
y_2	1	$\frac{2}{9}$	$\frac{2}{9}$	0
	2	$\frac{1}{9}$	0	0

b. $F(1, 0) = P(Y_1 \leq 1, Y_2 \leq 0) = p(0, 0) + p(1, 0) = \frac{1}{9} + \frac{2}{9} = \frac{1}{3}$

5.3 In this exercise Y_1 and Y_2 are both discrete random variables, and the joint distribution for Y_1 and Y_2 is given by

$$P(Y_1 = y_1, Y_2, y_2) = p(y_1, y_2)$$

We must calculate $p(y_1, y_2)$ for $y_1 = 0, 1, 2, 3$ and $y_2 = 0, 1, 2, 3$. The total number of ways of choosing 3 persons for the committee is $\binom{9}{3} = 84$. Now,

$$P(Y_1 = 0, Y_2 = 0) = P(3 \text{ divorced}) = 0$$

since there are only 2 divorced executives available. However,

$$P(Y_1 = 1, Y_2 = 0) = P(1 \text{ married, } 0 \text{ never married, } 2 \text{ divorced})$$

$$= \frac{\binom{4}{1}\binom{3}{0}\binom{2}{2}}{\binom{9}{3}} = \frac{4}{84}$$

Similar calculations, using an extension of the hypergeometric probability distribution discussed in Chapter 3, allow the student to obtain all 16 probabilities, and the joint probability distribution of Y_1 and Y_2 may be written in the form of a table.

$$p(2, 0) = \frac{\binom{4}{2}\binom{3}{0}\binom{2}{1}}{84} = \frac{12}{84} \qquad p(0, 2) = \frac{\binom{4}{0}\binom{3}{2}\binom{2}{1}}{84} = \frac{6}{84} \qquad p(1, 1) = \frac{\binom{4}{1}\binom{3}{1}\binom{2}{1}}{84} = \frac{24}{84}$$

$$p(3, 0) = \frac{\binom{4}{3}}{84} = \frac{4}{84} \qquad\qquad p(1, 2) = \frac{\binom{4}{1}\binom{3}{2}}{84} = \frac{12}{84} \qquad\qquad p(2, 1) = \frac{\binom{4}{2}\binom{3}{1}\binom{2}{0}}{84} = \frac{18}{84}$$

$$p(0, 3) = \frac{\binom{3}{3}}{84} = \frac{1}{84} \qquad\qquad p(0, 1) = \frac{\binom{4}{0}\binom{3}{1}\binom{2}{2}}{84} = \frac{3}{84}$$

$$p(3, 1) = p(2, 2) = p(3, 2) = p(3, 3) = p(1, 3) = p(2, 3) = 0$$

Note that $\displaystyle\sum_{y_1=0}^{3} \sum_{y_2=0}^{3} p(y_1, y_2) = 1.$

		y_2			
		0	1	2	3
	0	0	$\frac{3}{84}$	$\frac{6}{84}$	$\frac{1}{84}$
y_1	1	$\frac{4}{84}$	$\frac{24}{84}$	$\frac{12}{84}$	0
	2	$\frac{12}{84}$	$\frac{18}{84}$	0	0
	3	$\frac{4}{84}$	0	0	0

5.5   a.   Integrating the joint density function over the region indicated under the restriction that $y_1 \leq y_2$, we have

$$\int_0^1 \int_0^{y_2} K(1 - y_2) \, dy_1 \, dy_2$$

$$= \int_0^1 K\left(y_2 - y_2^2\right) dy_2 = \left(\frac{1}{2} - \frac{1}{3}\right)K$$

Hence $K = 6$. The region of integration is shown in Figure 5.1.

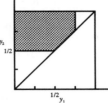

Figure 5.1

b. $P\left(Y_1 \le \frac{3}{4}, Y_2 \ge \frac{1}{2}\right) = \int\limits_{1/2}^{3/4} \int\limits_{0}^{y_2} 6(1 - y_2) \, dy_1 \, dy_2 + \int\limits_{3/4}^{1} \int\limits_{0}^{3/4} 6(1 - y_2) \, dy_1 \, dy_2$

$= \int\limits_{1/2}^{3/4} 6\left(y_2 - y_2^2\right) dy_2 + \int\limits_{3/4}^{1} \frac{9}{2}(1 - y_2) \, dy_2 = \left[3y^2 - 2y^3\right]_{1/2}^{3/4} + \frac{9}{2}\left[y - \frac{y^2}{2}\right]_{3/4}^{1}$

$= \frac{22}{64} + \frac{9}{64} = \frac{31}{64}$

5.7 The two lines that define the region shaded in the text's diagram are the lines $y_2 - y_1 = 1$ and $y_1 + y_2 = 1$. The triangle has area $\left(\frac{1}{2}\right)(1)(2) = 1$. Hence the joint density can be written as

$$f(y_1, y_2) = 1 \qquad\qquad y_2 - y_1 \le 1 \qquad \text{and} \qquad -1 \le y_1 \le 0$$
$$y_1 + y_2 \le 1 \qquad \text{and} \qquad 0 \le y_1 \le 1$$

a. The region of interest is the shaded area in Figure 5.3.

$P\left(Y_1 \le \frac{3}{4}, Y_2 \le \frac{3}{4}\right)$

$= \int\limits_{0}^{3/4} \int\limits_{y_2-1}^{1/4} dy_1 \, dy_2 + \int\limits_{1/4}^{3/4} \int\limits_{0}^{1-y_1} dy_2 \, dy_1$

$= \int\limits_{0}^{3/4} \left(\frac{5}{4} - y_2\right) dy_2 + \int\limits_{1/4}^{3/4} (1 - y_1) \, dy_1$

$= \left[\frac{5}{4}y_2 - \frac{y_2^2}{2}\right]_{0}^{3/4} + \left[y_1 - \frac{y_1^2}{2}\right]_{1/4}^{3/4} = \frac{29}{32}$

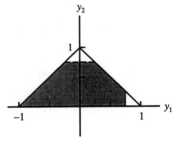

Figure 5.3

b.   Refer to Figure 5.4. The region of interest, which is the region on which $Y_1 \geq Y_2$, is the shaded area. Then

$$P(Y_1 - Y_2 \geq 0)$$

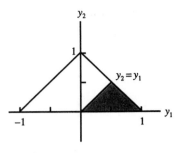

$$= P(Y_1 \geq Y_2) = \int_0^{1/2} \int_{y_2}^{1-y_2} dy_1 \, dy_2$$

$$= \int_0^{1/2} (1 - 2y_2) \, dy_2 = \left[y_2 - y_2^2 \right]_0^{1/2}$$

Figure 5.4

$$= \frac{1}{2} - \frac{1}{4} = \frac{1}{4}$$

This can be verified by calculating the area corresponding to the shaded region in Figure 5.4.

5.9   The region over which the density is positive is the area in the first quadrant ($y_1 \geq 0$, $y_2 \geq 0$) below the line $y_1 = y_2$. (See Figure 5.6).

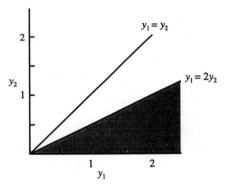

a. $$P(Y_1 < 2, Y_2 > 1) = \int_1^2 \int_{y_2}^2 e^{-y_1} \, dy_1 \, dy_2$$

$$= \int_1^2 -e^{-y_1} \Big]_{y_2}^2 \, dy_2 = \int_1^2 \left(e^{-y_2} - e^{-2} \right) dy_2$$

$$= \left[-e^{-y_2} - y_2 e^{-2} \right]_1^2$$

Figure 5.6

b.   Refer to Figure 5.6. Integrating over the shaded region, we obtain

$$P(Y_1 \geq 2Y_2) = \int_0^\infty \int_{2y_2}^\infty e^{-y_1} \, dy_1 \, dy_2 = \int_0^\infty e^{-2y_2} \, dy_2 = -\frac{1}{2} e^{-2y_2} \Big]_0^\infty = \frac{1}{2}$$

c.   Refer to Figure 5.7. Integrating over the shaded region, we obtain

$$P(Y_1 - Y_2 \geq 1) = \int_0^\infty \int_{1+y_2}^\infty e^{-y_1} \, dy_1 \, dy_2$$

$$= \int_0^\infty e^{-(1+y_2)} \, dy_2$$

$$= -e^{-(1+y_2)} \Big]_0^\infty = e^{-1}$$

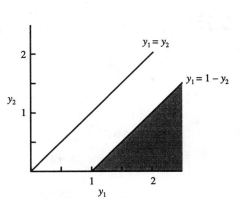

Figure 5.7

5.11 The region on which $Y_1 \leq Y_2$ is the shaded region
shown in Figure 5.9. The angle θ is $\theta = 45° = \frac{\pi}{4}$.
Transforming to polar coordinates, we have
$y_1 = r \cos \theta$ and $y_2 = r \sin \theta$. Also,
$dy_1 dy_2 = r \, dr \, d\theta$ (see Exercise 4.128). Then

$$P(Y_1 \leq Y_2) = \int \int_{y_1 \leq y_2} f(y_1, y_2) \, dy_1 \, dy_2$$

$$= \int_{\pi/4}^{5\pi/4} \int_0^1 \frac{r}{\pi} \, dr \, d\theta = \int_{\pi/4}^{5\pi/4} \frac{r^2}{2\pi} \Big]_0^1 \, d\theta$$

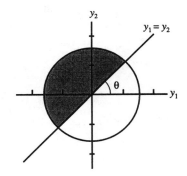

Figure 5.9

$$= \frac{\theta}{2\pi} \Big]_{\pi/4}^{5\pi/4} = \frac{\frac{4\pi}{4}}{2\pi} = \frac{1}{2}$$

This answer can be verified by calculating the area corresponding to the shaded
region given in Figure 5.9.

5.13 **a.**

y_1	0	1	2
$p(y_1)$	$\frac{4}{9}$	$\frac{4}{9}$	$\frac{1}{9}$

  b. No. Evaluating $f(y) = \binom{2}{y}\left(\frac{1}{3}\right)^y\left(\frac{2}{3}\right)^{2-y}$ for each value of Y_1 will result in the
same probabilities as those given in part (a).

5.15 a. By definition, $p_1(y_1) = \sum\limits_{y_2=0}^{3} p(y_1, y_2)$ for $y_1 = 0, 1, 2, 3$. These probabilities

may be obtained by summing across the rows in the table given in the solution to Exercise 5.3. Hence

$$P(Y_1 = 0) = 0 + \frac{3}{84} + \frac{6}{84} + \frac{1}{84} = \frac{10}{84}$$

Similarly,

$$P(Y_1 = 1) = \frac{40}{84} \qquad P(Y_1 = 2) = \frac{30}{84} \qquad P(Y_1 = 3) = \frac{4}{84}$$

The marginal distribution of Y_1 is shown in the following table.

y_1	$p_1(y_1)$
0	$\frac{10}{84}$
1	$\frac{40}{84}$
2	$\frac{30}{84}$
3	$\frac{4}{84}$

b. $P(Y_1 = 1 | Y_2 = 2) = \dfrac{P(Y_1 = 1, Y_2 = 2)}{P(Y_2 = 2)} = \dfrac{\dfrac{\binom{4}{1}\binom{3}{2}\binom{2}{0}}{\binom{9}{3}}}{\dfrac{\binom{3}{2}\binom{6}{1}}{\binom{9}{3}}} = \dfrac{12}{18} = \dfrac{2}{3}$

c. Again, use the hypergeometric and the extended hypergeometric distribution.

$$P(Y_3 = 1 | Y_2 = 1) = \frac{P(Y_2 = 1, Y_3 = 1)}{P(Y_2 = 1)} = \frac{\dfrac{\binom{3}{1}\binom{2}{1}\binom{4}{1}}{\binom{9}{3}}}{\dfrac{\binom{3}{1}\binom{6}{2}}{\binom{9}{3}}} = \frac{24}{45} = \frac{8}{15}$$

d. The probabilities are identical.

5.17 a. For this joint density function, $0 \le y_1 \le y_2 \le 1$. Integrating over y_2, we have

$$f_1(y_1) = \int_{y_1}^{1} 6(1 - y_2)\, dy_2 = 6\left(y_2 - \frac{y_2^2}{2}\right)\Big]_{y_1}^{1} = 3(1 - y_1)^2 \qquad \text{for } 0 \le y_1 \le 1$$

Similarly,

$$f_2(y_2) = \int_0^{y_2} 6(1 - y_2)\, dy_1 = 6y_2(1 - y_2) \qquad \text{for } 0 \le y_2 \le 1$$

b. Using part (a), calculate

$$P\left(Y_1 \le \tfrac{3}{4}\right) = \int_0^{3/4} 3(1 - y_1)^2\, dy_1$$

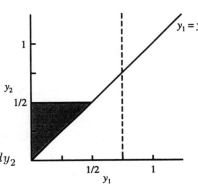

$$= 3\left[y_1 - y_1^2 + \frac{y_1^3}{3} \right]_0^{3/4} = \frac{63}{64}$$

Refer to Figure 5.10 and integrate the joint density over the shaded region to obtain

$$P\left(Y_2 \le \tfrac{1}{2},\, Y_1 \le \tfrac{3}{4}\right) = \int_0^{1/2} \int_0^{y_2} 6(1 - y_2)\, dy_1\, dy_2$$

$$= \int_0^{1/2} \left(6y_2 - 6y_2^2 \right) dy_2 = \frac{3}{4} - \frac{2}{8} = \frac{1}{2}$$

Figure 5.10

Then

$$P\left(Y_1 \le \tfrac{1}{2} \middle| Y_2 \ge \tfrac{3}{4}\right) = \frac{\left(\frac{1}{2}\right)}{\left(\frac{63}{64}\right)} = \frac{32}{63}$$

c. If $0 < y_2 < 1$,

$$f(y_1 | y_2) = \frac{f(y_1, y_2)}{f_2(y_2)} = \frac{6(1 - y_2)}{6y_2(1 - y_2)} = \frac{1}{y_2} \qquad \text{for } 0 \le y_1 \le y_2$$

d. If $0 < y_1 < 1$,

$$f(y_2 | y_1) = \frac{f(y_1, y_2)}{f_1(y_1)} = \frac{6(1 - y_2)}{3y_2(1 - y_1)^2} = \frac{2(1 - y_2)}{(1 - y_1)^2} \qquad \text{for } y_1 \le y_2 \le 1$$

e. Refer to part (d). Since $f\left(y_2 | y_1\right) = \tfrac{1}{2}\right) = \dfrac{2(1 - y_2)}{\left(\frac{1}{2}\right)^2} = 8(1 - y_2) \qquad \text{for } \tfrac{1}{2} \le y_2 \le 1,$

$$P\left(Y_2 \ge \tfrac{3}{4} \middle| Y_1 = \tfrac{1}{2}\right) = \int_{3/4}^1 8(1 - y_2)\, dy_2 = \left[8y_2 - 4y_2^2 \right]_{3/4}^1 = \frac{1}{4}$$

5.19 **a.** Refer to Exercise 5.7.

$$f_2(y_2) = \int_{y_2-1}^{1-y_2} 1 \, dy_1 = (1-y_2) - (y_2 - 1) = 2(1-y_2) \qquad 0 \le y_2 \le 1$$

In order to find $f_1(y_1)$, notice that the limits of integration are different for $0 \le y_1 \le 1$ and $-1 \le y_1 \le 0$. For the first case,

$$f_1(y_1) = \int_0^{1-y_1} dy_2 = 1 - y_1 \qquad \text{for } 0 \le y_1 \le 1$$

Then, for $-1 \le y_1 \le 0$,

$$f_1(y_1) = \int_0^{1+y_1} dy_2 = 1 + y_1$$

This can be written as $f_1(y_1) = 1 - |y_1|$ for $-1 \le y_1 \le 1$.

b. The conditional distribution of Y_2 given Y_1 is, for $-1 \le y_1 < 1$

$$f(y_2|y_1) = \frac{1}{1-|y_1|} \qquad \text{for } 0 \le y_2 \le 1 - |y_1|$$

Since $Y_1 = \frac{1}{4}$,

$$f\left(y_2|y_1 = \frac{1}{2}\right) = \frac{1}{\left(\frac{3}{4}\right)} = \frac{4}{3} \qquad \text{for } 0 \le y_2 \le \frac{3}{4}$$

Then

$$P\left(Y_2 > \frac{1}{2}|Y_1 = \frac{1}{4}\right) = \int_{1/2}^{3/4} \frac{4}{3} \, dy_1 = \frac{4}{3}\left(\frac{3}{4} - \frac{1}{2}\right) = \frac{1}{3}$$

5.21 Refer to Exercise 5.9. The probability of interest is $P(Y_2 < 1|Y_1 = 2)$. Calculate $f_1(y_1)$ as

$$f_1(y_1) = \int_0^{y_1} e^{-y_1} \, dy_2 = y_1 e^{-y_1} \qquad \text{for } 0 \le y_1 \le \infty$$

Then if $y_1 > 0$, $f(y_2|y_1) = \frac{1}{y_1}$ for $0 \le y_2 \le y_1$, which for $y_1 = 2$ is

$$f(y_2|Y_1 = 2) = \frac{1}{2} \qquad 0 \le y_2 \le 2$$

Finally,

$$P(Y_2 < 1|Y_1 = 2) = \int_0^1 \tfrac{1}{2}\, dy_2 = \tfrac{1}{2}$$

5.23 Calculate

$$f_2(y_2) = \int_0^\infty \frac{y_1}{8} e^{-(y_1+y_2)/2}\, dy_1 = \frac{e^{-y^2/2}}{8} \int_0^\infty \frac{y_1 e^{-y_1/2}}{4}\, dy_1 = \frac{e^{-y^2/2}}{2} \qquad y_2 > 0.$$

Notice that the preceding integral is that of a complete gamma density with $\alpha = 2$ and $\beta = 2$. Then

$$P(Y_2 > 2) = \int_2^\infty \frac{e^{-y^2/2}}{2}\, dy_2 = -e^{-y^2/2}\Big]_2^\infty = e^{-1}$$

5.25 By Definition 5.5, if $w = 0, 1, 2, \ldots$

$$P(Y_1 = y_1|W = w) = \frac{P(Y_1 = y_1, W = w)}{P(W = w)} = \frac{P(Y_1 = y_1, Y_1 + Y_2 = w)}{P(W = w)}$$

$$= \frac{P(Y_1 = y_1|Y_2 = w - y_1)}{P(W = w)} = \frac{P(Y_1 = y_1)P(Y_2 = 2 - y_2)}{P(W = w)}$$

(since Y_1 and Y_2 are independent)

If $y_1 = 0, 1, \ldots, w$,

$$= \frac{\left(\dfrac{\lambda_2^{y_1} e^{-\lambda_2}}{y_1!}\right)\left(\dfrac{\lambda_2^{w-y_1} e^{-\lambda_2}}{(w-y_1)!}\right)}{\left(\dfrac{(\lambda_1+\lambda_2)^w e^{-(\lambda_1+\lambda_2)}}{w!}\right)}$$

$$= \frac{w!}{y_1!(w-y_1)!} \frac{\lambda_1^{y_1}}{(\lambda_1+\lambda_2)^{y_1}} \frac{\lambda_2^{w-y_1}}{(\lambda_1+\lambda_2)^{w-y_1}}$$

$$= \binom{w}{y_1}\left(\frac{\lambda_1}{\lambda_1+\lambda_2}\right)^{y_1}\left(1 - \frac{\lambda_1}{\lambda_1+\lambda_2}\right)^{w-y_1} \qquad y_1 = 0, 1, \ldots, w$$

which is the probability mass function of a binomial random variable with $n = w$ and $p = \dfrac{\lambda_1}{\lambda_1 + \lambda_1}$.

5.27 Let Y be the number of defectives in a random selection of three items. Conditional on p, fixed, the probability distribution of Y is

$$P(Y = y|p) = \binom{3}{y} p^y q^{3-y} \qquad y = 0, 1, 2, 3$$

It is given that P is distributed uniformly on $(0, 1)$. That is, $f(p) = 1$ for $0 \le p \le 1$. We are interested in the unconditional marginal probability,

$$P(Y = 2) = \int_0^1 P(Y = 2, p)\, dp = \int_0^1 P(Y = 2|p)f(p) = \int_0^1 \binom{3}{2} p^2 (1-p)^1\, dp$$

$$= 3 \int_0^1 (p^2 - p^3)\, dp = 3\left(\tfrac{1}{3} - \tfrac{1}{4}\right) = \tfrac{3}{12} = \tfrac{1}{4}$$

5.29 No. For example, consider $P(Y_1 = 0, Y_2 = 0)$ and $p(Y_1 = 0)p(Y_2 = 0)$

$$p(0,0) = \tfrac{1}{9} \ne \left(\tfrac{4}{9}\right)\left(\tfrac{4}{9}\right) = p_1(0)p_2(0)$$

thus, not independent.

5.31 dependent

5.33 dependent

5.35 dependent

5.37 dependent

5.39 independent

5.41 **a.** Because of the independence of Y_1 and Y_2,

$$f(y_1, y_2) = f(y_1)f(y_2) = \tfrac{1}{9} e^{-(y_1+y_2)/3}$$

$$y_1 > 0,\ y_2 > 0$$

b. The probability of interest is the shaded area in Figure 5.11. Hence

$$P(Y_1 + Y_2 \le) = \int_0^1 \int_0^{1-y_2} f(y_1, y_2)\, dy_1\, dy_2$$

$$= \int_0^1 \left[1 - e^{-(1-y_2)/3}\right]\tfrac{1}{3} e^{-y_2/3}\, dy_2$$

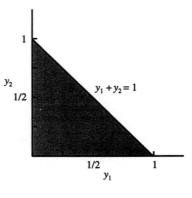

Figure 5.11

$$= \int_0^1 \left(\tfrac{1}{3} e^{-y_2/3} - \tfrac{1}{3} e^{-1/3} \right) dy_2 - e^{-y_2/3} \Big]_0^1 - \tfrac{1}{3} e^{-1/3} = 1 - \tfrac{4}{3} e^{-1/3}$$

5.43 Let $Y_1 = $ calling time to the switchboard of the first call

$$f(y_1) = 1; \qquad 0 \le y_1 \le 1$$

$Y_2 = $ calling time to the switchboard of the second call

$$f(y_2) = 1; \qquad 0 \le y_2 \le 1$$

$$f(y_1, y_2) = 1.$$

a. $P\left(Y_1 \le \tfrac{1}{2}, Y_2 \le \tfrac{1}{2}\right) = \left(\displaystyle\int_0^{1/2} 1 \, dy_1 \right)\left(\displaystyle\int_0^{1/2} 1 \, dy_2 \right)$

since Y_1 and Y_2 are independent

$$= \left(\tfrac{1}{2}\right)\left(\tfrac{1}{2}\right) = \tfrac{1}{4}.$$

b. Note that 5 minutes $= \tfrac{1}{12}$ of 1 hour.

$$P\left(|Y_1 = Y_2| < \tfrac{1}{12}\right) = \int_0^{1/12} \int_0^{y_1 + (1/12)} dy^2 \, dy_1 + \int_{1/12}^{11/12} \int_{y_1 - (1/12)}^{y_1 + (1/12)} dy_2 \, dy_1$$

$$+ \int_{11/12}^1 \left[\left(\tfrac{13}{12}\right) - y_1 \right] dy_1$$

$$= \left(\tfrac{y_1^2}{2}\right) + \tfrac{y_1}{12} \Big]_0^{1/12} + \tfrac{2y_1}{12} \Big]_{1/12}^{11/12} + \left(\tfrac{13y_1}{12}\right) - \tfrac{y_1^2}{2} \Big]_{11/12}^1 = \tfrac{46}{288} = \tfrac{23}{144}.$$

5.45 Refer to the solution for Exercise 5.15, where the marginal distribution for Y_1 is given.

$$E(Y_1) = \Sigma \, y_1 p_1(y_1) = 0\left(\tfrac{10}{84}\right) + 1\left(\tfrac{40}{84}\right) + 2\left(\tfrac{30}{84}\right) + 3\left(\tfrac{4}{84}\right) = \tfrac{40 + 60 + 12}{84} = \tfrac{112}{84} = \tfrac{4}{3}$$

5.47 Refer to Exercises 5.5 and 5.17.

a. $E(Y_1) = \displaystyle\int_0^1 3y_1(1 - y_1)^2 \, dy_1 = \int_0^1 \left(3y_1 - 6y_1^2 + 3y_1^3 \right) dy_1 = \tfrac{1}{4}$

$$E(Y_2) = \int_0^1 \left(6y_2^2 - 6y_2^3 \right) dy_2 = \tfrac{1}{2}$$

b. $E\left(Y_1^2\right) = \int\limits_0^1 \left(3y_1^2 - 6y_1^3 + 3y_1^4\right)dy_1 = \frac{1}{10}$ $\qquad\qquad V(Y_1) = \frac{1}{10} - \frac{1}{16} = \frac{3}{80}$

$\quad\ E\left(Y_2^2\right) = \int\limits_0^1 \left(6y_2^3 - 6y_2^4\right)dy_2 = \frac{6}{4} - \frac{6}{5} = \frac{3}{10}$ $\qquad\ V(Y_2) = \frac{3}{10} - \frac{1}{4} = \frac{1}{20}$

c. $E(Y_1 - 3Y_2) = \frac{1}{4} - 3\left(\frac{1}{2}\right) = -\frac{5}{4}$

5.49 Refer to Exercise 5.7. Integrating $y_1 y_2 f(y_1,\, y_2)$ over the two regions of integration, we have

$$E(Y_1 Y_2) = \int\limits_{-1}^0 \int\limits_0^{1+y_1} y_1 y_2 \, dy_1 dy_1 + \int\limits_0^1 \int\limits_0^{1-y_1} y_1 y_2 \, dy_2 \, dy_1$$

$$= \int\limits_{-1}^0 \frac{y_1(1+y_1)^2}{2} \, dy_1 + \int\limits_0^1 \frac{y_1(1-y_1)^2}{2} \, dy_1$$

$$= = \left[\frac{y_1^2}{4} + \frac{2y_1^3}{6} + \frac{y_1^4}{8}\right]_{-1}^0 + \left[\frac{y_1^2}{4} - \frac{2y_1^3}{6} + \frac{y_1^4}{8}\right]_0^1 = 0$$

5.51 Since Y_1 and Y_2 are independent, with $f_1(y_1) = \frac{1}{4}y_1 e^{-y_1/2}$ and $f_2(y_2) = \frac{1}{2}e^{-y_2/2}$,

$$E\left(\frac{Y_2}{Y_1}\right) = E\left(\frac{1}{Y_1}\right)E(Y_2) = \frac{1}{8}\int\limits_0^\infty e^{-y_1/2} \, dy_1 \int\limits_0^\infty y_2 e^{-y_2/2} \, dy_2$$

$$= \frac{1}{8}\left[-2e^{-y_1/2}\right]_0^\infty (4) = \frac{1}{4}(4) = 1$$

since the second integral is the variable factor of a gamma distribution with $\alpha = 2$, $\beta = 2$ and integrates to $\Gamma(2)2^2 = 4$.

5.53 Using $p(y)$ as derived in Exercise 5.28, we obtain

$$E(Y) = \sum_{y=0}^\infty y\left(\frac{1}{2}\right)^{y+1} = \frac{1}{4}\sum_{y=0}^\infty \frac{d}{da}a^y = \frac{1}{4}\frac{d}{da}\sum_{y=0}^\infty a^y$$

where $a = \frac{1}{2}$. The interchange of the summation sign and the derivative can be justified. Then

$$E(Y) = \frac{1}{4}\frac{d}{da}\left(\frac{1}{1-a}\right) = \frac{1}{4}\times\frac{1}{(1-a)^2} = \frac{1}{4}\times\frac{1}{\left(1-\frac{1}{2}\right)^2} = \frac{4}{4} = 1$$

5.55 $\text{Cov}(Y_1, Y_2) = E(Y_1 Y_2) - E(Y_1)E(Y_2).$

$$E(Y_1 Y_2) = \sum_{y_1} \sum_{y_2} y_1 y_2 p(y_1, y_2) = (0)(0)\left(\tfrac{1}{9}\right) + (1)(0)\left(\tfrac{2}{9}\right) + (2)(0)\left(\tfrac{1}{9}\right) + (0)(1)\left(\tfrac{2}{9}\right)$$
$$+ (1)(1)\left(\tfrac{2}{9}\right) + (0)(2)\left(\tfrac{1}{9}\right) = \tfrac{2}{9}.$$

Since Y_1 and Y_2 are both binomial with $n = 2$ and $p = \tfrac{1}{3}$,

$$E(Y_1) = E(Y_2) = 2\left(\tfrac{1}{3}\right) = \tfrac{2}{3}.$$

Thus, $\text{Cov}(Y_1, Y_2) = \left(\tfrac{2}{9}\right) - \left(\tfrac{2}{3}\right)\left(\tfrac{2}{3}\right) = -\tfrac{2}{9}.$

No, as value of Y_1 increases, value of Y_2 tends to decrease.

5.57 From Exercise 5.46; $E(Y_1) = E(Y_2) = \tfrac{2}{3}.$ Then

$$E(Y_1 Y_2) = \int_0^1 \int_0^1 4 y_1^2 y_2^2 \, dy_1 \, dy_2 = \int_0^1 \tfrac{4}{3} y_2^2 \, dy_2 = \tfrac{4}{9}$$

$$\text{Cov}(Y_1, Y_2) = \tfrac{4}{9} - \tfrac{4}{9} = 0.$$

No, Y_1 and Y_2 are independent.

5.59 From Exercise 5.49, $E(Y_1 Y_2) = 0.$ Calculate

$$E(Y_1) = \int_{-1}^0 \int_0^{1+y_1} y_1 \, dy_2 \, dy_1 = \int_0^1 \int_0^{1-y_1} y_1 \, dy_2 \, dy_1$$

$$= \int_{-1}^0 y_1(1 + y_1) \, dy_1 + \int_0^1 y_1(1 - y_1) \, dy_1 = -\tfrac{1}{2} + \tfrac{1}{3} + \tfrac{1}{2} - \tfrac{1}{3} = 0$$

Hence

$$\text{Cov}(Y_1, Y_2) = E(Y_1 Y_2) - E(Y_1)E(Y_2) = 0 - (0)E(Y_2) = 0$$

$$P = \frac{\text{Cov}(Y_1, Y_2)}{\sigma_1 \sigma_2} = 0$$

Note that $\text{Cov}(Y_1, Y_2) = 0$, even though Y_1 and Y_2 are dependent (c.f. Exercise 5.35).

5.61 The marginal distributions for Y_1 and Y_2 are shown in the accompanying tables.

y_1	$p_1(y_1)$		y_2	$p_2(y_2)$
-1	$\frac{1}{3}$		0	$\frac{2}{3}$
0	$\frac{1}{3}$		1	$\frac{1}{3}$
1	$\frac{1}{3}$			

Since, for example, $p(-1, 0) \neq p(-1)p(0)$, Y_1 and Y_2 are not independent. However,

$$E(Y_1) = -1\left(\frac{1}{3}\right) + 0\left(\frac{1}{3}\right) + 1\left(\frac{1}{3}\right) = 0$$

$$E(Y_1 Y_2) = (-1)(0)\left(\frac{1}{3}\right) + (0)(1)\left(\frac{1}{3}\right) + (1)(0)\left(\frac{1}{3}\right) = 0$$

so that $\text{Cov}(Y_1, Y_2) = 0$.

5.63 Refer to Theorem 5.12.

$$E(3Y_1 + 4Y_2 - 6Y_3) = 3(2) + 4(-1) - 6(4) = -22$$

$$V(3Y_1 + 4Y_2 - 6Y_3) = 9(4) + 16(6) + 36(8) + (2)(3)(4)(1) + (2)(3)(-6)(1)$$
$$+ 2(4)(-6)(0) = 480$$

5.65 $V(Y_1 - Y_2) = \frac{1}{18} + \frac{1}{18} - 2(0) = \frac{1}{9}$

(See Exercise 5.46 for $V(Y_1)$; $V(Y_2) = V(Y_2)$ by symmetry.)

5.67 Calculate

$$E(Y_1) = \int_0^1 \int_0^{1-y_2} 2y_1 \, dy_1 \, dy_2 = \int_0^1 (1-y_2)^2 \, dy_2 = \frac{1}{3}$$

$$E\left(Y_1^2\right) = \int_0^1 \int_0^{1-y_2} 2y_1^2 \, dy_1 \, dy_2 = \int_0^1 \frac{2}{3}(1-y_2)^3 \, dy_2 = \frac{1}{6}$$

$$V(Y_1) = \frac{1}{6} - \frac{1}{9} = \frac{1}{18}$$

By symmetry, $E(Y_2) = \frac{1}{3}$ and $V(Y_2) = \frac{1}{18}$.

$$E(Y_1 Y_2) = \int_0^1 \int_0^{1-y_2} 2y_1 y_2 \, dy_1 \, dy_2 = \int_0^1 (1-y_2)^2 \, dy_2 = \frac{1}{12}$$

and

$$\text{Cov}(Y_1, Y_2) = \tfrac{1}{12} - \tfrac{1}{9} = -\tfrac{1}{36}.$$

Then

$$E(Y_1 + Y_2) = \tfrac{2}{3}$$

and

$$V(Y_1 + Y_2) = V(Y_1) + V(Y_2) + 2\text{Cov}(Y_1, Y_2) = \tfrac{2}{18} - \tfrac{2}{36} = \tfrac{1}{18}.$$

5.69 Several intermediate results will be necessary.

(i) From Exercise 5.50, $E(Y_1) = \tfrac{7}{12}$ and $E(Y_2) = \tfrac{7}{12}.$

(ii) $$E(Y_1 Y_2) = \int_0^1 \int_0^1 (y_1 + y_2)\, y_1 y_2\, dy_1\, dy_2 = \int_0^1 \left[\frac{y_1^3 y_2}{3} + \frac{y_1^2 y_2^2}{2} \right]_0^1 dy_2$$

$$= \int_0^1 \left(\frac{y_2}{3} + \frac{y_2^2}{2} \right) dy_2 = \left[\frac{y_2^2}{6} + \frac{y_2^3}{6} \right]_0^1 = \tfrac{1}{3}$$

(iii) $$V(Y_1) = \int_0^1 \int_0^1 \left(y_1^3 + y_1^2 y_2 \right) dy_2\, dy_1 - [E(Y_1)]^2 = \int_0^1 \left(y_1^3 + \tfrac{1}{2} y_1^2 \right) dy_1 - \frac{49}{144}$$

$$= \left[\frac{y_1^4}{4} + \frac{y_1^3}{6} \right]_0^1 - \frac{49}{144} = \frac{11}{144}$$

and $V(Y_2) = V(Y_1) = \tfrac{11}{144}.$

(iv) $$\text{Cov}(Y_1, Y_2) = E(Y_1 Y_2) - E(Y_1)E(Y_2) = \tfrac{1}{3} - \left(\tfrac{7}{12} \right)\left(\tfrac{7}{12} \right) = -\tfrac{1}{144} = .0069$$

Thus

$$E(30Y_1 + 25Y_2) = 30E(Y_1) + 25E(Y_2) = 32.08$$

$$V(30Y_1 + 25Y_2) = 900\left(\tfrac{11}{144} \right) + 625\left(\tfrac{11}{144} \right) + 2(750)\left(-\tfrac{1}{144} \right) = 106.08$$

$$\sigma = \sqrt{V(30Y_1 + 25Y_2)} = 10.30$$

Using Tchebysheff's theorem with $k = 2$, the necessary interval is
$\mu \pm 2\sigma = 32.08 \pm 2(10.30) = 32.08 \pm 20.6$, or 11.48 to 52.68

5.71 It is known that X, the daily gain, is normally distributed with $E(X) = 50$ and $V(X) = 9$. Also, Y, the daily cost, is distributed as a gamma variable with $\alpha = 4$ and $\beta = 2$, so that $E(Y) = \alpha\beta = 8$ and $V(Y) = \alpha\beta^2 = 16$. The net daily gain is then $G = X - Y$. Since X and Y are independent,

$$E(X - Y) = E(X) - E(Y) = 50 - 8 = 42$$

and

$$V(X - Y) = V(X) + V(Y) = 9 + 16 = 25$$

Note that a gain of $G = 70$ lies 5.6 standard deviations away from the mean, $E(G) = 42$. We would not expect her gain to be higher than 70 since, even using the conservative Tchebysheff's theorem, at most $\left(\frac{1}{k^2}\right) = \left(\frac{1}{5.6^2}\right) = .032$ of the

measurements will fall beyond 5.6 standard deviations.

5.73 Refer to Example 5.29 in the text. The situation here is analogous to drawing n balls from an urn containing N balls, r_1 of which are red, r_2 of which are black, and $N - r_1 - r_2$ of which are of another color. Using the argument given there, we can deduce that

$$E(Y_1) = np_1 \qquad V(Y_1) = np_1(1 - p_1)\left(\frac{N - n}{N - 1}\right) \qquad \text{where} \qquad p_1 = \frac{r_1}{N}$$

$$E(Y_2) = np_2 \qquad V(Y_2) = np_2(1 - p_2)\left(\frac{N - n}{N - 1}\right) \qquad \text{where} \qquad p_2 = \frac{r_2}{N}$$

Define two new random variables for $i = 1, 2, \ldots, n$:

$$U_i = \begin{cases} 1, & \text{if alligator } i \text{ is male} \\ 0, & \text{if not} \end{cases} \qquad V_i = \begin{cases} 1, & \text{if alligator } i \text{ is female} \\ 0, & \text{if not} \end{cases}$$

Then $Y_1 = \sum\limits_{i=1}^{n} U_i$ and $Y_2 = \sum\limits_{i=1}^{n} V_i$. In order to find the variance of a linear form involving Y_1 and Y_2, it is necessary to find $\text{Cov}(Y_1, Y_2)$, which involves finding $E(Y_1 Y_2)$.

$$E(Y_1 Y_2) = E[(\Sigma U_i)(\Sigma V_i)] = E\left[\sum_{i=1}^{n} U_i V_i + \sum_{i \neq j} \sum U_i V_j\right]$$

$$= \sum_{i=1}^{n} E(U_i V_i) + \sum_{i \neq j} E(U_i V_j)$$

The only situation in which $U_i V_i$ would be nonzero is when $U_i = 1$ and $V_i = 1$, that is, when the ith alligator is both male and female. Since this event is impossible, $E(U_i V_i) = 0$ for all i. For $U_i V_j$ to be nonzero, we need $U_i = 1$ and $V_j = 1$, which happens with probability

$$P(U_i = 1, V_j = 1) = P(U_i = 1)P(V_j = 1 | U = 1) = \frac{r_1}{N} \times \frac{r_2}{N - 1} = p_1 \times \frac{N p_2}{N - 1}$$

$$= \frac{N}{N - 1} \times p_1 p_2$$

Since there are $n(n - 1)$ terms in $\sum\limits_{i \neq j} E(U_i V_j)$

$$E(Y_1 Y_2) = 0 + [n(n-1)] \times \frac{N}{N-1} \times p_1 p_2$$

$$\text{Cov}(Y_1, Y_2) = n(n-1) \times \frac{N}{N-1} \times p_1 p_2 - n^2 p_1 p_2 = \frac{n^2 p_1 p_2}{N-1} - \frac{nN}{M-1} \times p_1 p_2$$

$$= \frac{n(n-N)}{N-1} \times p_1 p_2$$

Then

$$E\left[\frac{Y_1}{n} - \frac{Y_2}{n}\right] = \frac{1}{n}(np_1 - np_2) = p_1 - p_1$$

$$V\left(\frac{Y_1}{n} - \frac{Y_2}{n}\right) = \frac{1}{n^2}V(Y_1 - Y_2) = \frac{1}{n^2}\left(V(Y_1) + V(Y_2) - 2\text{Cov}(Y_1 Y_2)\right)$$

$$= \frac{1}{n^2}\left(\frac{n(N-n)}{N-1}p_1(1-p_1) + \frac{n(N-n)}{N-1}p_2(1-p_2) - \frac{2n(n-N)}{N-1}p_1 p_2\right)$$

$$= \frac{N-n}{n(N-1)}\left[p_1(1-p_1) + p_2(1-p_2) + 2p_1 p_2\right]$$

$$= \frac{N-n}{n(N-1)}\left(p_1 + p_2 - (p_1 - p_2)^2\right)$$

5.75 **a.** Using the multinomial distribution with $p_1 = p_2 = p_3 = \frac{1}{3}$,

$$P(Y_1 = 4,\, Y_2 = 1,\, Y_3 = 2) = \frac{6!}{3!\,1!\,2!} = \left(\frac{1}{3}\right)^3 \left(\frac{1}{3}\right)^1 \left(\frac{1}{3}\right)^2 = 60\left(\frac{1}{3}\right)^6 = .0823$$

b. Refer to Theorem 5.13 in the text.

$$E(Y_1) = np_1 = \frac{n}{3} \qquad \text{and} \qquad V(Y_1) = np_1 q_1 = \frac{2n}{9}$$

c. Refer to Theorem 5.13 in the text.

$$\text{Cov}(Y_1, Y_2) = -np_2 p_3 = -n\left(\frac{1}{3}\right)\left(\frac{1}{3}\right) = -\frac{n}{9}$$

d. $E(Y_2 - Y_3) = \frac{n}{3} - \frac{n}{3} = 0 \qquad$ and $\qquad V(Y_2 - Y_3) = \frac{2n}{9} + \frac{2n}{9} - 2\left(-\frac{n}{9}\right) = \frac{6n}{9} = \frac{2n}{3}$

5.77 **a.** If N is large, the multinomial distribution is appropriate and

$$P(Y_1 = 2,\, Y_2 = 1) = \frac{5!}{2!\,1!\,2!} \times (.3)^2 (.1)^1 (.6)^2 = .0972$$

b. Using the multinomial means, variances, and covariances, we have

$$E\left[\frac{Y_1}{n} - \frac{Y_2}{n}\right] = p_1 - p_2 = .2$$

$$V\left[\frac{Y_1}{n} - \frac{Y_2}{n}\right] = \frac{p_1 q_2}{n} + \frac{p_2 q_2}{n} + 2\frac{p_1 p_2}{n} = \frac{.3(.7)}{5} + \frac{.1(.9)}{5} + \frac{2(.3)(.1)}{5} = .072$$

5.79 Let $Y_1 = \#$ of family home fires, $Y_2 = \#$ of apartment fires, $Y_3 = \#$ of fires in other types. The joint probability function for Y_1, Y_2, Y_3 is multinomial with $n = 4$, $p_1 = .73$, $p_2 = .2$, $p_3 = .07$.

$$P(Y_1 = 2, Y_2 = 1, Y_3 = 1) = \left(\frac{4!}{2!\,1!\,1!}\right)(.73)^2(.2)(.07) = .08953.$$

5.81 Let $Y_1 = \#$ of planes with no wing cracks, $Y_2 = \#$ of planes with detectable wing cracks, $Y_3 = \#$ of planes with critical wing cracks. The joint probability function for Y_1, Y_2, Y_3 is multinomial with $n = 5$, $p_1 = .7$, $p_2 = .25$, $p_3 = .05$.

a. $P(Y_1 = 2, Y_2 = 2, Y_3 = 1) = \left(\frac{5!}{2!\,2!\,1!}\right)(.7)^2(.25)^2(.05) = .046.$

b. The distribution of Y_3 is binomial with $n = 5$, $p = .05$. Thus,

$$P(Y_3 \geq 1) = 1 - P(Y_3 = 0) = 1 - (.95)^5 = 1 - .7738 = .2262.$$

5.83 Let $Y = \#$ of items with at least one defect. Y is binomial with $n = 10$, $p = .10 + .05 = .15$.

a. $P(Y = 2) = \binom{10}{2}(.15)^2(.85)^8 = .2759$

b. $P(Y \geq 1) = 1 - P(Y = 0) = 1 - (.85)^{10} = .8031.$

5.85 Similar to Exercise 5.84. It can be shown that

$$Q - \frac{(y_2 - \mu_2)^2}{\sigma_2^2} = \frac{\left[y_1 - \mu_1 - \left(\frac{p\sigma_1}{\sigma_2}\right)(y_2 - \mu_2)\right]^2}{\sigma_1^2(1 - p^2)}$$

Then, using the results in Exercise 5.84,

$$f(y_1|y_2) = \frac{f(y_1, y_2)}{f(y_2)} = \frac{\left(\dfrac{e^{-Q/2}}{2\pi\sigma_1\sigma_2\sqrt{1 - p^2}}\right)}{\left(\dfrac{e^{(-1/2)(y^2 - \mu_2)^2/\sigma_2^2}}{\sqrt{2\pi\sigma_2^2}}\right)} = \frac{\exp\left[-\frac{1}{2}\left(\dfrac{Q - (y_2 - \mu_2)^2}{\sigma_2^2}\right)\right]}{\sqrt{2\pi\sigma_1^2(1 - p^2)}}$$

$$= \frac{1}{\sqrt{2\pi\sigma_1^2(1 - p^2)}}\exp\left\{-\frac{1}{2}\frac{\left[y_1 - \left(\mu_1 + \left(\frac{p\sigma_1}{\sigma_2}\right)\right)(y_2 - \mu_2)\right]^2}{\sigma_1^2(1 - p^2)}\right\}$$

which is the density function of a normal distribution with mean

$$\mu_1 + p\left(\frac{\sigma_2}{\sigma_2}\right)(y_2 - \mu_2) \text{ and variance } \sigma_1^2(1 - p^2).$$

5.87 Refer to Exercise 5.17.

a. The conditional distribution of Y_1 given Y_2 is $f(y_1|y_2) = \frac{1}{y_2}$ for $0 \leq y_1 \leq y_2$.

Thus,

$$E(Y_1|Y_2 = 2) = \int_0^{y_2} y_1\left(\frac{1}{y_2}\right) dy_1 = \left[\frac{y_1^2}{2y_2}\right]_0^{y_2} = \frac{y_2}{2}$$

b. $E(Y_1) = E[E(Y_1|Y_2 = 2)] = \int_0^1 \frac{y_2}{2}\left(6y_2 - 6y_2^2\right) dy_2 = \left[\frac{3y_2^3}{3} - \frac{3}{4}y_2^4\right]_0^1 = \frac{1}{4}$

(same as answer to Exercise 5.47).

5.89 Refer to Exercise 5.27.

a. $E(Y|p) = \sum_{y=0}^{3} y p^y q^{3-y} = 3p$, the expected value of a binomial random variable.

Then, since $f(p) = 1$, $0 \leq p \leq 1$,

$$E(Y) = E[E(Y|p)] = \int_0^1 3p \; dp = \frac{3}{2}$$

b. $V(Y|p) = 3p(1-p)$ and from part (a), $E(Y|p) = 3p$.

$$V(Y) = E(3p(1-p)) + V(3p)$$

Since p is uniformly distributed on the interval $(0, 1)$, $E(p) = \frac{1}{2}$, $V(p) = \frac{1}{12}$,

and $E(p_2) = V(p) + [E(p)]^2 = \frac{1}{12} + \frac{1}{4} = \frac{1}{3}$. Then

$$V(Y) = 3\left[E(p) - E(p^2)\right] + 9V(p) = 3\left(\frac{1}{2} - \frac{1}{3}\right) + 9\left(\frac{1}{12}\right) = 1.25$$

5.91 Refer to Exercise 5.24. There we obtained $f(y_2|y_1) = \frac{1}{y_1}$ for $0 \leq y_2 \leq y_1$. For this

exercise we need $f\left(y_2|y_1 = \frac{3}{4}\right) = \frac{4}{3}$ for $0 \leq y_2 \leq \frac{3}{4}$. Then

$$E\left(Y_2|Y_1 = \frac{3}{4}\right) = \int_0^{3/4} \frac{4}{3}y_2 \; dy_2 = \frac{4}{6}y_2^2\Big|_0^{3/4} = \frac{4}{6} \times \frac{9}{16} = \frac{3}{8}$$

5.93 Consider the random variable $y_1 Y_2$ for a fixed value of Y_1. Then $y_1 Y_2$ has a normal distribution with mean 0 and variance y_1^2. Refer to Exercise 4.80. Hence

$$E\left(e^{t y_1 Y_2}|Y_1 = y_1\right) = e^{t^2 y_1^2/2}$$

Using Theorem 5.14, we have

$$E\left(e^{t Y_1 Y_2}\right) = m_{Y_1 Y_2}(t) = E\left(e^{t^2 Y_1^2/2}\right) = \int_{-\infty}^{\infty} \frac{1}{\sqrt{2\pi}} e^{t^2 y_1^2/2} e^{-y_1^2/2}\, dy_1$$

$$= \int_{-\infty}^{\infty} \frac{1}{\sqrt{2\pi}} e^{(-y_1^2/2)(1-t^2)}\, dy_1 = \left(1 - t^2\right)^{-1/2}$$

since the variable factor is that of a normal random variable with mean 0 and variance $\dfrac{1}{(1 - t^2)}$. Now

$$m'(0) = E(U) = t\left(1 - t^2\right)^{-3/2}\Big|_{t=0} = 0$$

$$m''(0) = E\left(U^2\right) = 3t^2 \left(1 - t^2\right)^{-5/2} + \left(1 - t^2\right)^{-3/2}\Big|_{t=0} = 1$$

Evaluating $E(U)$ and $V(U)$ directly and using the independence of Y_1 and Y_2, we have

$$E(U) = E(Y_1 Y_2) = E(Y_1)E(Y_2) = 0$$

$$E(U^2) = E\left(Y_1^2 Y_2^2\right) = E\left(Y_1^2\right)E\left(Y_2^2\right) = 1(1) = 1$$

$$V(U) = 1$$

5.95 Let Y_1 = arrival time for the first friend, $0 \le y_1 \le 1$, Y_2 = arrival time for the second friend, $0 \le y_2 \le 1$; $f(y_1, y_2) = 1$.

If friend 2 arrives $\frac{1}{6}$ hour (10 min.) before or $\frac{1}{6}$ hour after friend 1, they will meet. We can represent this as $|Y_1 - Y_2| < \frac{1}{3}$. We consider

$$P\left[|Y_1 - Y_2| < \tfrac{1}{3}\right] = \int_{0}^{1/6} \int_{0}^{y_1+(1/6)} dy_2\, dy_1 + \int_{1/6}^{5/6} \int_{y_1-(1/6)}^{y_1+(1/6)} dy_2\, dy_1$$

$$+ \int_{5/6}^{1} \int_{y_1-(1/6)}^{1} dy_2\, dy_1$$

$$= \int\limits_{0}^{1/6} \left(y_1 + \tfrac{1}{6}\right) dy_1 + \int\limits_{1/6}^{5/6} \left(\tfrac{1}{3}\right) dy_1 + \int\limits_{5/6}^{1} \left[\left(\tfrac{7}{6}\right) - y_1\right] dy_1$$

$$= \left(\frac{y_1^2}{2} + \frac{y_1}{6}\right)\Big]_{0}^{1/6} + \left(\frac{y_1}{3}\right)\Big]_{1/6}^{5/6} + \left(7y_1 - \frac{y_1^2}{2}\right)\Big]_{5/6}^{1} = \frac{22}{72} = \frac{11}{36}.$$

5.97 **a.** $f(y_1) = \int\limits_{0}^{y_1} 3y_1 \, dy_2 = 3y_1 y_2\Big]_{0}^{y_1} = 3y_1^2; \ 0 \le y_1 \le 1$

$$f(y_2) = \int\limits_{y_2}^{1} 3y_1 \, dy_1 = \frac{3y_1^2}{2}\Big]_{y_2}^{1} = \left(\tfrac{3}{2}\right)\left(1 - y_2^2\right); \ 0 \le y_2 \le 1$$

b. $P\left(Y_1 \le \tfrac{3}{4} \middle| Y_2 \le \tfrac{1}{2}\right) = \dfrac{P\left(Y_1 \le \tfrac{3}{4}, Y_2 \le \tfrac{1}{2}\right)}{P\left(Y_2 \le \tfrac{1}{2}\right)}$

$$= \frac{\displaystyle\int\limits_{0}^{1/2}\int\limits_{0}^{y_1} 3y_1 \, dy_2 \, dy_1 + \int\limits_{1/2}^{3/4}\int\limits_{0}^{1/2} 3y_1 \, dy_2 \, dy_1}{\displaystyle\int\limits_{0}^{1/2} \left(\tfrac{3}{2}\right)\left(1 - y_2^2\right) dy_2}$$

$$= \frac{\tfrac{1}{8} + \tfrac{15}{64}}{\tfrac{11}{16}} = \frac{23}{44}$$

c. If $0 \le y_2 < 1$, $f(y_1 | y_2) = \dfrac{f(y_1, y_2)}{f(y_2)} = \dfrac{3y_1}{\left(\tfrac{3}{2}\right)\left(1 - y_2^2\right)} = \dfrac{2y_1}{\left(1 - y_2^2\right)} \quad y_2 \le y_1 \le 1$

d. $P\left(Y_1 \le \tfrac{3}{4} \middle| Y_2 = \tfrac{1}{2}\right) = \int\limits_{1/4}^{3/4} f\left(y_1 \middle| y_2 = \tfrac{1}{2}\right) dy_1 = \int\limits_{1/4}^{3/4} \dfrac{2y_1}{\left(1 - \tfrac{1}{4}\right)} dy_1 = \dfrac{8y_1^2}{6}\Big]_{1/4}^{3/4} = \dfrac{2}{3}$

5.99 **a.** Since Y_1 and Y_2 are independent,

$$f(y_1, y_2) = f(y_1) f(y_2) = \left(\frac{e^{-y_1/\beta}}{\beta}\right)\left(\frac{e^{-y_2/\beta}}{\beta}\right)$$

$$= \left(\frac{1}{\beta^2}\right)e^{-(y_1+y_2)/\beta}; \ y_1 > 0, \ y_2 > 0.$$

Note the similarity with Exercise 5.41.

b. $P(Y_1 + Y_2 \leq a) = \displaystyle\int_0^a \int_0^{a-y_2} \left(\frac{1}{\beta^2}\right)e^{-(y_1+y_2)/\beta} \, dy_1 \, dy_2$

$$= \left(\frac{1}{\beta}\right)\int_0^a \left[1 - e^{-(a-y_2)/\beta}\right]e^{-y_2/\beta} \, dy_2$$

$$= -e^{-y_2/\beta}\Big]_0^a - \left(\frac{a}{\beta}\right)e^{-a/\beta}$$

$$= 1 - \left[1 + \left(\frac{\alpha}{\beta}\right)\right]e^{-a/\beta} \quad a > 0$$

5.101 a.   Let X be the number of eggs laid by an insect, and let Y be the number of eggs hatched. Given that X eggs were laid, Y has a binomial distribution with $p = P(\text{egg hatched})$. Hence $E(Y|x) = px$. Since X has a Poisson distribution with parameter λ, $E(Y) = E(pX) = pE(X) = p\lambda$.

b. $V(Y|x) = xp(1-p)$ and, from part (a), $E(Y|x) = xp$.

$V(Y) = E[V[Y|X]] + V[E(Y \mid X)] = E[Xp(1-p)] + V[Xp]$.

$E(x) = \lambda$ and $V(x) = \lambda$. Then

$V(Y) = p(1-p)E(X) + p^2 V(X) = p\lambda - p^2\lambda + p^2\lambda = p\lambda$.

5.103 $f(y) = \displaystyle\int_0^\infty p(y|\lambda)f(\lambda) \, d\lambda = \int_0^\infty \frac{\lambda^y e^{-\lambda}}{y!} \times \frac{1}{\Gamma(a)\beta^\alpha} e^{-\lambda/\beta} \lambda^{\alpha-1} \, d\lambda$

$$= \int_0^\infty \frac{\lambda^{y+\alpha-1}e^{-\lambda[(\beta+1)/\beta]}}{\Gamma(y+1)\Gamma(\alpha)\beta^\alpha} \, d\lambda \quad y = 0, 1, 2, \ldots$$

$$= \frac{\Gamma(y+\alpha)\left(\frac{\beta}{\beta+1}\right)^{y+\alpha}}{\beta^\alpha \Gamma(y+1)\Gamma(\alpha)} = \binom{y+\alpha-1}{y}\left(\frac{\beta}{\beta+1}\right)^y \left(\frac{\beta}{\beta+1}\right)^\alpha \left(\frac{1}{\beta}\right)^\alpha$$

$$= \binom{y+\alpha-1}{y}\left(\frac{\beta}{\beta+1}\right)^y \left(\frac{1}{\beta+1}\right)^\alpha = \frac{\Gamma(y+\alpha)}{\Gamma(y+1)\Gamma(\alpha)}\left(\frac{\beta}{\beta+1}\right)^y \left(\frac{1}{\beta+1}\right)^\alpha,$$

$$y = 0, 1, 2, \ldots$$

5.105 The mean and variance of the geometric random variable W_i are given in Table 3.2 as

$$E(W_i) = \frac{1}{p} \qquad \text{and} \qquad V(W_i) = \frac{q}{p^2}$$

Since W_i and W_j are independent, $\text{Cov}(W_i, W_j) = 0$ for $i \neq j$, and

$$E(Y) = E\left(\sum_{i=1}^{r} W_i\right) = \sum_{i=1}^{r} E(W_i) = \frac{r}{p}$$

$$V(Y) = E\left(\sum_{i=1}^{r} W_i\right) = \sum_{i=1}^{r} V(W_i) = \frac{rq}{p^2}$$

5.107 $E(\overline{Y} - \overline{X}) = E\left(\dfrac{\sum Y_i}{n} - \dfrac{\sum X_i}{m}\right) = \mu_1 - \mu_2$

$$V(\overline{Y} - \overline{X}) = V(\overline{Y}) + V(\overline{X}) = \frac{n\sigma_1^2}{n^2} + \frac{m\sigma_2^2}{m^2} = \frac{\sigma_1^2}{n} + \frac{\sigma_2^2}{m} \text{ since } \overline{Y} \text{ and } \overline{X} \text{ are independent.}$$

5.109 **a.** $m(t_1, t_2, t_3) = \sum_{x_1} \sum_{x_2} \sum_{x_3} \dfrac{n!}{x_1! \, x_2! \, x_3!} \, e^{t_1 x_1 + t_2 x_2 + t_3 x_3} \, p_1^{x_1} p_2^{x_2} p_3^{x_3}$

$$= \sum_{x_1} \sum_{x_2} \sum_{x_3} \frac{n!}{x_1! \, x_2! \, x_3!} \left(p_1 e^{t_1}\right)^{x_1} \left(p_2 e^{t_2}\right)^{x_2} \left(p_3 e^{t_3}\right)^{x_3}$$

$$= \left(p_1 e^{t_1} + p_2 e^{t_2} + p_3 e^{t_3}\right)^n$$

$\left(\sum x_i = n\right)$, using the multinomial theorem.

b. The moment-generating function of X_1 is given by

$m(t, 0, 0) = \left(p_1 e^t + p_2 + p_3\right)^n$, which is the moment-generating function of a binomial random variable with $p = p_1$ and $q = p_2 + p_3$. Hence, by Theorem 6.1, X_1 must be binomial with parameter p_1.

c. $E(X_1) = \dfrac{\partial m(t_1, t_2, t_3)}{\partial t_1}\bigg|_{t_1 = t_2 = t_3 = 0} = n\left(p_1 e^{t_1} + p_2 e^{t_2} + p_3 e^{t_3}\right)^{n-1}\left(p_1 e^{t_1}\right)\bigg|_{t_i = 0}$

$$= np_1$$

Similarly,

$$E(X_2) = \frac{\partial m(t_1, t_2, t_3)}{\partial t_2}\bigg|_{t_1 = 0} = n\left(\sum_{i=1}^{3} p_i e^{t_i}\right)^{n-1}\left(p_2 e^{t_2}\right)\bigg|_{t_i = 0} = np_2$$

Also,

$$E(X_1 X_2) = \frac{\partial^2 m(t_1,\, t_2,\, t_3)}{\partial m_1 \partial t_2}\Bigg|_{t_1=0} = n(n-1)\left(\sum_{i=1}^{3} p_i e^{t_i}\right)^{n-2} p_1 e^{t_1} p_2 e^{t_2}\Bigg|_{t_i=0}$$

$$= n(n-1)p_1 p_2$$

so that $\text{Cov}(X_1,\, X_2) = -n p_1 p_2$.

5.111 a. For this exercise a quadratic form of interest is

$$At^2 + Bt + C = E\left(Y_1^2\right)t^2 + [-2E(Y_1 Y_2)]t + E(Y_2)^2$$

Since $E\left[(ty_1 - Y_2)^2\right]$ is the integral of a nonnegative quantity, it must be nonnegative, so that we must have $At^2 + Bt + C \geq 0$. In order to satisfy this inequality, the quadratic form must not dip below the horizontal axis in Figure 5.14. That is, the two roots of the equation $At^2 + Bt + C$ must either be imaginary (a) or equal (b). In terms of the discriminant, then,

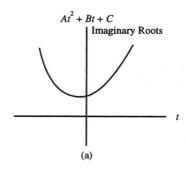

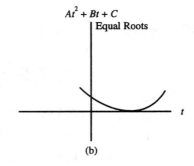

(a)

(b)

Figure 5.14a   Figure 5.14b

$$B^2 - 4AC \leq 0$$

$$[-2E(Y_1 Y_2)]^2 - 4E\left(Y_1^2\right)E\left(Y_2^2\right) \leq 0$$

and thus,

$$[E(Y_1 Y_2)]^2 \leq E\left(Y_1^2\right)E\left(Y_2^2\right)$$

b. Consider

$$\rho^2 = \frac{[E(Y_1 - \mu_1)(Y_2 - \mu_2)]^2}{E(Y_1 - \mu_1)^2 E(Y_2 - \mu_2)^2} = \frac{[E(Z_1 Z_2)]^2}{E\left(Z_1^2\right)E\left(Z_2^2\right)} \quad \text{where } Z_i = Y_i - \mu_i \;\; i = 1,\, 2$$

Using part (a), we know

$$[E(Z_1 Z_2)]^1 \leq E\left(Z_1^2\right)E\left(Z_2^2\right) \qquad \text{or} \qquad \rho^2 = \frac{[E(Z_1 Z_2)]^2}{E\left(Z_1^2\right)E\left(Z_2^2\right)} \leq 1$$

CHAPTER 6 FUNCTIONS OF RANDOM VARIABLES

6.1 **a.** Using the distribution function approach, we write

$$F_{U_1}(u) = P(U_1 \leq u) = P(2Y - 1 \leq u) = P\left(Y \leq \frac{u+1}{2}\right) = F_Y\left(\frac{u+1}{2}\right)$$

Now

$$F_Y(y) = \int_0^y (2 - 2t)\, dt = 2y - y^2, \qquad \text{for } 0 \leq y \leq 1$$

$$F_Y(y) = 0, \qquad \text{for } y < 0$$

$$F_Y(y) = 1, \qquad \text{for } y > 1$$

Since $F_{U_1}(u) = F_Y\left(\frac{u+1}{2}\right)$, the distribution function can be written as

$$F_{U_1}(u) = \begin{cases} 0, & \frac{u+1}{2} < 0 \\ 2\left(\frac{u+1}{2}\right) - \left(\frac{u+1}{2}\right)^2, & 0 \leq \frac{u+1}{2} \leq 1 \\ 1, & \frac{u+1}{2} > 1 \end{cases}$$

The density function can be obtained by differentiating $F_{U_1}(u)$ with respect to u. Thus,

$$f_{U_1}(u) = -\frac{u}{2} + \frac{1}{2} = \frac{1-u}{2}, \qquad -1 \leq u \leq 1$$

$$= 0, \qquad\qquad\qquad \text{elsewhere}$$

b. Similar to part (a).

$$F_{U_2}(u) = P(1 - 2Y \leq u) = P\left(Y \geq \frac{1-u}{2}\right) = 1 - F_Y\left(\frac{1-u}{2}\right)$$

$F_Y(y)$ was given in part (a), so that

$$F_{U_1}(u) = \begin{cases} 1 - 0, & \frac{1-u}{2} \leq 0 \\ 1 - 2\left(\frac{1-u}{2}\right) + \left(\frac{1-u}{2}\right)^2, & 0 \leq \frac{1-u}{2} \leq 1 \\ 1 - 1, & \frac{1-u}{2} > 1 \end{cases}$$

or

$$F_{U_2}(u) = \begin{cases} 0, & u < -1 \\ \dfrac{1 + 2u + u2}{4}, & -1 \le u \le 1 \\ 1, & u > 1 \end{cases}$$

By differentiating $F_{U_2}(u)$, we obtain

$$f_{U_2}(u) = \tfrac{1}{2} + \tfrac{u}{2} = \tfrac{u+1}{2}, \qquad\qquad -1 \le u \le 1$$

$$= 0, \qquad\qquad\qquad \text{elsewhere}$$

c. Write

$$F_{U_3}(u) = P(U \le u) = P(Y^2 \le u) = P(Y \le \sqrt{u}) = F_Y(\sqrt{u})$$

since Y is defined only on the range $0 \le y \le 1$. Hence

$$F_{U_3}(u) = \begin{cases} 0, & u < 0 \\ 2\sqrt{u} - u, & 0 \le u \le 1 \\ 1, & u > 1 \end{cases}$$

and

$$f_{U_3}(u) = \frac{1}{\sqrt{u}} - 1 = \frac{1 - \sqrt{u}}{\sqrt{u}}, \qquad\qquad 0 \le u \le 1$$

$$= 0, \qquad\qquad\qquad \text{elsewhere}$$

d. Using the derived density functions, we have

$$E(U_1) = \int_{-\infty}^{\infty} u f_{U_1}(u)\, du = \int_{-1}^{1} u\left(\frac{1-u}{2}\right) du = \left[\tfrac{1}{4}u^2 - \tfrac{1}{5}u^3\right]_{-1}^{1}$$

$$= \tfrac{1}{4} - \tfrac{1}{6} - \tfrac{1}{4} - \tfrac{1}{6} = -\tfrac{1}{3}$$

$$E(U_2) = \int_{-\infty}^{\infty} u f_{U_2}(u)\, du = \int_{-1}^{1} \left(\frac{u^2}{2} + \tfrac{1}{2}u\right) du = \left[\frac{u^3}{6} + \tfrac{1}{4}u^2\right]_{-1}^{1}$$

$$= \tfrac{1}{6} + \tfrac{1}{4} + \tfrac{1}{6} - \tfrac{1}{4} = \tfrac{1}{3}$$

$$E(U_3) = \int_{-\infty}^{\infty} u f_{U_3}(u) \ du = \int_0^1 \left(\frac{u - u\sqrt{u}}{\sqrt{u}} \right) du = \int_0^1 (\sqrt{u} - u) \ du$$

$$= \left[\tfrac{2}{3} u^{3/2} - \tfrac{1}{2} u^2 \right]_0^1$$

$$= \tfrac{1}{6}$$

e.   Calculate

$$E(Y) = \int_0^1 (2 - 2y)y \ dy = \left[y^2 - \tfrac{2}{3} y^3 \right]_0^1 = \tfrac{1}{3}$$

$$E(Y^2) = \int_0^1 (2 - 2y)y^2 \ dy = \left[\tfrac{2}{3} y^3 - \tfrac{2}{4} y^4 \right]_0^1 = \tfrac{2}{12} = \tfrac{1}{6}$$

Then

$$E(U_1) = 2E(Y) - 1 = 2\left(\tfrac{1}{3}\right) - 1 = -\tfrac{1}{3}$$

$$E(U_2) = 1 - 2E(Y) = 1 - 2\left(\tfrac{1}{3}\right) = \tfrac{1}{3}$$

$$E(U_3) = E(Y^2) = \tfrac{1}{6}$$

6.3   Similar to Exercise 6.2. Calculate

$$F_Y(y) = \begin{cases} \dfrac{y^2}{2}, & 0 \le y \le 1 \\[2mm] \left(\dfrac{1}{2}\right) + (y - 1), & 1 \le y \le 1.5 \\[2mm] 1, & y \ge 1.5 \end{cases}$$

a. $F_U(u) = P(10Y - 4 \le u) = P\left(Y \le \dfrac{u+4}{10}\right) = F_Y\left(\dfrac{u+4}{10}\right)$

Hence

$$F_U(u) = \begin{cases} \dfrac{(u+4)^2}{200}, & -4 \le u \le 6 \\[2mm] \left(\dfrac{1}{2}\right) + \dfrac{u-6}{10}, & 6 \le u \le 11 \\[2mm] 1, & u > 11 \end{cases}$$

Differentiating with respect to u, we have

$$f_U(u) = \begin{cases} \dfrac{u+4}{100}, & -4 \le u \le 6 \\[2mm] \dfrac{1}{10}, & 6 \le u \le 11 \\[2mm] 0, & \text{elsewhere} \end{cases}$$

b. $E(U) = \displaystyle\int_{-4}^{6} \frac{u^2 + 4u}{100}\,du + \int_{6}^{11} \frac{u}{10}\,du = \left[\frac{u^3}{300} + \frac{u^2}{50}\right]_{-4}^{6} + \left[\frac{u^2}{20}\right]_{6}^{11} = 5.583$

c.   Using an alternative procedure, calculate

$$E(Y) = \int_{0}^{1} y^2\,dy + \int_{1}^{1.5} y\,dy = \frac{1}{3} + \left[\frac{y^2}{2}\right]_{1}^{1.5} = \frac{23}{24}$$

Then $E(U) = 10\left(\frac{23}{24}\right) - 4 = 5.58$.

6.5   It is given that $f(y) = \frac{1}{4}$ for $1 \le y \le 5$. Then

$F_Y(y) = \dfrac{y}{4}$ $\qquad\qquad\qquad\qquad\qquad$ for $1 \le y \le 5$

$F_U(u) = P\left(2Y^2 + 3 \le u\right) = P\left[Y \le \sqrt{\dfrac{u-3}{2}}\right]$ $\qquad$ for $1 \le \sqrt{\dfrac{u-3}{2}} \le 5$

$\qquad = \dfrac{1}{4}\sqrt{\dfrac{u-3}{2}}$ $\qquad\qquad\qquad\qquad\qquad$ for $5 \le u \le 53$

Differentiating, we have

$$f_U(u) = \begin{cases} \dfrac{1}{16}\left(\dfrac{u-3}{2}\right)^{-1/2} = \dfrac{1}{8\sqrt{2(u-3)}}, & 5 \le u \le 53 \\[3mm] 0, & \text{elsewhere} \end{cases}$$

6.7   The region on which the density is positive is shown in Figure 6.2. This triangle has area $\frac{1}{2}$, so that the uniform density will be

$f(y_1, y_2) = 2$ $\qquad$ $0 \le y_1 \le 1,\ 0 \le y_2 \le 1,\ 0 \le y_1 + y_2 \ge 1$

a.   Let $Y_1 + Y_2 = u$, where $0 \le u \le 1$.

$\qquad F_U(u) = P(Y_1 + Y_2 \le u)$

$$= \int_0^u \int_0^{u-y_2} 2 \, dy_1 \, dy_2$$

$$= \int_0^u 2(u - y_2) \, dy_2$$

$$= u^2 \qquad 0 \le u \le 1$$

Differentiating, we have

$$f_U(u) = \begin{cases} 2u, & 0 \le u \le 1 \\ 0, & \text{elsewhere} \end{cases}$$

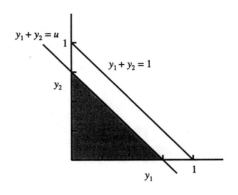

Figure 6.2

b. $E(U) = \int_0^1 2u^2 \, du = = \frac{2}{3} u^3 \Big]_0^1 = \frac{2}{3}$

c. We must first obtain the marginal densities of Y_1 and Y_2. Integrating over the ranges shown in Figure 6.2, we have

$$f_1(y_1) = \int_0^{1-y_1} 2 \, dy_2 = 2(1 - y_1) \qquad\qquad 0 \le y_1 \le 1$$

$$f_2(y_2) = \int_0^{1-y_2} 2 \, dy_1 = 2(1 - y_2) \qquad\qquad 0 \le y_2 \le 1$$

The marginal densities are identical. Then

$$E(Y_i) = \int_0^1 2y_i(1 - y_i) \, dy_i = \left[y_i^2 - \frac{2}{3} y_i^3 \right]_0^1 = \frac{1}{3}$$

and $E(Y_1 + Y_2) + E(Y_1) + E(Y_2) = \frac{2}{3}$.

6.9 It is given that $f(y_i) = e^{-y_i}$ for $i = 1, 2$.

Since Y_1 and Y_2 are independent,

$f(y_1, y_2) = e^{-(y_1 + y_2)}$ for $y_1 \ge 0$, $y_2 \ge 0$.

a.   Let $U = \dfrac{Y_1 + Y_2}{2}$. Then (see Figure 6.4)

$$F_U(u) = P\left(\frac{Y_1 + Y_2}{2} \le u\right)$$

$$= P(Y_1 + Y_2 \le 2u)$$

$$= \int_0^{2u} \int_0^{2u - y_2} e^{-(y_1 + y_2)} \, dy_1 \, dy_2$$

$$= \int_0^{2u} \left(e^{-y_2} - e^{-2u}\right) dy_2$$

$$= 1 - e^{-2u} - 2ue^{-2u} \quad \text{for } u \ge 0$$

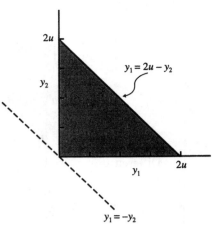

Figure 6.4

Differentiating with respect to u, we have

$$f_U(u) = 2e^{-2u} + 4ue^{-2u} - 2e^{-2u} = 4ue^{-2u} \quad \text{for } \mu \ge 0$$

b.   Since U has a gamma distribution with $\alpha = 2$, $\beta = \frac{1}{2}$, then $E(U) = \alpha\beta = 1$ and $V(U) = \alpha\beta^2 = \frac{1}{2}$. Using Theorem 5.12 with $E(Y_i) = V(Y_i) = 1$, we obtain

$$E(U) = \tfrac{1}{2}(1) + \tfrac{1}{2}(1) = 1 \qquad\qquad V(U) = \tfrac{1}{4}(1+1) = \tfrac{1}{2}$$

6.11   From Exercise 4.4, $F_Y(y) = 1 - e^{-y^2}$, $y \ge 0$. Suppose that U has a uniform distribution on $[0, 1]$, in which case $P(U \le u) = u$, $0 < u < 1$. We want to find $Y = G(U)$ such that $F_Y(y) = 1 - e^{-y^2}$, $w \ge 0$. Now, $G^{-1}(U) = Y$, and so

$$F_Y(y) = P(Y \le y) = P(G(U) \le y) = P\left(U \le G^{-1}(y)\right) = P(U \le u) = u, \quad 0 < u < 1$$

since U is uniform on $[0, 1]$. Thus, we want $u = 1 - e^{-y^2}$ so $-y^2 = \ln(1-u)$ and $y = [-\ln(1-u)]^{1/2}$. So $G(U) = [-\ln(1-U)]^{1/2}$.

(NOTE: $F_Y(y) = P\left([-\ln(1-u)]^{1/2} \le y\right) = P\left(1 - u \ge e^{-y^2}\right) = P\left(u \le 1 - e^{-y^2}\right)$
$= 1 - e^{-y^2}$, $y > 0$.)

6.13 **a.**   Taking the derivative of F, we have

$$f(y) = \frac{\alpha y^{\alpha - 1}}{\theta^\alpha}, \ 0 \le y \le \theta$$

b.   This is similar to Exercise 6.11. Let $W = G(U)$. We need $F_Y(y) = u = \left(\dfrac{y}{\theta}\right)^\alpha$ or $y = \theta u^{1/\alpha}$. So, $G(U) = \theta U^{1/\alpha}$.

c. From part (b), $y = 4\sqrt{u}$. The values are $4\sqrt{.2700} = 2.0785$, $4\sqrt{.6901} = 3.229$, $4\sqrt{.1413} = 1.5036$, $4\sqrt{.1523} = 1.5610$, $4\sqrt{.3609} = 2.4030$.

6.15 From Exercise 6.14, Y is always positive, so that $X = \frac{1}{Y}$ is always positive. Thus, $F_X(x) = 0$ if $x < 0$. Consider $x \geq 0$.

$$F_X(x) = P\left(\frac{1}{Y} \leq x\right) = P\left(Y \geq \frac{1}{x}\right) = 1 - F_Y\left(\frac{1}{x}\right)$$

$$= 1 - \{1 - (\beta x)^\alpha\} \qquad \text{if } \frac{1}{x} \geq \beta$$

$$= 1 \qquad\qquad\qquad \text{if } \frac{1}{x} < \beta$$

Thus,

$$F_X(x) = \begin{cases} 0, & x < 0 \\ \left(\dfrac{x}{\beta^{-1}}\right)^\alpha, & 0 \leq x \leq \beta^{-1} \\ 1, & x > \beta^{-1} \end{cases}$$

so the result is shown.

6.17 By definition,

$$P(X = i) = P[F(i-1) < U \leq F(i)] = P[U \leq F(i)] - P[U \leq F(i-1)]$$

$$= F(i) - F(i-1), \; i = 1, 2, 3, \ldots$$

since $P(U \leq a) = a$ for any $0 \leq a \leq 1$. From Exercise 4.1

$$P(Y = i) = F(i) - F(i-1), \; i = 1, 2, 3, \ldots.$$

Therefore X has the same distribution as Y.

6.19 a. If $U = 2Y - 1$, then $Y = \dfrac{U+1}{2}$. Differentiating, we have $\dfrac{dy}{du} = \dfrac{1}{2}$ and

$$f_U(u) = \frac{1}{2}(2)\left(1 - \frac{u+1}{2}\right) = \frac{1-u}{2} \quad \text{for} \quad 0 \leq \frac{u+1}{2} \leq 1 \quad \text{or} \quad -1 \leq u \leq 1$$

b. If $U = 1 - 2Y$, then $Y = \dfrac{1-U}{2}$, and $\dfrac{dy}{du} = \dfrac{1}{2}$. Then

$$f_U(u) = \left(1 - \frac{1-u}{2}\right) = \frac{1+u}{2} \quad \text{for} \quad 0 \leq \frac{1-u}{2} \leq 1 \quad \text{or} \quad -1 \leq u \leq 1$$

c. If $U = Y^2$, then $Y = \sqrt{U}$ and $\dfrac{dy}{du} = \dfrac{1}{2\sqrt{U}}$. Then (since $u = y^2$ is increasing for $y > 0$)

$$f_U(u) = \frac{1}{2\sqrt{u}} \times 2(1 - \sqrt{u}) = \frac{1 - \sqrt{u}}{\sqrt{u}} \quad \text{for} \quad 0 \le \sqrt{u} \le 1 \quad \text{or} \quad 0 \le u \le 1$$

6.21 The variable of interest is $U = \dfrac{Y_1 + Y_2}{2}$. Fix $Y_2 = y_2$. Then $Y_1 = 2u - y_2$ and

$\dfrac{dy_1}{du} = 2$. The joint density for U and Y_2 is $g(u, y_2) = 2e^{-2u}$, where $u \ge 0$, $y_2 \ge 0$,

and $2u - y_2 \ge 0$, or $y_2 \le 2u$. Integrating over all possible values of y_2, we have

$$f_U(u) = \int_0^{2u} 2e^{-2u} \, dy_2 = 4ue^{-2u} \qquad \text{for } u \ge 0$$

6.23 $f_Y = \left(\dfrac{1}{\beta}\right)e^{-y/\beta}$

   a. $F_w(w) = P(W \le w) = P(Y \le w^2)$

$$f_w(w) = f_Y(w^2)(2w) = \left(\frac{2w}{\beta}\right)e^{-w^2/\beta} \qquad \text{for } w > 0$$

      which is Weibull with $\alpha = \beta$ and $m = 2$.

   b. In Exercise 6.22, part (b), it was shown that

$$E(Y^k) = \Gamma\left(\frac{1+k}{m}\right)\alpha^{k/m}.$$

      Thus,

$$E(Y^{k/2}) = E(W^k) = \Gamma\left(\frac{1+k}{2}\right)\beta^{k/2}.$$

6.25 **a.** If $w = \dfrac{mV^2}{2}$, then $V = \sqrt{\dfrac{2W}{m}}$ and $\left|\dfrac{dv}{dw}\right| = \left(\dfrac{1}{2}\right)\sqrt{\dfrac{2}{m}}(w^{-1/2}) = \dfrac{1}{\sqrt{2mw}}$. Then

$$f_W(w) = \frac{a\left(\frac{2w}{m}\right)}{\sqrt{2mw}} \times e^{-2bw/m} = \frac{a\sqrt{2w}}{m^{3/2}} \times e^{-w/kT} = \frac{a\sqrt{2}}{m^{3/2}} \times w^{1/2} \times e^{-w/kT}$$

Since the above density must integrate to 1 and since the variable part of the density is that of a gamma variable with $\alpha = \frac{3}{2}$ and $\beta = kT$, a must be chosen so that

$$\frac{a\sqrt{2}}{m^{3/2}} = \frac{1}{\Gamma\left(\frac{3}{2}\right)(kT)^{3/2}}$$

and the density is

$$f_W(w) = \frac{1}{\Gamma\left(\frac{3}{2}\right)(kT)^{3/2}} \times w^{1/2} \times e^{-w/kT} \qquad \text{for } w \ge 0$$

b. For a gamma-type variable, $E(W) = \alpha\beta = \left(\frac{3}{2}\right)kT$.

6.27 Similar to Exercise 6.21. Fix $Y_1 = y_1$. Then if $U = \frac{Y_2}{y_1}$, $Y_2 = y_1 U$ and $\left|\frac{dy_2}{du}\right| = y_1$. The joint density of Y_1 and U is

$$f(y_1, u) = \frac{1}{8} y_1^2 e^{-(y_1 + uy_1)/2} = \frac{1}{8} y_1^2 e^{-y_1(1+u)/2} \qquad \text{for } y_1 \geq 0, \ u \geq 0$$

Integrating over all possible values of y_1, we have

$$f_U(u) = \frac{1}{8} \int_0^\infty y_1^2 e^{-y_1(1+u)/2} \, dy_1 = \frac{\Gamma(3)\left[\frac{2}{(1+u)}\right]^3}{8} = \frac{2}{(1+u)^3} \qquad u \geq 0$$

since the variable part of the integrand is that of a gamma variable with $\alpha = 3$, $\beta = \frac{2}{1+u}$.

6.29 If $U = 5 - \left(\frac{Y}{2}\right)$, then $Y = 2(5 - U)$ and $\left|\frac{dy}{du}\right| = 2$. Then

$$f_U(u) = 2\left[\left(\frac{3}{2}\right)(4)(5-u)^2 + 2(5-u)\right] = 4\left(80 - 31u + 3u^2\right)$$
$$\text{for} \quad 0 \leq 2(5-u) \quad \text{or} \quad 4.5 \leq u \leq 5$$
$$f_U(u) = 0 \quad \text{elsewhere}$$

6.31 Since Y_i has an exponential distribution, $m_{Y_i}(t) = \frac{1}{1-t}$, for $i = 1, 2$. Using Theorem 6.2, we have

$$m_U(t) = m_{Y_1}\left(\frac{t}{2}\right) m_{Y_2}\left(\frac{t}{2}\right) = \frac{1}{\left[1 - \left(\frac{t}{2}\right)\right]^2}$$

which is the moment-generating function for a gamma random variable with $\alpha = 2$, $\beta = \frac{1}{2}$. Hence $f_U(u) = 4ue^{-2u}$ for $u \geq 0$.

6.33 The moment-generating function of any of the random variables Y_i, $i = 1, 2, \ldots, n$, is given in Appendix II as

$$m_{Y_i}(t) = e^{\mu t + \left(t^2 \sigma^2/2\right)}$$

Then

$$m_{a_i Y_i}(t) = m_{Y_i}(a_i t) = e^{a_i \mu t + \left(a_i^2 t^2 \sigma^2 / 2 \right)}$$

Using Theorem 6.2, we have

$$m_U(t) = \prod_{i=1}^{n} m_{a_i Y_i}(t) = \prod_{i=1}^{n} e^{a_i \mu t + \left(a_i^2 t^2 \sigma^2 / 2 \right)} = e^{\mu t \Sigma a_i + \left(t^2 \sigma^2 / 2 \right) \Sigma a_i^2}$$

which is the moment-generating function of a normal random variable with mean $\mu \Sigma a_i$ and variance $\sigma^2 \Sigma a_i^2$. Hence U is distributed normally with

$$E(U) = \mu \sum_{i=1}^{n} a_i \qquad\qquad V(U) = \sigma^2 \sum_{i=1}^{n} a_i^2$$

6.35 Assuming that Y_1 and Y_2 are independent, Theorem 6.3 can be used to find the distribution of U, which will be normal with $E(U) = 100 + 7(10) + 3(4) = 182$ and $V(U) = 49(.5)^2 + 9(.2)^2 = 12.61$. It is necessary to find a value c such that $P(U > c) = .01$.   Now $P(U > c) = P\left[Z < \dfrac{c - 182}{\sqrt{12.61}} \right]$. The value of the standard normal random variable that satisfies the requirement is $z = 2.33$.   Hence

$$\frac{c - 182}{\sqrt{12.61}} = 2.33 \qquad\qquad \text{and} \qquad\qquad c = \$190.27$$

6.37 Looking in the table under $n = 2\alpha = 7$ degrees of freedom, we find

$$P(Y > 33.627) = P\left(\frac{2Y}{\beta} > 16.0128 \right) = .025.$$

6.39 Using the moment-generating function approach, we have (see Exercise 3.86 or Appendix II)

$$m_{Y_1}(t) = \left(pe^t + q \right)^{n_1} \qquad\qquad m_{Y_2}(t) = \left(pe^t + q \right)^{n_2}$$

And since Y_1 and Y_2 are independent, we have

$$m_{Y_1 + Y_2}(t) = m_{Y_1}(t)\, m_{y_2}(t) = \left(pe^t + q \right)^{n_1 + n_2}$$

which is the moment-generating function of the binomial random variable with parameters $n_1 + n_2$ and p. By Theorem 6.1,

$$P(Y_1 + Y_2 = k) = \binom{n_1 + n_2}{k} p^k q^{n_1 + n_2 - k} \qquad k = 0, 1, 2, \ldots, n_1 + n_2$$

6.41 Let $Y = \#$ of customers that arrive in a two-hour period. Using the result in Exercise 6.40(a), Y is Poisson with $\mu = 14$.

$$P(Y > 20) = 1 - P(Y \le 19) = 1 - .923 = .077.$$

6.43 $m_{Y_i}(t) = (1 - \beta t)^{-\alpha_i}$

$$m_Y(t) = m_{Y_1}(t) m_{Y_2}(t) \cdots m_{Y_n}(t)$$

$$= (1 - \beta t)^{-\alpha_1} (1 - \beta t)^{-\alpha_2} \cdots (1 - \beta t)^{-\alpha_n}$$

$$= (1 - \beta t)^{-(\alpha_1 + \alpha_2 + \ldots + \alpha_n)}$$

which is the gamma distribution with parameters $(\alpha_1 + \ldots + \alpha_n)$ and β.

6.45 From Example 4.13 with $\alpha = \frac{\nu_i}{2}$ and $\beta = 2$, $m_{Y_i}(t) = (1 - 2t)^{\nu_i/2}$. From Theorem 6.2,

$$m_U(t) = m_{Y_1}(t) m_{Y_2}(t) = (1 - 2t)^{-(\nu_1 + \nu_2)/2}$$

which is the moment-generating function of a chi-square random variable with $\nu_1 + \nu_2$ degrees of freedom. The result follows from Theorem 6.1.

6.47 Y_1 and Y_2 are independently and identically distributed with density function $f(y) = 1$ for $0 \le y \le 1$ and cumulative density function

$$F(Y) = \begin{cases} 0, & y < 0 \\ y, & 0 \le y \le 1 \\ 1, & y > 1 \end{cases}$$

a. Referring to Section 6.6, with $n = 2$, we have

$$g_1(u) = 2[1 - F(u)]^{2-1} f(u) = 2(1 - u) \qquad 0 \le u \le 1$$

b. $E(u_1) = \displaystyle\int_0^1 u[2(1 - u)] \, du = \frac{1}{3}.$

$$E\left(u_1^2\right) = \int_0^1 u^2[2(1 - u)] \, du = \frac{1}{6}.$$

Then

$$V(u_1) = E\left(u_1^2\right) - [E(u_1)]^2 = \frac{1}{6} - \frac{1}{9} = \frac{1}{18}$$

c. Similarly, $g_2(u) = 2[F(u)]^{2-1} f(u) = 2u$ \qquad for $0 \le u \le 1$.

d. $E(u_2) = \displaystyle\int_0^1 u[2u]\, du = \frac{2}{3}.$

$$E\left(u_2^2\right) = \int_0^1 u^2[2u]\, du = \frac{1}{2}.$$

Then

$$V(u_2) = E\left(u_2^2\right) - [E(u_2)]^2 = \frac{1}{2} - \frac{4}{9} = \frac{1}{18}.$$

6.49 $P(Y < 10) = F(10) = \left(\frac{10}{15}\right)^5 = \left(\frac{2}{3}\right)^5.$

6.51   **a.** $f_{(j)(k)}(y_j, y_k) = \dfrac{n!}{(j-1)!\,(k-1-j)!\,(n-k)!}\left(\dfrac{y_j}{\theta}\right)^{j-1}$

$$\times \left[\frac{y_k}{\theta} - \frac{y_j}{\theta}\right]^{k-1-j}\left[1 - \frac{y_k}{\theta}\right]^{n-k}\left(\frac{1}{\theta}\right)^2 \qquad 0 \le y_j \le y_k \le \theta$$

b. $\mathrm{Cov}\left(Y_{(j)}, Y_{(k)}\right) = E\left(Y_{(j)}Y_{(k)}\right) - E\left(Y_{(j)}\right)E\left(Y_{(k)}\right).$

To simplify calculations, let $u = \dfrac{Y}{\theta}$.   Then

$$f(u_j, u_k) = \frac{n!}{(j-1)!\,(k-1-j)!\,(n-k)!}\, u_j^{j-1}[u_k - u_j]^{k-1-j}[1 - u_k]^{n-k}$$

$$0 \le u_j \le u_k \le 1.$$

c. $E\left[u_{(j)}u_{(k)}\right] = L \displaystyle\int_0^1 \int_0^{u_k} u_j^{j}[u_k - u_j]^{k-1-j} u_k[1 - u_k]^{n-k}\, du_j\, du_k$

(for the appropriate constant L)

Let $w = \dfrac{u_j}{u_k}$, $u_j = w u_k$, and $du_j = u_k\, dw$.   The expected value becomes

$$= L \int_0^1 \int_0^1 (wu_k)^j\, u_k^{k-1-j}[1 - w]^{k-1-j}\, u_k[1 - u_k]^{n-k}\, u_k\, dw\, du_k$$

$$= L \int_0^1 u_k^{k+1}[1 - u_k]^{n-k} \int_0^1 w^j(1 - w)^{k-1-j}\, dw\, du_k$$

$$= L B(k+2,\, n-k+1)\, B(j+1,\, k-j)$$

$$= \frac{n!}{(j-1)!\,(k-1-j)!\,(n-k)!}\ \frac{(k+1)!\,(n-k)!}{(n+2)!}\ \frac{j!\,(k-j-1)!}{k!} = \frac{(k+1)j}{(n+2)(n+1)}.$$

From Exercise 6.50, $E\left[u_{(j)}\right] = \frac{j}{n+1}$. Then

$$\mathrm{Cov}\left(U_{(j)}U_{(k)}\right) = \frac{(k+1)j}{(n+2)(n+1)} - \left(\frac{k}{n+1}\right)\left(\frac{j}{n+1}\right)$$

$$= \frac{(n+1)(nk+j) - (n+2)jk}{(n+2)(n+1)^2} = \frac{(n-k+1)j}{(n+1)^2(n+2)}.$$

Finally, $\mathrm{Cov}\left(Y_{(j)},\, Y_{(k)}\right) = \frac{(n-k+1)}{(n+1)2(n+2)}\,\theta^2.$

d. $V\left(Y_{(k)}Y_{(j)}\right) = V\left(Y_{(k)}\right) + V\left(Y_{(j)}\right) - 2\mathrm{Cov}\left(Y_{(k)},\, Y_{(j)}\right)$

$$= \left[\frac{k(n-k+1)}{(n+1)^2(n+2)} + \frac{j(n-k+1)}{(n+1)^2(n+2)} - \frac{2(n-k+1)j}{(n+1)^2(n+2)}\right]\theta^2$$

$$= \left[\frac{(k+j-2j)(n-k+1) + j(k-j)}{(n+1)^2(n+2)}\right]\theta^2$$

$$= \frac{(k-j)(n-k+j+1)}{(n+1)^2(n+2)}\,\theta^2.$$

6.53 a. $f(y) = \left(\frac{1}{\beta}\right)e^{-y/\beta};\ F(y) = 1 - e^{-y/\beta}$ if $y > 0$

$$f_{Y_{(1)}}(y) = n\left[1 - F(y)^{n-1}\right]f(y) = n\left[\left(e^{-y/\beta}\right)^{n-1}\right]\left[\left(\frac{1}{\beta}\right)e^{-y/\beta}\right]$$

$$= \left(\frac{n}{\beta}\right)e^{-ny/\beta}, \text{ which is exponential with } \mu = \frac{\beta}{n}.$$

b. $f_{Y_{(1)}}(y) = 2.5e^{-y/.4}$ $y > 0$

$F(3.6) = 1 - e^{-(3.6)/(.4)} = 1 - e^{-9}$

6.55 a. This is similar to Exercise 6.47. The density function of $Y_{(1)}$ is

$g_1(u) = n[1 - F(u)]^{n-1}f(u)$. Now

$$1 - F(u) = \int_u^\infty \frac{1}{2}e^{-(1/2)(t-4)}\,dt = \int_{u-4}^\infty \frac{1}{2}e^{-(1/2)s}\,ds = \left[-e^{-(1/2)s}\right]_{u-4}^\infty$$

$$= e^{-(1/2(u-4))}$$

so that

$$g_1(u) = 2\left[e^{-(1/2)(u-4)}\right]^{2-1} \times \frac{1}{2}e^{-(1/2)(u-4)} = e^{-(u-4)} \qquad \text{for } u \geq 4$$

b. $E(u) = \displaystyle\int_4^\infty ue^{-(u-4)}\,du$

Let $y = u - 4$, so that $dy = du$ and

$$E(u) = \int_0^\infty (y+4)e^{-y}\,dy = \int_0^\infty ye^{-y}\,dy + \int_0^\infty 4e^{-y}\,dy = 1 + 4 = 5$$

6.57   Since $f(y) = 1$ and $F(y) = y$ for $0 \leq y \leq 1$, Theorem 6.5 gives

$$f_{(1)(n)}(y_1, y_n) = \frac{n!}{(n-2)!}[y_n - y_1]^{n-2} \qquad 0 \leq y_1 \leq y_n \leq 1$$

Now use the method of transformations to get the joint distribution of

$R = Y_{(n)} - Y_{(1)}$ and $Y_{(1)}$ by letting $Y_{(1)} = y_1$ so that $Y_{(n)} = R + y_1$, $\dfrac{dy_n}{dR} = 1$, and

$$g(r, y_1) = \frac{n!}{(n-2)!}r^{n-2} = n(n-1)\,r^{n-2}$$

The ranges of $\left(Y_{(1)},\, Y_{(n)}\right)$ and $(R,\, Y_{(1)})$ are shown in Figure 6.6.

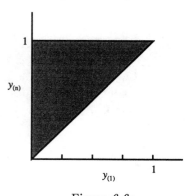

Figure 6.6a

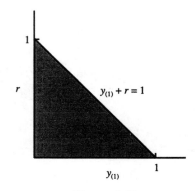

Figure 6.6b

Since $Y_{(1)} \leq Y_{(n)}$, $0 \leq r \leq 1 - y_1$, or $0 \leq y_1 \leq 1 - r$. Hence the marginal density of R is

$$g(r) = \int\limits_0^{1-r} g(r, y_1)\, dy_1 = \int\limits_0^{1-r} n(n-1)\, r^{n-2}\, dy_1 = n(n-1)\, r^{n-2}(1-r) \quad 0 \le r \le 1$$

$$g(r) = 0 \qquad\qquad \text{elsewhere}$$

6.59 a. It is given that $f(y) = \left(\frac{1}{\theta}\right) e^{-y/\theta}$ for $y \ge 0$, $\theta \ge 0$. Then $F(y) = 1 - e^{-y/\theta}$ for $y \ge 0$. In order to solve the exercise, we need the joint distribution of the two order statistics, W_{j-1} and W_j. Using Theorem 6.5, we obtain

$$f(w_j, w_{j-1}) = \frac{n!}{(j-2)!\,(n-j)!}\left(1 - e^{-w_{j-1}/\theta}\right)^{j-2}\left(e^{-w_j/\theta}\right)^{n-j}\left(\frac{1}{\theta^2}\right)$$

$$\times \left(e^{-(w_j + w_{j-1})/\theta}\right) \qquad 0 \le w_{j-1} \le w_j < \infty$$

Fix $W_{j-1} = w_{j-1}$ and let $T_j = W_j - w_{j-1}$. Then $\dfrac{dw_j}{dY_j} = 1$ and the joint distribution of T_j and w_{j-1} is

$$f(t_j, w_{j-1}) = \frac{n!}{(j-2)!\,(n-j)!}\left(1 - e^{-w_{j-1}/\theta}\right)^{j-2}\left(e^{-(t_j + w_{j-1}/\theta)}\right)^{n-j}\left(\frac{1}{\theta^2}\right)$$

$$\times \left(e^{-(t_j + 2w_{j-1})/\theta}\right) \qquad 0 \le t_j;\ 0 < w_{j-1}$$

Integrating over all possible values of W_{j-1}, we obtain

$$f(t_j) = \frac{n!}{(j-2)!\,(n-j)!}\exp\left[\frac{-t_j(n-j) - t_j}{\theta}\right] \times \frac{1}{\theta^2}\int\limits_0^{\infty}\left(1 - e^{-w_{j-1}/\theta}\right)^{j-2}$$

$$\times \left(e^{-w_{j-1}/\theta}\right)^{n-j+2} dw_{j-1}$$

Let $u = e^{-w_{j-1}/\theta}$ so that $w_{j-1} = -\theta \ln u$ and $\dfrac{dw_{j-1}}{du} = -\dfrac{\theta}{u}$. Then

$$f(t_j) = \frac{1}{\theta^2} e^{-t_j(n-j+1)/\theta}\int\limits_0^1 \frac{n!}{(j-2)!\,(n-j)!}(1-u)^{j-2}\, u^{n-j+2}\left(\frac{\theta}{u}\right) du$$

$$= \frac{n-j+1}{\theta} e^{-t_j(n-j+1)/\theta}\int\limits_0^1 \frac{n!}{(j-2)!\,(n-j+1)!}(1-u)^{j-2}\, u^{n-j+1}\, du$$

$$= \frac{n-j+1}{\theta} e^{-t_j(n-j+1)/\theta} \qquad 0 < t_j$$

since the integrand is that of a complete beta random variable with $\alpha = j - 1$, $\beta = n - j + 2$. Hence T_j has an exponential distribution with mean $\dfrac{\theta}{n-j+1}$.

b. Look at

$$\sum_{j=1}^{r} (n-j+1)T_j = nW_1 + (n-1)(W_2 - W_1) + (n-2)(W_3 - W_2)$$
$$+ (n-3)(W_4 - W_3) + \ldots + (n-r+2)(W_{r-1} - W_{r-2})$$
$$+ (n-r+1)(W_r - W_{r-1})$$
$$= W_1 + W_2 + \ldots + W_{r-1} + (n-r+1)W_r$$
$$= \sum_{j=1}^{r} W_j + (n-r)W_r = U_r$$

Hence

$$E(U_r) = \sum_{j=1}^{r} (n-j+1)E(T_j) = \sum_{j=1}^{r} \frac{(n-j+1)\theta}{n-j+1} = r\theta$$

6.61 Since I and R are independent,

$$f(i, r) = f(i)f(r) = 2r \qquad \text{for } 0 \le r \le 1, \, 0 \le i \le 1$$

Fix $R = r$. Then $W = I^2 r$ and $I = \sqrt{\frac{W}{r}}$ so that $\left|\frac{di}{dw}\right| = \left(\frac{1}{2r}\right)\left(\frac{W}{r}\right)^{-1/2}$ for $0 \le w \le r \le 1$. Notice that since $W = I^2 r$ and I is in the interval $(0, 1)$, w will always be less than r. Then $f(w, r) = \sqrt{\frac{r}{w}}$ and

$$f(w) = \int_{w}^{1} \sqrt{\frac{r}{w}}\, dr = \frac{1}{\sqrt{w}}\left(\frac{2}{3}\right)r^{3/2}\Big]_{w}^{1} = \frac{2}{3}\left(\frac{1}{\sqrt{w}} - 2\right) \qquad \text{for } 0 \le w \le 1$$

$$f(w) = 0 \qquad \qquad \text{otherwise}$$

6.63 The joint density of Y_1 and Y_2 can be written as

$$f(y_1, y_2) = 1 \qquad 0 \le y_1 \le 1, \, 0 \le y_2 \le 1$$

a. On the region defined above, the function $U_1 = Y_1^2$ is an increasing function of Y. Hence

$$f_{U_1}(u) = f_{Y_1}(\sqrt{u})\left|\frac{dy_1}{du}\right| = 1 \times \frac{1}{2\sqrt{u}} = \frac{1}{2\sqrt{u}} \qquad 0 \le u \le 1$$

$$f_{U_1}(u) = 0 \qquad \text{elsewhere}$$

Note that we used the marginal density of Y_1, $f_{Y_1}(y_1) = 1$ for $0 \le y_1 \le 1$.

b. Parts (b), (c), and (d) involve functions of both Y_1 and Y_2. However, the transformation approach is still valid. For a more general approach to transformations involving functions of two variables, the reader should check the references at the end of the text's Chapter 6. Consider the joint distribution of

$U = \dfrac{Y_1}{Y_2}$ and Y_2. Letting Y_2 be fixed at y_2, we can write $U = \dfrac{Y_1}{y_2}$. Then

$Y_1 = y_2 U$ and $\dfrac{dy_1}{du} = y_2$, so that the joint density of Y_2 and U is $g(y_2, u) = y_2$.

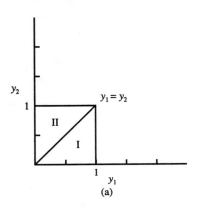

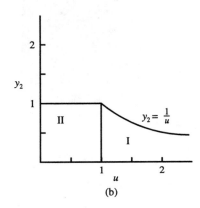

| Figure 6.7a | Figure 6.7b |

In region II of Figure 6.7(a) we have $0 \le y_1 < y_2 \le 1$, so that $U = \dfrac{Y_1}{Y_2}$ is in the interval $0 \le u \le 1$ and $0 \le y_2 \le 1$. However, region I, where $0 \le y_2 \le y_1 \le 1$, U has minimum value $u = \dfrac{y_2}{y_2} = 1$ and maximum value $u = \dfrac{1}{y_2}$, so that

$$1 \le u \le \tfrac{1}{y_2} \qquad 0 \le y_2 \le 1$$

Written in another form, the ranges are $0 \le y_2 \le \tfrac{1}{u}$, $u > 1$. The marginal density of U is given by

$$f_U(u) = \int_{-\infty}^{\infty} g(y_2, u)\, dy_2 = \int_0^1 y_2\, dy_2 = \left. \dfrac{y_2^2}{2} \right]_0^1 = \tfrac{1}{2} \qquad \text{if } 0 \le u \le 1$$

and

$$f_U(u) = \int_0^{1/u} y_2\, dy_2 = \left. \dfrac{y_2^2}{2} \right]_0^{1/u} = \dfrac{1}{2u^2} \qquad \text{if } u > 1$$

c. Consider the joint distribution of $U = -\ln Y_1 Y_2$ and Y_1. Letting $Y_1 = y_1$, we have $U = -\ln y_1 Y_2$, so that

$$Y_2 = \dfrac{e^{-U}}{y_1} \qquad \text{and} \qquad \dfrac{dy_2}{du} = \dfrac{-1}{y_1} e^{-u}$$

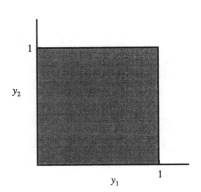

Figure 6.8a   Figure 6.8b

The joint density of Y_1 and U is then $g(y_1, u) = \frac{1}{y_1}e^{-u}$. The ranges of Y_1 and U are shown in Figure 6.8. When $0 \le y_2 \le 1$, $U = -\ln Y_1 Y_2$ ranges in the interval $-\ln y_1 \le u \le \infty$ and $0 \le y_1 \le 1$. Written in another form, the ranges are $e^{-u} \le y_1 \le 1$, $0 \le u \le \infty$. The marginal density of U is then

$$f_U(u) = \int_{e^{-u}}^{1} \frac{1}{y_1}e^{-u}\, dy_1 = \left[\ln y_1\right]_{e^{-u}}^{1} = ue^{-u} \qquad \text{for } 0 \le u \le \infty.$$

d. Consider the joint distribution of $U = Y_1 Y_2$ and Y_1. For $Y_1 = y_1$ we have

$$Y_2 = \frac{U}{y_1} \qquad\qquad \frac{dy_2}{du} = \frac{1}{y_1}$$

and

$$g(y_1, u) = \frac{1}{y_1}$$

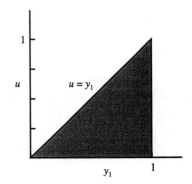

Figure 6.9a   Figure 6.9b

The ranges of Y_1 and U are $0 \le y_1 \le 1$ and $0 \le u \le y_1$, or $u \le y_1 \le 1$ and $0 \le u \le 1$, as shown in Figure 6.9. Then

$$f_U(u) = \int_u^1 \frac{1}{y_1} \, dy_1 = \ln y_1 \big]_u^1 = -\ln u \qquad 0 \le u \le 1$$

6.65 If we let $(A, B) = (-1, 1)$ and $T = 0$, then the density function for X, the landing point, is $f(x) = \frac{1}{2}$ for $-1 \le x \le 1$. It is necessary to obtain the probability density function for $U = |X|$. Write

$$F_U(u) = P(U \le u) = P(|X| \le u) = P(-u \le X \le u) = F_X(u) - F_X(-u)$$

On differentiating with respect to u, we obtain

$$f_U(u) = f_X(u) + f_X(-u) = \frac{1}{2} + \frac{1}{2} = 1 \qquad 0 \le u \le 1$$

Note that since $U = |X|$, U has positive density only on the range $0 \le u \le 1$.

6.67 The joint distribution of Y_1 and Y_2 is

$$f(y_1, y_2) = \frac{1}{2\pi} \exp\left[-\frac{1}{2}\left(y_1^2 + y_2^2\right)\right] \qquad -\infty \le y_1 \le \infty,\ -\infty \le y_2 \le \infty$$

Consider the joint distribution of $U = \dfrac{Y_1}{Y_2}$ and Y_2 by letting $Y_2 = y_2$ so that

$$Y_1 = Uy_2, \left|\frac{dy_1}{du}\right| = |y_2|, \text{ and } g(y_2, u) = \frac{|y_2|}{2\pi} \exp\left\{-\frac{1}{2}\left[y_2^2(1 + u^2)\right]\right\}$$

Now

$$g(u) = \int_{-\infty}^{\infty} \frac{|y_2|}{2\pi} \exp\left\{-\frac{1}{2}\left[y_2^2(1 + u^2)\right]\right\} dy_2 = \frac{2}{2\pi} \int_0^{\infty} y_2 \exp\left\{-\frac{1}{2}\left[y_2^2(1 + u^2)\right]\right\} dy_2$$

since the integrand is an even function. Making transformation $y_2^2 = z$, we have

$$g(u) = \left(\frac{1}{\pi}\right)\frac{1}{(1 + u^2)} \int_0^{\infty} (1 + u^2) \exp\left[-\frac{1}{2}z(1 + u^2)\right] dz = \frac{1}{\pi(1 + u^2)} \qquad -\infty \le u \le \infty$$

since the integral is that of an exponential variable with $\beta = \dfrac{2}{(1 + u^2)}$.

6.69 If Y is distributed uniformly on the interval $(-1, 3)$, then

$$f(y) = \begin{cases} \frac{1}{4}, & -1 \le y \le 3 \\ 0, & \text{elsewhere} \end{cases}$$

Now, if $U = Y^2$, then

$$g(u) = \frac{1}{2\sqrt{u}}[f(\sqrt{u}) + f(-\sqrt{u})]$$

as in Example 6.4. If $-1 \leq y \leq 3$, then $0 \leq u \leq 9$; however, if $1 \leq u \leq 9$, $f(-\sqrt{u})$ is not positive. That is,

$$g(u) = \frac{1}{2\sqrt{u}}\left(\frac{1}{4} + \frac{1}{4}\right) = \frac{1}{4\sqrt{u}} \qquad \text{if } 0 \leq u \leq 1$$

but

$$g(u) = \frac{1}{2\sqrt{u}}\left(\frac{1}{4} + 0\right) = \frac{1}{8\sqrt{u}} \qquad \text{if } 1 \leq u \leq 9$$

6.71 Let C_3 be production cost. Then the profit function (per gallon) is

$$U = \begin{cases} C_1 - C_3, & \frac{1}{3} \leq y \leq \frac{2}{3} \\ \\ C_2 - C_3, & 0 \leq y \leq \frac{1}{3} \text{ or } \frac{2}{3} \leq y \leq 1 \end{cases}$$

Since profit can only take on two different values, depending on the region into which y falls, the profit distribution is a discrete random variable with

$$P(U = C_1 - C_3) = \int_{1/3}^{2/3} (20y^3 - 20y^4)\, dy = \left[5y^4 - 4y^5\right]_{1/3}^{2/3} = .4156$$

$$P(U = C_2 - C_3) = 1 - .4156 = .5844$$

6.73 $P(\text{largest} > \text{median}) = 1 - P(\text{largest} < \text{median}) = 1 - P(\text{all } Y\text{'s} < \text{median})$

$$= 1 - \left(\frac{1}{2}\right)^4.$$

6.75 a. Let $U = \ln Y$ so that $\frac{dU}{dy} = \frac{1}{y}$. Since the density function of U is known, $f_Y(y)$ can be derived as

$$f_Y(y) = f_U(\ln y)\left|\frac{du}{dy}\right| = \frac{1}{y\sigma\sqrt{2\pi}} \exp\left[\frac{-(\ln y - \mu)^2}{2\sigma^2}\right] \qquad \text{for } y \geq 0$$

b. $E(Y) = \int_0^\infty \frac{1}{\sigma\sqrt{2\pi}} \exp\left[\frac{-(\ln y - \mu)^2}{2\sigma^2}\right] dy.$

Let $z = \ln y$. Then $-\infty \leq z \leq \infty$ and $dz = \left(\frac{1}{y}\right) dy$. Thus,

$$E(Y) = \int_{-\infty}^{\infty} \frac{e^z}{\sigma\sqrt{2\pi}} \exp\left[\frac{-(z-\mu)^2}{2\sigma^2}\right] dz.$$

Let $x = z - \mu$ so that $z = x + \mu$ and $dz = dx$. Then

$$E(Y) = e^\mu \int_{-\infty}^{\infty} \frac{1}{\sigma\sqrt{2\pi}} \exp\left[-\frac{x^2 - 2\sigma^2 x}{2\sigma^2}\right] dx$$

$$= e^{\mu + (\sigma^2/2)} \int_{-\infty}^{\infty} \frac{1}{\sigma\sqrt{2\pi}} \exp\left[\frac{-(x-\sigma^2)^2}{2\sigma^2}\right] dx = e^{\mu + (\sigma^2/2)}$$

6.77 **a.** Let R be the distance from a randomly chosen point to the nearest particle. Consider

$$P(R > r) = P(\text{no particles in the sphere of radius } r)$$
$$= P\left[Y = 0 \text{ in a volume of } \left(\tfrac{4}{3}\right)\pi r^3\right]$$

Since the number of particles in a volume V has a Poisson distribution with mean λV,

$$P(R > r) = \frac{\left[\left(\tfrac{4}{3}\right)\pi r^3 \lambda\right]^0 e^{-(4/3)\pi r^3 \lambda}}{0!} = e^{-(4/3)\pi r^3 \lambda}$$

But $P(R > r) = 1 - F(r)$, so that $F(r) = 1 - e^{-(4/3)\pi r^3 \lambda}$ and $f(r) = 4\pi r^2 \lambda e^{-(4/3)\pi r^3 \lambda}$ for $r > 0$.

b. Let $U = R^3$ so that $R = U^{1/3}$ and $\frac{dr}{du} = \left(\tfrac{1}{3}\right)u^{-2/3}$. Then

$$f(u) = 4\pi\lambda u^{2/3} e^{-(4/3)\pi\lambda u}\left(\tfrac{1}{3}u^{-2/3}\right) = \frac{4\lambda\pi}{3} e^{-(4\pi\lambda/3)u}$$

which is an exponential distribution with $\beta = \left(\tfrac{3}{4}\right)\pi\lambda$.

CHAPTER 7 SAMPLING DISTRIBUTIONS AND THE CENTRAL LIMIT THEOREM

7.1 Since the distribution of basal areas is normally distributed with mean μ and variance $\sigma^2 = 16$, the sample mean will also be normally distributed, from Theorem 7.1. Then

$$P\left(\left|\overline{Y} - \mu\right| \le 2\right) = P\left[-2 \le \left(\overline{Y} - \mu\right) \le 1\right] = P\left(\frac{-2}{\left(\frac{\sigma}{\sqrt{n}}\right)} \le \frac{Y - \mu}{\left(\frac{\sigma}{\sqrt{n}}\right)} \le \frac{2}{\left(\frac{\sigma}{\sqrt{n}}\right)}\right)$$

$$= P(-1.5 \le z \le 1.5)$$

$$= 1 - 2P(Z > 1.5) = 1 - 2(.0668) = .8664$$

7.3 Similar to Exercise 7.1. It is necessary to calculate

$$P\left(\left|\overline{Y} - \mu\right| \le .5\right) = P\left(\frac{\left|\overline{Y} - \mu\right|}{\left(\frac{\sigma}{\sqrt{n}}\right)} \le \frac{.5}{\left(\frac{\sigma}{\sqrt{n}}\right)}\right) = P(|x| \le 2.5) = 1 - 2(.0062) = .9876$$

7.5 Use Theorems 6.3 and 7.1 together,

a. $E(\overline{X} - \overline{Y}) = E(\overline{X}) - E(\overline{Y}) = \mu_1 - \mu_2$

b. $V(\overline{X} - \overline{Y}) = V(\overline{X})\text{V} + (\overline{Y}) = \frac{\sigma_1^2}{m} + \frac{\sigma_2^2}{n}$

c. It is necessary that

$$P\left(\left|(\overline{X} - \overline{Y}) - (\mu_1 - \mu_2)\right| \le 1\right) = .95$$

or

$$P\left(\frac{\left|(\overline{X} - \overline{Y}) - (\mu_1 - \mu_2)\right|}{\sqrt{\frac{2}{n} + \frac{2.5}{n}}} \le \frac{1}{\sqrt{\frac{2}{n} + \frac{2.5}{n}}}\right) = .95$$

That is, $\dfrac{1}{\sqrt{\frac{4.5}{n}}} = 1.96$, or $n = 17.29$. Thus two samples of size $m = n = 18$ should be drawn.

7.7   We are given that $s^2 = .065$ and $n = 10$. Suppose that $\sigma^2 = .04$. Then from Theorem 7.3, $\dfrac{9S^2}{.04}$ has a χ^2 distribution with 9 degrees of freedom. Thus,

$$P\left(S^2 > .065\right) = P\left(\frac{9S^2}{.04} > \frac{9(.065)}{.04}\right) = P\left(\chi_9^2 > 14.625\right) = .10$$

There is some suggestion that $\sigma > .2$.

7.9   Since $P\left(a \le S^2 \le b\right) = P\left[\dfrac{(n-1)a}{\sigma^2} \le \dfrac{(n-1)S^2}{\sigma^2} \le \dfrac{(n-1)b}{\sigma^2}\right]$, we can take

$$\frac{(n-1)a}{\sigma^2} = \chi_{.975}^2 \qquad \text{and} \qquad \frac{(n-1)b}{\sigma^2} = \chi_{.025}^2$$

to obtain the desired result. Indexing $\nu = n - 1 = 19$ with $\sigma^2 = 1.4$, we have $\chi_{.975}^2 = 8.90655$ and $\chi_{.025}^2 = 32.8523$. Then

$$a = \frac{1.4(8.90655)}{19} = .656 \qquad \text{and} \qquad b = \frac{1.4(32.8523)}{19} = 2.42$$

7.11   Since Y has an F distribution, $Y = \dfrac{\left(\dfrac{\chi_1^2}{\nu_1}\right)}{\left(\dfrac{\chi_2^2}{\nu_2}\right)}$.

$$U = \frac{1}{Y} = \frac{\left(\dfrac{\chi_1^2}{\nu_2}\right)}{\left(\dfrac{\chi_2^2}{\nu_2}\right)} = \frac{\left(\dfrac{\chi_2^2}{\nu_2}\right)}{\left(\dfrac{\chi_1^2}{\nu_1}\right)}$$

which has an F distribution with ν_2 numerator and ν_1 denominator degrees of freedom.

7.13   It was shown that $T = \dfrac{Z}{\sqrt{\dfrac{Y}{\nu}}}$ has a t distribution with ν degrees of freedom where Z has a standard normal distribution and Y is an independent χ^2 random variable. Then

$$U = T^2 = \frac{\left(\dfrac{Z^2}{1}\right)}{\left(\dfrac{Y}{\nu}\right)}.$$

Theorem 7.2 shows that Z^2 has a χ^2 distribution with 1 degree of freedom (and still independent of Y). Finally, by Definition 7.3, U has an F distribution with 1 numerator degree of freedom and ν denominator degrees of freedom.

7.15 **a.** Using the result in Exercise 7.14, part (a)

$$E(F) = \frac{\nu_2}{\nu_2 - 2} = \frac{70}{68} = 1.029$$

b. Using the result in Exercise 7.14, part (b),

$$V(F) = \frac{2\nu_2^2(\nu_1 + \nu_2 - 2)}{\nu_1(\nu_2 - 2)^2(\nu_2 - 4)} = \frac{2(70)^2(118)}{50(68)^2(66)} = .076$$

c. $3 = \mu + 7.15\sigma$. It is not likely that F will exceed 3.

7.17 **a.** χ^2 with 5 degrees of freedom, by Theorem 7.2

b. χ^2 with 4 degrees of freedom, by Theorem 7.3. Note that $\sigma^2 = 1$.

c. χ^2 with 5 degrees of freedom. Let $R = \sum_{i=1}^{5} (Y_i - \overline{Y})^2 + Y_6^2 = U + Y_6^2$, which is $\chi^2(4 + 1)$.

7.19 The Central Limit Theorem (Theorem 7.4) states that $Y_n = \frac{\sqrt{n}(\overline{X} - \mu)}{\sigma}$ converges in distribution to a standard normal random variable, which is denoted by Z. For this exercise, $n = 100$, $\sigma = 2.5$, and the approximation is

$$P\left(|\overline{X} - \mu| \le .5\right) = P\left(-.5 \le \overline{X} - \mu \le .5\right) = P\left[\frac{-.5(10)}{2.5} \le Z \le \frac{.5(10)}{2.5}\right]$$

$$= P(-2 \le Z \le 2) = 1 - 2(.0228) = .9544$$

7.21 Similar to Exercise 7.19. for this exercise, $\mu = 5.00$, $\sigma = .50$, $n = 64$, and the is approximation

$$P(\overline{X} \le 4.90) = P\left[Z \le \frac{\sqrt{64}\,(4.90 - 5.00)}{.5}\right] = P(Z \le -1.6) = .0548$$

7.23 Refer to Exercise 7.22. It is necessary to have

$$P\left(|\overline{Y} - \mu| \le .1\right) = P\left[|Z| \le \frac{\sqrt{n}(.1)}{.75}\right] = .90$$

Hence, take $\frac{\sqrt{n}(.1)}{.75} = 1.645$, or $n = 152.21$. At least 153 core samples should be taken.

7.25 We are given that $\mu = 1.4$ and $\sigma = .7$. Although X, the service time on one automobile, may not have a normal distribution, $\overline{X}$ will. Thus,

$$P(\overline{X} > 1.6) = P\!\left(Z > \frac{\sqrt{50}(1.6 - 1.4)}{.7}\right) = P(Z > 2.02) = .0217$$

7.27 Let X_i be the length of life for the ith heat lamp, $i = 1, 2, \ldots, 25$. It is given that the X_i's are independent, each with mean 50 and standard deviation 4. Then by the Central Limit Theorem, the random variable

$$Y_n = \frac{\sqrt{n}(X - \mu)}{\sigma} = \frac{n(\overline{X} - \mu)}{\left(\dfrac{\sigma}{\sqrt{n}}\right)} = \frac{\Sigma X_i - n\mu}{\left(\dfrac{\sigma}{\sqrt{n}}\right)}$$

converges in distribution to a standard normal random variable. Hence, since the lifetime of the lamp system is represented by $V = \displaystyle\sum_{i=1}^{25} X_i$, the probability of interest is

$$P\!\left(\sum_{i=1}^{25} X_1 \ge 1300\right) = P\!\left(\frac{\Sigma X_i - n\mu}{\left(\dfrac{\sigma}{\sqrt{n}}\right)} \ge \frac{1300 - 1250}{\sqrt{400}}\right) = P(Z > 2.5) = .0062$$

7.29 Use the results of Exercise 7.28. It is given that $n = 50$, $\sigma_1 = \sigma_2 = 2$, and $\mu_1 = \mu_2$. Let $\overline{X}$ be the mean for operator B and $\overline{Y}$ be the mean for operator A. Then the probability of interest is

$$P(\overline{X} - \overline{Y} > 1) = P\!\left[\frac{(\overline{X} - \overline{Y}) - (\mu_1 - \mu_2)}{\sqrt{\dfrac{\sigma_1^2 + \sigma_2^2}{n}}} > \frac{1 - 0}{\sqrt{\dfrac{4 + 4}{50}}}\right] = P(Z > 2.5) = .0062$$

7.31 It is necessary to have

$$P\!\left[\left|(\overline{X} - \overline{Y}) - (\mu_1 - \mu_2)\right| \le .04\right] = P\!\left(|Z| \le \frac{.04}{\sqrt{\dfrac{.01 + .02}{n}}}\right) = .90$$

Hence $1.645 = \dfrac{.04}{\sqrt{\dfrac{.03}{n}}}$, or $n = 50.74$. Each sample must be at least of size $n = 51$.

7.33 We need to consider $P\!\left(\displaystyle\sum_{i=1}^{n} X_i < 120\right) = .1$.

$$P\!\left(\sum_{i=1}^{n} x_i < 120\right) = P\!\left(\overline{X} < \frac{120}{n}\right) = P\!\left[z < \left(\frac{\dfrac{120}{n} - 1.5}{\dfrac{1}{\sqrt{n}}}\right)\right] = P(z < z_0) = .1.$$

Going to the normal table, we find that $z_0 \cong -1.28$. Thus,

$$\frac{\frac{120}{n} - 1.5}{\frac{1}{\sqrt{n}}} = -1.28.$$

Solving for n,

$$\sqrt{n}\left(\frac{120}{n} - 1.5\right) = -1.28.$$

$$1.5n - 1.28\left(\sqrt{n}\right) - 120 = 0.$$

Using the quadratic formula, where $a = 1.5$, $b = 1.28$, $c = -120$,

$$\sqrt{n} = \frac{1.28 \pm \sqrt{(-1.28(2) - 4(1.5)(-120)}}{2(1.5)} = \frac{1.28 \pm 26.86}{3} = 9.38 \text{ or } -8.5266.$$

Taking the positive value, we find $n = 87.98 \cong 88$.

7.35 The random variable of interest is X, the number of persons not showing up for a given flight. This is a binomial random variable with

$$n = 160$$

$$p = P(\text{person does not show up}) = .05.$$

If there is to be a seat available for every person planning to fly, then there must be at least five persons not showing up. Hence, the probability of interest is $P(X \geq 5)$. Calculate

$$\mu = np = 160(.05) = 8,$$

$$\sigma = \sqrt{npq} = \sqrt{160(.05)(.95)} = \sqrt{7.6} = 2.76.$$

A correction for continuity is made to include the entire area under the rectangle associated with the value $X = 5$, and the approximation becomes $P(X \geq 4.5)$. The Z value corresponding to $X = 4.5$ is

$$Z = \frac{4.5 - 8}{2.76} = 1.27,$$

so that

$$P(X \geq 4.5) = \mathrm{P}(Z \geq -1.27) = 1 - P(Z \leq 1.27) = 1 - .1020 = .8980.$$

7.37 Similar to previous exercises. With $p = .20$ and $n = 64$, the probability of interest is

$$P\left(\left|\frac{Y}{n} - p\right| \leq .06\right) = P\left(-.06 \leq \frac{Y}{n} - p \leq .06\right) = P\left(\frac{-.06}{\sqrt{\frac{(.2)(.8)}{64}}} \leq Z \leq \frac{.06}{\sqrt{\frac{(.2)(.8)}{64}}}\right)$$

$$= P(-1.2 \leq Z \leq 1.2) = 1 - 2(.1151) = .7698$$

7.39 Similar to Exercise 7.38. It is necessary to solve for n in the following equality:

$$\frac{.15\sqrt{n}}{\sqrt{pq}} = 2.33. \quad \text{Using } p = \frac{1}{2} \text{ to approximate } p, \text{ we have}$$

$$\sqrt{n} = \frac{2.33(.5)}{.15} \qquad \text{or} \qquad n = 60.32$$

Hence $n = 61$ customers should be sampled.

7.41 Similar to Example 7.10 in the text.

a. To use the normal approximation to the binomial, treat Y as having approximately the same distribution as W, where W is normally distributed with mean np and variance $np(1-p)$. Then

$$P(Y \geq 2) = P(W \geq 1.5) = P\left(Z \geq \frac{1.5 - 2.5}{\sqrt{25(.1)(.9)}}\right) = P(Z > -.67) = .7486$$

b. Indexing $n = 25$ and $p = .1$ in Table 1, Appendix III, we have

$$P(Y \geq 2) = 1 - P(Y \leq 1) = 1 - .271 = .729$$

which is quite close to the approximate probability.

7.43 It is given that $p_1 = .1$ and $p_2 = .2$. Using the results of Exercise 7.42, we obtain

$$P\left[\left|\left(\frac{Y_1}{n_1} - \frac{Y_2}{n_2}\right) - (p_1 - p_2)\right| \leq .1\right] = P\left(|Z| \leq \frac{.1}{\sqrt{\frac{(.1)(.9)}{50} + \frac{(.2)(.8)}{50}}}\right) = P(|Z| \leq 1.41)$$

$$= 1 - 2(.0793) = .8414$$

7.45 Let $X = $ waiting time over a 2-day period. X is exponential with $\beta = 10$ minutes. Let $Y = \#$ of customers whose waiting time is greater than 10 minutes. Y is binomial with $n = 100$ and

$$p = P(X > 10) = \int_{10}^{\infty} \left(\frac{1}{10}\right) e^{-x/10}\, dx = e^{-1} = .3679.$$

$$P(Y \geq 50) \approx P\left(Z \geq \frac{49.5 - 36.79}{\sqrt{100(.3679)(.6321)}}\right) = P(Z \geq 2.636) \approx .0041.$$

7.47 It is necessary that

$$P\left[\overline{X} > 10\right] = P\left[Z > \frac{10 - \mu}{\frac{.5}{\sqrt{8}}}\right] = .8$$

But, $P[Z > -.84] = .8$. Thus, $-.84 = \dfrac{10 - \mu}{\frac{.5}{\sqrt{8}}}$ or $\mu = 10.15$.

7.49 Since X, Y, and Z are normally distributed, $\overline{X}$, $\overline{Y}$, and $\overline{Z}$ are normally distributed, and Theorem 6.3 states that

$$E(U) = .4\mu_1 + .2\mu_2 + .4\mu_3 \qquad\qquad V(U) = .16\left(\frac{\sigma_1^2}{n_1}\right) + .04\left(\frac{\sigma_2^2}{n_2}\right) + .16\left(\frac{\sigma_3^2}{n_3}\right)$$

7.51 Since Y has an exponential distribution with mean θ, $f(y) = \left(\frac{1}{\theta}\right)e^{-y/\theta}$ for $y > 0$. Let $u = \frac{2Y}{\theta}$. Then $Y = \left(\frac{\theta}{2}\right)U$ and $\left|\frac{dy}{du}\right| = \frac{\theta}{2}$. Finally,

$$f(u) = \tfrac{1}{\theta}e^{-u/2}\,\tfrac{\theta}{2} = \tfrac{1}{2}e^{-u/2} \qquad \text{for } u > 0$$

Hence U has a gamma distribution with $\alpha = 1$, $\beta = 2$ or, equivalently, a χ^2 distribution with $\alpha = \frac{\nu}{2} = 1$ or $\nu = 2$, or equivalently, an exponential distribution with $\beta = 2$.

7.53 **a.–b.** Since $\mu = 0, T = \dfrac{\overline{Y}}{\left(\frac{S}{\sqrt{n}}\right)}$ (see the discussion following Definition 7.2) and

$T^2 = \dfrac{n\overline{Y}^2}{S^2}$ has an F distribution with 1 and $(n-1)$ degrees of freedom. By Definition 7.3, $\dfrac{1}{T^2} = \dfrac{S^2}{n\overline{Y}^2}$ also has an F distribution with $(n-1)$ and 1 degrees of freedom.

c. Consider the probability statement given at the beginning of the exercise.

$$P\left(-c \le \frac{S}{\overline{Y}} \le c\right) = P\left(0 \le \frac{S^2}{\overline{Y}^2} \le c^2\right) = P\left(0 \le \frac{S^2}{10\overline{Y}^2} \le \frac{c^2}{10}\right) = .95$$

The constant $\frac{c^2}{10}$ is thus the same as $F_{.05}$ for an F distribution with 9 and 1 degrees of freedom. Then $\frac{c^2}{10} = 240.5$, or $c = 49.04$.

7.55 **a.** Note that $E(X_i) = 1$ and $V(X_i) = 2$. By the Central Limit Theorem,

$$\overline{X} \sim N\left(1, \sqrt{\tfrac{2}{n}}\right)$$

or

$$\frac{\frac{Y}{n}-1}{\sqrt{\frac{2}{n}}}=\frac{Y-n}{\sqrt{2n}}\sim N(0,\ 1).$$

b. It is given that Y has a normal distribution with mean $\mu = 6$ and variance $\sigma^2 = .2$. Let C_i be the cost for a single rod, $i = 1, 2, \ldots, 50$. That is, $C_i = 4(Y_i - \mu)^2$. The total cost for the day, then, will be

$$\sum_{i=1}^{50} C_i = 4 \sum_{i=1}^{50} (y_i - \mu)^2$$

where the Y_i are independent and distributed as Y above. Recall from Chapter 6 that

$$\frac{\sum_{i=1}^{50}(Y_i-\mu)^2}{\sigma^2}=\sum_{i=1}^{50}\left(\frac{Y_i-\mu}{\sigma}\right)^2=\sum_{i=1}^{50}Z_i^2$$

has a chi-square distribution with 50 degrees of freedom, since each Z_i is standard normal, and hence Z_i^2 has a chi-square distribution with 1 degree of freedom, $i = 1, 2, \ldots, 50$. The probability of interest, using the results of part (a), is

$$P\left(\Sigma C_i > 48\right) = P\left[\Sigma(Y_i-\mu)^2 > 12\right] = P\left[\frac{\Sigma(Y_i-\mu)^2}{.2} > 60\right] = P(X > 60)$$

where X is a chi-square random variable with $n = 50$ degrees of freedom. Hence the approximation is

$$P(X > 60) = P\left(\frac{X-n}{\sqrt{2n}} > \frac{60-50}{\sqrt{100}}\right) = P(Z > 1) = .1587$$

7.57 a. Similar to Exercise 7.56. For fixed W_2, let $F = \frac{W_1}{c}$, where $c = \frac{w_2 \nu_1}{\nu_2}$. Since W_1 has a chi-square distribution, we have

$$m_{W_1}(t) = (1-2t)^{-\nu_1/2} \qquad\qquad m_{W_1/c}(t) = m_{W_1}\left(\frac{t}{c}\right) = \left(1-\frac{2}{c}t\right)^{-\nu_1/2}$$

Hence the distribution of F, conditional on $W_2 = w_2$, is that of a gamma random variable with parameters $\alpha = \frac{\nu_1}{2}$ and $\beta = \frac{2}{c} = \frac{2\nu_2}{w_2\nu_1}$.

b. Now, since W_2 has a chi-square distribution, we can write

$$g(f, w_2) = g(f|w_2)g(w_2) = \frac{f^{(\nu_1/2)-1}e^{-fw_2\nu_1/2\nu_2}w_2^{(\nu_2/2)-1}e^{-w_2/2}}{\Gamma\left(\frac{\nu_1}{2}\right)\left(\frac{2\nu_2}{w_2\nu_1}\right)^{\nu_1/2}\Gamma\left(\frac{\nu_2}{2}\right)2^{\nu_2/2}}$$

$$= \frac{f^{(\nu_1/2)-1}e^{-(w_2/2)[((f\nu_1/\nu_2)+1]}w_2^{(\nu_2/2)-1}w_2^{\nu_1/2}}{\Gamma\left(\frac{\nu_1}{2}\right)\left(\frac{2\nu_2}{\nu_1}\right)^{\nu_1/2}\Gamma\left(\frac{\nu_2}{2}\right)2^{\nu_2/2}} \quad \text{for } f \geq 0,\ w_2 \geq 0$$

c. Finally,

$$g(f) = \int_0^\infty g(f,\ w_2)\ dw_2$$

$$= \frac{f^{(\nu_1/2)-1}\left(\frac{\nu_1}{\nu_2}\right)^{\nu_1/2}}{\Gamma\left(\frac{\nu_1}{2}\right)\Gamma\left(\frac{\nu_2}{2}\right)2^{(\nu_1+\nu_2)/2}}\int_0^\infty w_2^{[(\nu_1+\nu_2)/2]-1}e^{-(w_2/2)[(f\nu_1/\nu_2)+1]}\ dw_2$$

$$= \frac{f^{(\nu_1/2)-1}\left(\frac{\nu_1}{\nu_2}\right)^{\nu_1/2}\Gamma\left(\frac{\nu_1+\nu_2}{2}\right)\left(1+\frac{f\nu_1}{\nu_2}\right)^{-(\nu_1+\nu_2)/2}}{\Gamma\left(\frac{\nu_1}{2}\right)\Gamma\left(\frac{\nu_2}{2}\right)} \quad f \geq 0$$

7.59 Using the result of Exercise 7.58, we have

$$P(X \leq 110) = P\left(\frac{X-\lambda}{\sqrt{\lambda}} \leq \frac{110-\lambda}{\sqrt{\lambda}}\right) = P\left(Z \leq \frac{110-100}{\sqrt{100}}\right) = P(Z \leq 1)$$

$$= 1 - .1587 = .8413$$

7.61 From Exercise 7.58, X converges in distribution to a normal distribution with mean λ_1 and variance λ_1, while Y converges in distribution to a normal distribution with mean λ_2 and variance λ_2. It can be shown that, as in Exercise 7.30, the quantity

$$\frac{(X-Y)-(\lambda_1-\lambda_2)}{\sqrt{\lambda_1+\lambda_2}}$$

will also converge in distribution to a standard normal random variable. Using this result, the approximation is

$$P(X-Y > 10) = P\left(\frac{X-Y}{\sqrt{\lambda_1+\lambda_2}} > \frac{100}{\sqrt{100}}\right) = P(Z > 1) = .1587$$

7.63 Let $Y = \#$ of people that suffer an adverse reaction. Y is binomial with $n = 1000$, $p = .001$. Using the result of Exercise 7.62, we let

$$\lambda = 1000(.001) = 1.$$

Using the Poisson distribution, we evaluate

$$P(Y \geq 2) = 1 - P(Y \leq 1) = 1 - .736 = .264.$$

CHAPTER 8 ESTIMATION

8.1 Write

$$\text{MSE}(\hat{\theta}) = E[(\hat{\theta} - \theta)]^2 = E[(\hat{\theta} - E(\hat{\theta}) + B^2] = E\{[\hat{\theta} - E(\hat{\theta})]^2\} + E(B^2)$$
$$+ 2B\{E[\hat{\theta} - E(\hat{\theta})]\}$$
$$= V(\hat{\theta}) + B^2 + 2B[E(\hat{\theta}) - E(\hat{\theta})] = V(\hat{\theta}) + B^2$$

8.3 Given that $E(\hat{\theta}_1) = E(\hat{\theta}_2) = \theta$; $V(\hat{\theta}_1) = \sigma_1^2$; $V(\hat{\theta}_2) = \sigma_2^2$;

$\hat{\theta}_3 = a\hat{\theta}_1 + (1 - a)\hat{\theta}_2$; $\text{cov}(\hat{\theta}_1, \hat{\theta}_2) = c$.

$$V(\hat{\theta}_3) = V[a\hat{\theta}_1 + (1 - a)\hat{\theta}_2]$$
$$= a^2 V(\hat{\theta}_1) + (1 - a)^2 V(\hat{\theta}_2) + 2a(1 - a)\,\text{cov}(\hat{\theta}_1, \hat{\theta}_2)$$
$$= a^2 \sigma_1^2 + (1 - a)\sigma_2^2 + 2a(1 - a)c.$$

To choose a value of a so that $V(\hat{\theta}_3)$ is minimized, consider

$$\frac{d}{da} V(\hat{\theta}_3) = 2a\sigma_1^2 - 2(1 - a)\sigma_2^2 + 2c(1 - 2a).$$

Setting this equal to 0 and solving for a,

$$a\sigma_1^2 - (1 - a)\sigma_2^2 + c(1 - 2a) = 0$$
$$a\sigma_1^2 + a\sigma_2^2 - 2ac = \sigma_2^2 - c$$
$$a(\sigma_1^2 + \sigma_2^2 - 2c) = \sigma_2^2 - c$$
$$a = \frac{\sigma_2^2 - c}{\sigma_1^2 + \sigma_2^2 - 2c}$$

8.5 Since Y has an exponential distribution with mean $\theta + 1$, $E(Y) = \theta + 1$ and $E(\overline{Y}) = \theta + 1$. Hence if we use $\hat{\theta} = \overline{Y} - 1$, $E(\hat{\theta}) = \theta$ and we have constructed an unbiased estimator.

8.7 The third central moment is defined to be

$$E[(Y - \mu)^3] = E[(Y - 3)^3]$$
$$= E[Y^3 - 9Y^2 + 27Y - 27]$$

$$= E[Y^3] - 9E[Y^2] + 27E[Y] - 27$$
$$= E[Y^3] - 9E[Y^2] + 27(3) - 27$$
$$= E[Y^3] - 9E[Y^2] + 54.$$

Given that $\hat{\theta}_3$ is an unbiased estimator of $E[Y^3]$ and $\hat{\theta}_2$ is an unbiased estimator of $E[Y^2]$, we consider $\hat{\theta}_3 - 9\hat{\theta}_2 + 54$ as an estimator. By the definition of unbiasness,

$$E[\hat{\theta}_3 - 9\hat{\theta}_2 + 54] = E[\hat{\theta}_3] - 9E[\hat{\theta}_2] + 54 = E[Y^3] - 9E[Y^2] + 54 = E[(Y-3)^3].$$

8.9   **a.**   For a binomial random variable Y, $E(Y) = np$ and $E(Y^2) = V(Y) + n^2p^2$
$= npq + n^2p^2$. Hence

$$E\left\{n\left(\frac{Y}{n}\right)\left[1 - \left(\frac{Y}{n}\right)\right]\right\} = E(Y) - \frac{1}{n}E(Y_2) = np - pq - np^2 = np(1-p) - pq$$
$$= (n-1)pq$$

b.   An unbiased estimator $\hat{\theta}$ has expected value npq. Hence we can use

$$\left(\frac{n}{n-1}\right)n \times \frac{Y}{n}\left(1 - \frac{Y}{n}\right) = \frac{n^2}{n-1}\left(\frac{Y}{n}\right)\left(1 - \frac{Y}{n}\right)$$

8.11   We consider the following:

$$f(y) = 3\beta^3 y^{-4}, \qquad \beta \le y$$

$$F(y) = \int_{\beta}^{y} 3\beta^3 y^{-4}\, dy = 1 - \left(\frac{\beta}{y}\right)^3$$

$$E(Y) = \frac{\alpha\beta}{\alpha - 1}$$

$$E(Y^2) = \frac{\alpha\beta^2}{\alpha - 2}$$

$$F_{Y(1)}(y) = 1 - \left(\frac{\beta}{y}\right)^{n\alpha}$$

$$f_{Y(1)}(y) = -n\alpha\left(\frac{\beta}{y}\right)^{n\alpha-1}(-\beta y^{-2}) = n\alpha\beta^{n\alpha}y^{-n\alpha-1}$$

$$E(Y_{(1)}) = \frac{n\alpha\beta}{n\alpha - 1} = \frac{3n\beta}{3n - 1}$$

$$E(Y_{(1)}^2) = \frac{3n\beta^2}{3n - 2}$$

Using the above information,

a. Bias $= \dfrac{3n\beta}{3n-1} - \beta$

$\qquad = \dfrac{3n\beta - 3n\beta + \beta}{3n-1}$

$\qquad = \dfrac{\beta}{3n-1}$

b. $\text{MSE}(\hat{\beta}) = E(\hat{\beta}^2) - 2\beta E(\hat{\beta}) + \beta^2$

$\qquad = \dfrac{3n\beta^2}{3n-2} - 2\beta\left(\dfrac{3n\beta}{3n-1}\right) + \beta^2$

$\qquad = \dfrac{2\beta^2}{(3n-1)(3n-2)}$

8.13 Note that

$$E(\hat{p}_1) = E\left(\frac{Y}{n}\right) = \left(\frac{1}{n}\right)(np) = p$$

$$E(\hat{p}_2) = E\left(\frac{Y+1}{n+2}\right) = \frac{1}{(n+2)}(np+1) = \frac{np+1}{n+2}$$

a. Bias $= \dfrac{np+1}{n+2} - p = \dfrac{np+1-np-2p}{n+2} = \dfrac{1-2p}{n+2}.$

b. $\text{MSE}(\hat{p}_1) = V(\hat{p}_1) + B^2 = V\left(\dfrac{Y}{n}\right) + 0 = \left(\dfrac{1}{n^2}\right)np(1-p) = \dfrac{p(1-p)}{n}.$

$$\text{MSE}(\hat{p}_2) = V(\hat{p}_2) + B^2 = V\left(\frac{Y+1}{n+2}\right) + \left(\frac{1-2p}{n+2}\right)^2$$

$$= \left[\frac{1}{(n+2)^2}\right]V(Y+1) + \frac{(1-2p)^2}{(n+2)^2}$$

$$= \frac{np(1-p) + (1-2p)^2}{(n+2)^2}.$$

c. We need to consider $\text{MSE}(\hat{p}_2) < \text{MSE}(\hat{p}_1)$.

$$\frac{np(1-p) + (1-2p)^2}{(n+2)^2} < \frac{p(1-p)}{n}$$

$$n^2 p(1-p) + n(1-2p)^2 - p(1-p)(n+2)^2 < 0$$

This simplifies to

$$(8n+4)p^2 - (8n+4)p + n < 0$$

One can proceed to use the quadratic formula to solve for p and will see that p will be near the value of $\frac{1}{2}$.

8.15 $E(Y_{(1)}) = \frac{\beta}{n}$.

$$E(\hat{\theta}) = E[nY_{(1)}] = nE(Y_{(1)}) = n\left(\frac{\beta}{n}\right) = \beta = \theta.$$

Thus, $nY_{(1)}$ is an unbiased estimator for θ.

$$MSE(\hat{\theta}) = V(\hat{\theta}) + B^2$$

$$= Y(nY_{(1)}) = n^2V(Y_{(1)}) = n^2\left(\frac{\beta}{n}\right)^2 = \beta^2.$$

8.17 The point estimate of μ is $\bar{y} = 39.8°$, and the bound on the error of estimation with $s = 17.2$ and $n = 50$ is

$$2\sigma_{\bar{y}} = 2\frac{\sigma}{\sqrt{n}} = 2\frac{s}{\sqrt{n}} = \frac{2(17.2)}{\sqrt{50}} = 4.86$$

8.19 **a.** The point estimate of μ is $\bar{y} = 11.3$, and the bound on the error of estimation is $2\sigma_{\bar{y}}$. With $n = 467$ and $s = 16.6$, this bound is

$$2\sigma_{\bar{y}} = 2\frac{\sigma}{\sqrt{n}} = 2\frac{s}{\sqrt{n}} = \frac{2(16.6)}{\sqrt{467}} = 1.54$$

b. The point estimate of $\mu_R - \mu_C$ is $\bar{y}_R - \bar{y}_C = 46.4 - 45.1 = 1.3$. The bound on the error of estimation is

$$2\sqrt{\frac{\sigma_R^2}{n_R} + \frac{\sigma_C^2}{n_C}} = 2\sqrt{\frac{s_R^2}{n_R} + \frac{s_C^2}{n_C}} = 2\sqrt{\frac{(9.8)^2}{191} + \frac{(10.2)^2}{467}} = 1.7$$

c. The point estimate of $p_C - p_R$ is $\hat{p}_C - \hat{p}_R = .78 - .61 = .17$. The bound on the error of estimation is

$$2\sqrt{\frac{p_C q_C}{n_C} + \frac{p_R q_R}{n_R}} = 2\sqrt{\frac{\hat{p}_C \hat{q}_C}{n_C} + \frac{\hat{p}_R \hat{q}_R}{n_R}} = 2\sqrt{\frac{(.78)(.23)}{467} + \frac{(.61)(.39)}{191}} = .08$$

8.21 We estimate the difference to be

$$\bar{y}_1 - \bar{y}_2 = 2.4 - 3.1 = -.7.$$

A bound for the error of estimation is

$$b = 2\sqrt{\frac{1.44 + 2.64}{100}} = 2\sqrt{.0408} = .404.$$

8.23 a. The bound is $2\sqrt{\dfrac{\hat{p}(1-\hat{p})}{n}} = 2\sqrt{\dfrac{(.67)(.33)}{308,007}} = .0017$.

b. The bound is $2\sqrt{\dfrac{\hat{p}(1-\hat{p})}{n}} = 2\sqrt{\dfrac{(.71)(.29)}{308,007}} = .0016$.

c. No, ± 2 percentage points is too large for the margin of error. The bound on the margin of error is closer to $\pm .2$ percentage points.

8.25 a. Let p_1 = proportion of Americans who ate the recommended amount of fibrous foods in 1983 and p_2 = proportion of Americans who ate the recommended amount of fibrous foods in 1992. Then $n_1 = 1250$, $n_2 = 1251$, $\hat{p}_1 = .59$, and $\hat{p}_2 = .53$. The point estimator for the difference in proportions is

$$\hat{p}_1 - \hat{p}_2 = .59 - .53 = .06.$$

The bound on the error of estimation is

$$2\sqrt{\dfrac{\hat{p}_1(1-\hat{p}_1)}{n_1} + \dfrac{\hat{p}_2(1-\hat{p}_2)}{n_2}} = 2\sqrt{\dfrac{(.59)(.41)}{1250} + \dfrac{(.53)(.47)}{1251}} = .04$$

b. Since $.06 - .04 > 0$, we can conclude that there has been a demonstrable decrease in the proportion of Americans who eat the recommended amount of fibrous foods.

8.27 Let p = proportion of individuals who regularly used seat belts in 1992. Then the point estimator is $\hat{p} = .70$ and the bound is

$$2\sqrt{\dfrac{(.7)(.3)}{1251}} = .026$$

It is not likely that the estimate is off by as much as 10%. The bound on the margin of error is only 2.6%.

8.29 The point estimate is $\hat{p} = .3$. A bound on the error of estimation is

$$2\sqrt{\dfrac{\hat{p}\hat{q}}{n}} = \sqrt{\dfrac{(.3)(.7)}{20}} = .205$$

It is fairly likely that the proportion in compliance exceeds .80 since the value .20 is well within the margin of error from the estimation of the proportion not in compliance.

8.31 We will use the result of Exercise 8.30.

a. The point estimate is $\hat{\lambda}_A = \bar{y} = 20$. A bound on the error of estimation is

$$2\sqrt{\dfrac{\bar{y}}{n}} = 2\sqrt{\dfrac{20}{50}} = 1.265$$

b. The point estimate is $\hat{\lambda}_A - \hat{\lambda}_B = \bar{y}_A - \bar{y}_B = 20 - 23 = -3$. A bound on the error of estimation is

$$2\sqrt{\frac{\bar{y}_A}{n_A} + \frac{\bar{y}_B}{n_B}} = \sqrt{\frac{20}{50} + \frac{23}{50}} = 1.855$$

If the two regimes are identical, then $|\lambda_A - \lambda_B| = 0$. The value $|-3|$ is well beyond the bound of the error of estimation. Thus, we would say the regime B tends to produce a larger mean number of nucleation sites.

8.33 Refer to Exercise 8.32. An estimate of θ is $\hat{\theta} = \bar{y} = 1020$, and the bound on error is approximately

$$2\hat{\sigma}_{\bar{Y}} = 2\,\frac{\bar{y}}{\sqrt{n}} = 2\left(\frac{1020}{\sqrt{10}}\right) = 645.10$$

8.35 Using Table 6 and indexing $\nu = 4$ degrees of freedom, we can write

$$P\left(\chi^2_{.95} \le X \le \chi^2_{.05}\right) = .90$$

$$P\left(.71072) < \frac{2Y}{\beta} < 9.48773\right) = .90$$

$$P\left(\frac{2Y}{9.48773} < \beta < \frac{2Y}{.710721}\right) = .90$$

Hence the interval $\left(\frac{2Y}{9.48773}, \frac{2Y}{.710721}\right)$ forms a 90% confidence interval for β.

8.37 **a.** $.95 = P\left(\chi^2_{.975} \le \frac{Y^2}{\sigma^2} \le \chi^2_{.025}\right) = P\left(.000982) \le \frac{Y^2}{\sigma^2} \le 5.02389\right)$

$$= P\left(\frac{Y^2}{5.02389} \le \sigma^2 \le \frac{Y^2}{.0009821}\right)$$

b. $.95 = P\left(\chi^2_{.95} \le \frac{Y^2}{\sigma^2}\right) = P\left(\sigma^2 \le \frac{Y^2}{.0039321}\right)$

c. $.95 = P\left(\frac{Y^2}{\sigma^2} \le \chi^2_{.05}\right) = P\left(\sigma^2 \ge \frac{Y^2}{3.84146}\right)$

8.39 $f(y) = \frac{1}{\theta}$; $F(y) = \frac{y}{\theta}$

a. $F_u(u) = P(U \le u) = P(Y_{(n)} \le \theta u) = F_{Y_{(n)}}(\theta u)$.

$$F_{Y_{(n)}}(y) = n[F(y)]^{n-1}f(y) = \left(\frac{y}{\theta}\right)^n$$

$$F_{Y(n)}(\theta u) = \left(\frac{\theta u}{\theta}\right)^n = u^n.$$

Thus,

$$F(u) = \begin{cases} 0, & u < 0 \\ u^n, & 0 \le u < 1 \\ 1, & u > 1 \end{cases}$$

b. $P\left[\left(\frac{Y_n}{\theta}\right) < a\right] = .95$

$a^n = .95$

$a = (.95)^{1/n}.$

Thus, a 95% lower confidence bound for θ is

$$\frac{Y_{(n)}}{\left[(.95)^{1/n}\right]}.$$

8.41 $\hat{p} = \frac{1912}{2374} = .805;\ \alpha = .01,\ Z_{\alpha/2} = Z_{.005} = 2.576$

$$\hat{p} \pm Z_{\alpha/2}\sqrt{\frac{\hat{p}(1-\hat{p})}{n}} = .805 \pm 2.576\sqrt{\frac{(.805)(1-.805)}{2374}} = .805 \pm .021 = (.784, .826)$$

Thus, at the 99% confidence level, the proportion of adults in the continental United States registered to vote is between .784 and .826.

8.43 **a.** We are given that $n = 224$, $\hat{p} = \frac{2}{3}$, $\alpha = .10$, and $z_{.05} = 1.645$. Then a 90% confidence interval for the proportion of children aged 9 to 17 who would like to experience space travel is

$$\hat{p} \pm z_{\alpha/2}\sqrt{\frac{\hat{p}(1-\hat{p})}{n}} = \frac{2}{3} \pm 1.645\sqrt{\frac{\left(\frac{2}{3}\right)\left(\frac{1}{3}\right)}{224}} = .667 \pm .052 = (.615, .719)$$

b. Since the entire interval is greater than one-half, it is believable that most of the children in the group think that they would like to experience space travel.

8.45 For a 95% confidence interval, $z = 1.96$. We consider

$$(167.1 - 140.9) \pm (1.96)\sqrt{\frac{(23.4)^2}{30} + \frac{(17.6)^2}{30}} = 26.2 \pm 10.477 = (15.723, 36.677).$$

8.47 **a.** A 95% confidence interval for p is approximately

$$\hat{p} \pm 1.96\sqrt{\frac{\hat{p}\hat{q}}{n}}$$

$$.73 \pm 1.96 \sqrt{\frac{(.73)(.27)}{1506}}$$

$$.73 \pm .022 \quad \text{or} \quad .708 < p < .752$$

b.   The sampling must be random, and hence each trial (person) independent.

8.49   Let p_1 = proportion of Americans who used seat belts in 1983 and p_2 = proportion of Americans who used seat belts in 1992. Then, $n_1 = 1250$, $n_2 = 1251$, $\hat{p}_1 = .19$, $\hat{p}_2 = .70$, $\alpha = .10$, and $z_{\alpha/2} = z_{.05} = 1.645$.

$$(\hat{p}_1 - \hat{p}_2) \pm z_{\alpha/2} \sqrt{\frac{\hat{p}_1(1 - \hat{p}_1)}{n_1} + \frac{\hat{p}_2(1 - \hat{p}_2)}{n_2}}$$

$$= (.19 - .70) \pm 1.645 \sqrt{\frac{(.19)(.81)}{1250} + \frac{(.70)(.30)}{1251}} = -.51 \pm .028$$

$$= (-.538, -.482)$$

Yes, it is believable, at the 90% confidence level, that the proportion of Americans that used seat belts in 1992 is greater than the proportion in 1983 by as little as .482 or as much as .538.

8.51   This is similar to Example 8.8 in the text. The 98% confidence interval is

$$(\hat{p}_1 - \hat{p}_2) \pm 2.33 \sqrt{\frac{\hat{p}_1(1 - \hat{p}_1)}{n_1} + \frac{\hat{p}_2(1 - \hat{p}_2)}{n_2}}$$

or

$$(.18 - .12) \pm 2.33 \sqrt{\frac{(.18)(.82)}{100} + \frac{(.12)(.88)}{100}} \quad \text{or} \quad .06 \pm .12 \quad \text{or} \quad -.05 \text{ to } .18$$

8.53   **a.**   The parameter to be estimated is μ, the mean number of ships passing within ten miles of the proposed power site location per day. The 95% confidence interval is approximately

$$\bar{y} \pm 1.96 \left(\frac{s}{\sqrt{n}} \right) \quad \text{or} \quad 7.2 \pm 1.96 \sqrt{\frac{8.8}{60}} \quad \text{or} \quad 7.2 \pm .751$$

b.   Now we are interested in the difference between means, $\mu_1 - \mu_2$, for the summer versus the winter months. The interval estimator for the difference between two means is

$$(\bar{y}_1 - \bar{y}_2) \pm z_{\alpha/2} \sqrt{\frac{\sigma_1^2}{n_1} + \frac{\sigma_2^2}{n_2}},$$

which is approximated as

$$(\bar{y}_1 - \bar{y}_2) \pm 1.645\sqrt{\frac{s_1^2}{n_1} + \frac{s_2^2}{n_2}} \quad \text{or} \quad (7.2 - 4.7) \pm 1.645\sqrt{\frac{8.8}{60} + \frac{4.9}{90}}$$

or

$$2.5 \pm .738$$

c. The population used is the difference between the daily mean number of ships sighted in summer months and the mean number sighted in winter months for all summers and winters. One possible problem with the sample of parts (a) and (b) is that the months were not chosen independently or randomly and all of the months were chosen in the same year. Practically, it would be nearly impossible to choose them independently and randomly.

8.55 Since the $\hat{p}_i$ are independent, $V\left[\hat{p}_3 - \hat{p}_1) - (\hat{p}_4 - \hat{p}_2)\right] = \sum\limits_{i=1}^{4} \frac{p_i q_i}{n_i}$. If $\hat{p}_i$ is used to estimate p_i, the approximate 95% confidence interval is

$$\left[(\hat{p}_3 - \hat{p}_1) - (\hat{p}_4 - \hat{p}_2)\right] \pm 1.96\sqrt{V\left[(\hat{p}_3 - \hat{p}_1) - (\hat{p}_4 - \hat{p}_2)\right]}$$

$$[(.69 - .65) - (.25 - .43)] \pm 1.96 = \sqrt{\frac{(.65)(.35)}{31} + \frac{(.43)(.57)}{30} + \frac{(.69)(.31)}{26} + \frac{(.25)(.75)}{28}}$$

$$.22 \pm .34 \quad \text{or} \quad -.12 \text{ to } .56$$

8.57 $B = 2$, $\sigma = 10$.

$$n = \frac{4\sigma^2}{B^2} = \frac{4(10)^2}{4} = 100.$$

8.59 From 8.43, $\hat{p} = \frac{2}{3}$. It is given that $B = .02$, $\alpha = .01$, and $z_{\alpha/2} = z_{.005} = 2.576$. Then

$$z_{\alpha/2}\sqrt{\frac{\hat{p}(1 - \hat{p})}{n}} = B$$

$$2.576\sqrt{\frac{\left(\frac{2}{3}\right)\left(\frac{1}{3}\right)}{n}} = .02$$

$$n = \frac{(2.576)^2\left(\frac{2}{3}\right)\left(\frac{1}{3}\right)}{(.02)^2} = 3686.54.$$

Use $n = 3687$.

8.61 There are now two populations of interest and the parameter to be estimated is $\mu_1 - \mu_2$. The bound on the error of estimation is $B = .1$, $\sigma_1^2 = \sigma_2^2 = .25$, and $n_1 = n_2 = n$. With $(1 - \alpha) = .90$, $z_{.05} = 1.645$. Hence,

$$1.645\sqrt{\frac{\sigma_1^2}{n_1} + \frac{\sigma_2^2}{n_2}} \le .1$$

$$1.645\sqrt{\frac{.25 + .25}{n}} \le .1$$

$$\frac{1.645\sqrt{.5}}{.1} \le \sqrt{n}$$

$$n \ge 135.30$$

or $n = 136$ samples should be selected at each location.

8.63 Using our estimates for p_1 and p_2, $\hat{p}_1 = .7$ and $\hat{p}_2 = .54$, we consider

$$1.645\sqrt{\frac{(.7)(.3) + (.54)(.46)}{n}} = .05$$

$$n = 496.17 \cong 497.$$

8.65 **a.** Assuming equal sample sizes, the estimate of the standard error of the difference is

$$\hat{\sigma}_{\overline{x}_1 - \overline{x}_2} = \sqrt{\frac{(2.10)^2 + (2.26)^2}{n}} = \frac{3.085}{\sqrt{n}}.$$

Then $z_{.005} = 2.576$ and $B = 1$, so that we have

$$2.576\left(\frac{3.085}{\sqrt{n}}\right) = 1$$

or

$$n = [2.576(3.085)]^2 = 63.16.$$

Use 64.

b. For hobbies,

$$\hat{\sigma}_{\overline{x}_1 - \overline{x}_2} = \sqrt{\frac{(6.47)^2 + (7.46)^2}{n}} = \frac{9.874}{\sqrt{n}}.$$

Then

$$2.576\left(\frac{9.874}{\sqrt{n}}\right) = 1$$

or

$$n = [2.576(9.874)]^2 = 646.96.$$

Collect 647 each of males and females.

c. The activity scores for hobbies have the largest standard deviation of all the activities. Any sample size that meets the criterion for this activity would do better for the other activities. Therefore, a sample of size 647 each of males and females would ensure that 99% confidence intervals to compare males and females in all 7 activities would have widths no larger than 2 units.

8.67 Calculate $\Sigma y_i = 608$, $\Sigma y_i^2 = 37{,}538$, $n = 10$. Then

$$\overline{y} = \frac{608}{10} = 60.8$$

$$s^2 = \frac{37{,}538 - \dfrac{(608)^2}{10}}{9} = 63.5111$$

The 95% confidence interval is

$$\overline{Y} \pm t_{.025}\left(\frac{s}{\sqrt{n}}\right)$$

$$60.8 \pm 2.262\sqrt{\frac{63.5111}{10}}$$

$$60.8 \pm 5.701 \qquad \text{or} \qquad 55.099 < \mu < 66.501$$

8.69 a. Let μ_1 = mean compartment pressure for all runners under resting condition and μ_2 = mean compartment pressure for all cyclists under resting condition. Then the small-sample 95% confidence interval for $\mu_1 - \mu_2$ is

$$\overline{x}_1 - \overline{x}_2 \pm t_{\alpha/2}\, S_p \sqrt{\frac{1}{n_1} + \frac{1}{n_2}}$$

where $t_{.025} = 2.101$ with $n_1 + n_2 - 2 = 18$ degrees of freedom. Also

$$S_p^2 = \frac{(n_1 - 1)S_1^2 + (n_2 - 1)S_2^2}{n_1 + n_2 - 2} = \frac{9(3.92)^2 + 9(3.98)^2}{18} = 15.6034.$$

Then, the interval is

$$14.5 - 11.1 \pm 2.101\sqrt{15.6034\left(\frac{1}{10} + \frac{1}{10}\right)}$$

$$3.4 \pm 3.7$$

$$(-.3,\ 7.1)$$

b. Similar to part (a). Let μ_1 = mean compartment pressure for all runners who exercise at 80% of maximal oxygen consumption and μ_2 = mean compartment pressure for all cyclists who exercise at 80% of maximal oxygen consumption. Then calculate

$$S_p^2 = \frac{(n_1 - 1)S_1^2 + (n_2 - 1)S_2^2}{n_1 + n_2 - 2} = \frac{9(3.49)^2 + 9(4.95)^2}{18} = 18.3413.$$

A 90% confidence interval for $\mu_1 - \mu_2$ is

$$(\bar{x}_1 - \bar{x}_2) \pm t_{.025}\sqrt{S_p^2\left(\frac{1}{n_1} + \frac{1}{n_2}\right)} = 12.2 - 11.5 \pm 1.734\sqrt{18.3413\left(\frac{1}{10} + \frac{1}{10}\right)}$$

$$= .7 \pm 3.32 \qquad \text{or} \qquad (-2.62, 4.02)$$

c. Because both intervals contain 0, we cannot conclude that a difference exists in mean compartment pressures between runners and cyclists in either condition.

8.71 This is similar to Exercise 8.69. Calculate

$$s^2 = \frac{15(6)^2 + 19(8)^2}{34} = 51.647$$

The 95% confidence interval for $\mu_1 - \mu_2$ is

$$(\bar{y}_1 - \bar{y}_2) \pm t_{.025}\sqrt{s^2\left(\frac{1}{n_1} + \frac{1}{n_2}\right)} \qquad \text{or} \qquad (11 - 12) \pm 1.96\sqrt{51.647\left(\frac{1}{16} + \frac{1}{20}\right)}$$

or

$$-1 \pm 4.72 \qquad \text{or} \qquad -5.72 \text{ to } 3.72.$$

8.73 Refer to Exercise 8.72, where $\bar{y}_1 = 9$ and $(n_1 - 1)s_1^2 = \Sigma y_1^2 - \left[\frac{(\Sigma y_1)^2}{n}\right] = 454$. From this,

$$\Sigma y_2 = 10.7 \qquad\qquad\qquad \Sigma y_2^2 = 65.09$$

$$\Sigma y_2^2 - \frac{(\Sigma y_2)^2}{n_2} = 26.92667 \qquad\qquad s_2 = \frac{26.92667}{2} = 13.4633$$

a. The 90% confidence interval for μ_2 is

$$\bar{y}_2 \pm t_{.05}\left(\frac{s_2}{\sqrt{n_2}}\right) \qquad\qquad \text{or} \qquad\qquad 3.57 \pm 2.92\sqrt{\frac{13.4633}{3}}$$

or

$$3.57 \pm 6.19 \qquad\qquad \text{or} \qquad\qquad -2.62 \text{ to } 9.76$$

b. Calculate

$$s^2 = \frac{454 + 26.92667}{13} = 36.9944$$

The 90% confidence interval for $\mu_1 - \mu_2$ is

$$(\bar{y}_1 - \bar{y}_2) \pm t_{.05}\sqrt{s^2\left(\frac{1}{n_1} + \frac{1}{n_2}\right)} \qquad \text{or} \qquad (9 - 3.57) \pm 1.771\sqrt{36.9944\left(\frac{1}{12} + \frac{1}{3}\right)}$$

or

$$5.43 \pm 6.95 \qquad\qquad \text{or} \qquad\qquad -1.52 \text{ to } 12.38$$

We must assume that the LC50 measurements are normally distributed and independent and that $\sigma_1^2 = \sigma_2^2$.

8.75 Assume two independent random samples from normal populations with $\sigma_1^2 = \sigma_2^2$. The following calculations are necessary:

	Spring	Summer
Σy_i	78.1	289.1
Σy_i^2	1612.15	22,641.47
$\overline{y}$	15.62	72.275
$\Sigma (y_i - \overline{y})^2$	392.228	1746.7675
n	5	4

Calculate
$$s^2 = \frac{392.228 + 1746.7675}{7} = 305.57079$$
Then the 95% confidence interval for $\mu_1 - \mu_2$ is

$$(15.62 - 72.275) \pm 2.365 \sqrt{s^2 \left(\tfrac{1}{5} + \tfrac{1}{4}\right)} \quad \text{or} \quad -56.66 \pm 27.73 \quad \text{or} \quad -84.39 \text{ to } -28.93$$

8.77 **a.** Since X and Y are both normally distributed with given means and variances, $2\overline{X} + \overline{Y}$ is normally distributed, with mean $2\mu_1 + \mu_2$ and variance $\frac{4\sigma^2}{n} + \frac{3\sigma^2}{m}$

$= \sigma^2 \left(\frac{4}{n} + \frac{3}{m}\right)$. If σ^2 is known, then $2\overline{X} + \overline{Y} \pm 1.96\sigma \sqrt{\frac{4}{n} + \frac{3}{m}}$ is a 95% confidence interval for $2\mu_1 + \mu_2$.

b. Since $\Sigma \dfrac{(Y_i - \overline{Y})^2}{\sigma^2}$ has a χ^2 distribution with $(n - 1)$ degrees of freedom and

$\Sigma \dfrac{(X_i - \overline{X})^2}{3\sigma^2}$ has a distribution with $(m - 1)$ degrees of freedom, the sum

$$\frac{\Sigma (Y_i - \overline{Y})^2 + \frac{1}{3}\Sigma (X_i - \overline{X})^2}{\sigma^2}$$

has a χ^2 distribution with $(n + m - 2)$ degrees of freedom. Then using Definition 7.2,

$$t = \frac{(2\overline{X} + \overline{Y}) - (2\mu_1 + \mu_2)}{\sqrt{\hat{\sigma}\left(\frac{4}{n} + \frac{3}{m}\right)}}$$

where

$$\hat{\sigma}^2 = \frac{\Sigma(Y_i - \overline{Y})^2 + \left(\frac{1}{3}\right)\Sigma(X_i - \overline{X})^2}{n + m - 2}$$

Then, the 95% confidence interval is

$$(2\overline{X} + \overline{Y}) \pm t_{.025}\sqrt{\hat{\sigma}^2\left(\frac{4}{n} + \frac{3}{m}\right)}$$

8.79 It is desired to place a 90% confidence interval on σ^2, the variance of the truck noise emission readings. Indexing $\chi^2_{.05}$ and $\chi^2_{.95}$ with $(n-1) = 5$ degrees of freedom in Table 5 yields

$$\chi^2_{.95} = 1.145476 \qquad \text{and} \qquad \chi^2_{.05} = 11.0705.$$

Calculate $\Sigma y_i = 514.4$, $\Sigma y_i^2 = 44{,}103.74$,

$$s^2 = \frac{44{,}103.74 - \dfrac{(514.4)^2}{6}}{5} = .502667.$$

Then the 90% confidence interval for σ^2 is

$$\frac{5(.502667)}{11.0705} < \sigma^2 < \frac{5(.502667)}{1.145476}$$

$$.227 < \sigma^2 \le 2.194.$$

Intervals constructed in this manner will enclose σ^2 90% of the time in repeated sampling. Hence, we are fairly certain that σ^2 is between .227 and 2.194.

8.81 **a.** Find a number χ^2_α with $(n-1)$ degrees of freedom such that $P\left(\dfrac{(n-1)s^2}{\sigma^2} \ge \chi^2_\alpha\right)$
$= 1 - \alpha$. Then

$$1 - \alpha = P\left(\frac{(n-1)s^2}{\sigma^2} \le \chi^2_\alpha\right) = P\left(\frac{(n-1)s^2}{\chi^2_\alpha} \ge \sigma^2\right).$$

Then

$$\frac{(n-1)s^2}{\chi^2_\alpha} \text{ is a } 100(1-\alpha)\% \text{ upper confidence bound for } \sigma^2.$$

b. Similar to part (a). Find a number $\chi^2_{1-\alpha}$ such that

$$P\left(\frac{(n-1)s^2}{\sigma^2} \le \chi^2_{1-\alpha}\right) = 1 - \alpha.$$

Then

$$1 - \alpha = P\left(\frac{(n-1)s^2}{\sigma^2} \leq \chi^2_{1-\alpha}\right) = P\left(\frac{(n-1)s^2}{\chi^2_{1-\alpha}} \leq \sigma^2\right).$$

Therefore,

$\dfrac{(n-1)s^2}{\chi^2_{1-\alpha}}$ is a $100(1-\alpha)\%$ lower confidence bound for σ^2.

8.83 Similar to Exercise 8.81.

    a. A $100(1-\alpha)\%$ upper confidence bound for σ is

$$\sqrt{\frac{(n-1)s^2}{\chi^2_\alpha}}$$

    where χ^2_α is chosen such that

$$P\left(\frac{(n-1)s^2}{\sigma^2} \geq \chi^2_\alpha\right) = 1 - \alpha.$$

    b. A $100(1-\alpha)\%$ lower confidence bound for σ is

$$\sqrt{\frac{(n-1)s^2}{\chi^2_{1-\alpha}}}$$

    where $\chi^2_{1-\alpha}$ is chosen so that

$$P\left(\frac{(n-1)s^2}{\sigma^2} \leq \chi^2_{1-\alpha}\right) = 1 - \alpha.$$

8.85 Calculate

$$\Sigma y_i = 56.91 \qquad \Sigma y_i^2 = 539.9341 \qquad \Sigma y_i^2 - \frac{(y_i)^2}{n} = .14275$$

Thus, $s^2 = .02855$. Then the 90% confidence interval for σ^2 is

$$\frac{(n-1)s^2}{\chi^2_{.05}} < \sigma^2 < \frac{(n-1)s^2}{\chi^2_{.95}} \quad \text{or} \quad \frac{.14275}{11.0705} < \sigma^2 < \frac{.14275}{1.145476} \quad \text{or} \quad (.0129, .1246)$$

8.87 Assume that the measurements are normally distributed. The 90% confidence interval for σ^2 is

$$\frac{(n-1)s^2}{X_U^2} < \sigma^2 < \frac{(n-1)s^2}{X_L^2} \qquad \text{or} \qquad \frac{11}{7.81473} < \sigma^2 < \frac{11}{.351846}$$

or

$$1.407 < \sigma^2 < 31.264$$

The guarantee is not reasonable since a range of ± 2 units implies

$$\sigma = \frac{\text{range}}{4} \qquad \text{or} \qquad \sigma = \frac{4}{4} = 1$$

which does not lie in the above confidence interval.

8.89 We are given that $B = .03$ and $\alpha = .05$. Using the sample statistics from the previous study, we have

$$z_{\alpha/2}\sqrt{\frac{\hat{p}_1\hat{q}_1}{n_1} + \frac{\hat{p}_2\hat{q}_2}{n_2}} = B$$

as

$$1.96\sqrt{\frac{(.903)(.097)}{n} + \frac{(.898)(.102)}{n}} = .03$$

$$\frac{(1.96)^2}{(.03)^2}(.179) = n$$

$$764.05 = n.$$

Use $n = 765$.

8.91 a. With $y = 25$ and $n = 400$, the best estimate of p, the proportion of unemployed workers, is

$$\hat{p} = \frac{25}{400} = .0625$$

and a bound on error

$$1.96\sqrt{\frac{pq}{n}} = 1.96\sqrt{\frac{\hat{p}\hat{q}}{n}} = 1.96\sqrt{\frac{(.0625)(.9375)}{400}} = .0237.$$

b. Estimating p by $\hat{p}$ and assuming that the bound on error is .02, we have

$$1.96\sqrt{\frac{\hat{p}\hat{q}}{n}} \le .02 \qquad \frac{(1.96)(.24206)}{\sqrt{n}} \le .02 \qquad \sqrt{n} \ge 23.722 \qquad n \ge 562.73$$

so $n = 563$.

8.93 Assume $p = .5$ and the desired bound is .005. Then solving for n in the correct inequality, we have

$$1.96\sigma_{\hat{\theta}} \le B \quad \text{or} \quad 1.96\sqrt{\frac{pq}{n}} \le .005 \quad \text{or} \quad 1.96\sqrt{\frac{(.5)(.5)}{n}} \le .005$$

or $n = 38{,}916$.

8.95 Assume that $p = p_1 = p_2 = .6$, $n_1 = n_2 = n$, and that the desired bound is .05. Then

$$2\sqrt{\frac{p_1 q_1}{n_1} + \frac{p_2 q_2}{n_2}} \text{ will be } 2\sqrt{\frac{2pq}{n}}$$

and n is calculated as follows:

$$\sqrt{\frac{2(.6)(.4)}{n}} = .05 \qquad \text{or} \qquad n = 768$$

8.97 Refer to Exercise 8.96. The 90% confidence interval for σ^2 is

$$\frac{(n-1)s^2}{\chi^2_{.05}} < \sigma^2 < \frac{(n-1)s^2}{\chi^2_{.95}} \qquad \text{or} \qquad \frac{278}{9.48773} < \sigma^2 < \frac{278}{.710721}$$

or

$$29.30 < \sigma^2 < 391.15$$

8.99 The 90% confidence interval is

$$\bar{y} \pm y_{.05,\,16}\left(\frac{s}{\sqrt{n}}\right) \qquad \text{or} \qquad 11.3 \pm 1.746\left(\frac{3.4}{\sqrt{17}}\right) \qquad \text{or} \qquad 11.3 \pm 1.440$$

or

$$9.860 < \mu < 12.740$$

8.101 Since n_1 and n_2 are large, the 95% confidence interval for $\mu_1 - \mu_2$ is

$$(\bar{y}_1 - \bar{y}_2) \pm 1.96\sqrt{\frac{s_1^2}{n_1} + \frac{s_2^2}{n_2}} \quad \text{or} \quad (75 - 72) \pm 1.96\sqrt{\frac{(10)^2}{50} + \frac{(8)^2}{45}} \quad \text{or} \quad 3 \pm 3.63$$

or

$$-.63 \text{ to } 6.63.$$

8.103 Assume the reaction times are normally distributed with $\sigma_1^2 = \sigma_2^2$. Calculate

$$\bar{y}_1 = \frac{\sum_j y_{1j}}{n_1} = \frac{15}{8} = 1.875 \qquad\qquad \bar{y}_2 = \frac{21}{8} = 2.625$$

$$s^2 = \frac{33 - \left[\frac{(15)^2}{8}\right] + 61 - \left[\frac{(21)^2}{8}\right]}{14} = \frac{94 - \left(\frac{666}{8}\right)}{14} = \frac{10.75}{14} = .7678$$

0% confidence interval for $\mu_1 - \mu_2$ is

$$(\bar{y}_1 - \bar{y}_2) \pm t_{.05,14}\sqrt{s^2\left(\frac{1}{n_1} + \frac{1}{n_2}\right)} \quad \text{or} \quad (1.875 - 2.625) \pm 1.761\sqrt{.7678\left(\frac{1}{8} + \frac{1}{8}\right)}$$

which yields $-.75 \pm .77$, or $-1.52 < \mu_1 - \mu_2 < .02$.

8.105 It is given that $n = 2300$ and $y = 1914$. Thus, $\hat{p} = \frac{1914}{2300} = .832$ and a 95% confidence interval for p is

$$\hat{p} \pm 1.96\sqrt{\frac{\hat{p}\hat{q}}{n}} \quad \text{or} \quad .832 \pm 1.96\sqrt{\frac{(.832)(.168)}{2300}} \quad \text{or} \quad .832 \pm .0153$$

8.107 a. From Definition 7.3, the quantity

$$F = \frac{\left[\dfrac{\dfrac{(n_1-1)S_1^2}{\sigma^2}}{(n_1-1)}\right]}{\left[\dfrac{\dfrac{(n_2-1)S_2^2}{\sigma^2}}{(n_2-1)}\right]} = \frac{\left(\dfrac{S_1^2}{\sigma_1^2}\right)}{\left(\dfrac{S_2^2}{\sigma_2^2}\right)} = \frac{S_1^2}{S_2^2} \times \frac{\sigma_2^2}{\sigma_1^2}$$

has an F distribution with $(n_1 - 1)$ and $(n_2 - 1)$ degrees of freedom.

b. Similar to Exercise 7.16. We want

$$P(F_L \le F \le F_U) = 1 - \alpha$$

where F_L and F_U are chosen so that $P(F > F_U) = P(F < F_L) = \frac{\alpha}{2}$. From Table 7, $F_U = F_{\nu_1, \nu_2, \alpha/2}$, where $\nu_i = n_i - 1$, $i = 1, 2$. Then, as in Exercise 7.16 and using Exercise 7.11,

$$F_L = \frac{1}{F_{\nu_2, \nu_1, \alpha/2}}.$$

Now, we have

$$P\left(\frac{1}{F_{\nu_2, \nu_1, \alpha/2}} \le \frac{S_1^2}{S_2^2} \times \frac{\sigma_2^2}{\sigma_1^2} \le F_{\nu_1, \nu_2, \alpha/2}\right)$$

$$= P\left(\frac{S_2^2}{S_1^2 F_{\nu_2, \nu_1, \alpha/2}} \le \frac{\sigma_2^2}{\sigma_1^2} \le \frac{S_2^2}{S_1^2} F_{\nu_1, \nu_2, \alpha/2}\right) = 1 - \alpha.$$

Therefore, the $100(1-\alpha)\%$ confidence interval for $\dfrac{\sigma_2^2}{\sigma_1^2}$ will be

$$\left(\frac{1}{F_{\nu_2,\nu_1,\alpha/2}}\left(\frac{S_2^2}{S_1^2}\right),\ F_{\nu_1,\nu_2,\alpha/2}\left(\frac{S_2^2}{S_1^2}\right)\right).$$

8.109 Refer to Exercise 7.8.

a. Notice that

$$S'^2 = \frac{\Sigma(Y_i - \overline{Y})^2}{n} = \frac{(n-1)S^2}{n}$$

Hence

$$V(S'^2) = \frac{(n-1)^2}{n^2}V(S^2) = \frac{2(n-1)\sigma^4}{n^2}$$

b. Since

$$\frac{V(S^2)}{V(S^{2\prime})} = \frac{2\sigma^4 n^2}{2(n-1)^2\sigma^4} = \left(\frac{n}{n-1}\right)^2 > 1$$

we can say that

$$V(S'^2) < V(S^2)$$

8.111 From Exercise 7.8 we know that

$$E(S_i^2) = \sigma^2 \qquad \text{and} \qquad V(S_1^2) = \frac{2\sigma^4}{n_i - 1}$$

a. $$E(S_p^2) = \frac{(n_1 - 1)E(S_1^2) + (n_2 - 1)E(S_2^2)}{n_1 + n_2 - 2} = \frac{n_1 + n_2 - 2}{n_1 + n_2 - 2}(\sigma^2) = \sigma^2$$

so that S_p^2 is an unbiased estimator of σ^2.

b. $$V(S_p^2) = \frac{(n_1 - 1)^2 V(S_1^2) + (n_2 - 1)^2 V(S_2^2)}{(n_1 + n_2 - 2)^2} = \frac{2\sigma^4(n_1 - 1) + 2\sigma^4(n_2 - 1)}{(n_1 + n_2 - 2)^2}$$

$$= \frac{2\sigma^4}{n_1 + n_2 - 2}$$

Finally, the $100(1-\alpha)^5$ confidence interval for $\dfrac{\sigma_1^2}{\sigma_2^2}$ will be

$$\frac{1}{F_{\nu_1,\nu_2,\alpha/2}} \times \frac{S_1^2}{S_2^2} \le \frac{\sigma_1^2}{\sigma_2^2} \le F_{\nu_2,\nu_1,\alpha/2} \times \frac{S_1^2}{S_2^2}$$

8.113 The $100(1 - \alpha)\%$ confidence interval for σ^2 is given as

$$\frac{(n-1)S^2}{X_U^2} \leq \sigma^2 \frac{(n-1)S^2}{X_L^2}$$

so that the midpoint of this interval will be

$$M = \frac{1}{2}\left[\frac{(n-1)S^2}{X_U^2} + \frac{(n-1)S^2}{X_L^2}\right]$$

Now

$$E(M) = \left(\frac{n-1}{2X_U^2} + \frac{n-1}{2X_L^2}\right)E(S^2) = \frac{(n-1)\sigma^2}{2}\left(\frac{1}{X_U^2} + \frac{1}{X_L^2}\right) \neq \sigma^2$$

CHAPTER 9 PROPERTIES OF POINT ESTIMATORS AND METHODS OF ESTIMATION

9.1 Refer to Exercise 8.4, where the four necessary variances were calculated. The efficiencies are as follows:

$\hat{\theta}_1$ relative to $\hat{\theta}_5$: $\dfrac{V(\hat{\theta}_5)}{V(\hat{\theta}_1)} = \dfrac{\left(\dfrac{\theta^2}{3}\right)}{\theta^2} = \dfrac{1}{3}$
$\qquad$
$\hat{\theta}_2$ relative to $\hat{\theta}_5$: $\dfrac{V(\hat{\theta}_5)}{V(\hat{\theta}_2)} = \dfrac{\left(\dfrac{\theta^2}{3}\right)}{\left(\dfrac{\theta^2}{2}\right)} = \dfrac{2}{3}$

$\hat{\theta}_3$ relative to $\hat{\theta}_5$: $\dfrac{V(\hat{\theta}_5)}{V(\hat{\theta}_3)} = \dfrac{\left(\dfrac{\theta^2}{3}\right)}{\left(\dfrac{5\theta^2}{9}\right)} = \dfrac{3}{5}$

9.3 For the given uniform distribution, $E(Y_i) = \theta + \frac{1}{2}$ and $V(Y_i) = \frac{1}{12}$.

(1) For $\hat{\theta}_1$,

$$E(\hat{\theta}_1) = E\left(\overline{Y} - \frac{1}{2}\right) = \left(\theta + \frac{1}{2}\right) - \frac{1}{2} = \theta \quad \text{and} \quad V(\hat{\theta}_1) = V(\overline{Y}) = \frac{\left(\frac{1}{12}\right)}{n} = \frac{1}{12n}$$

(2) The distribution of $Y_{(n)}$ is given by $g_n(y) = n[F(y)]^{n-1} f(y)$, where

$$F(y) = \int_{\theta}^{y} dt = (y - \theta) \qquad \text{and} \qquad f(y) = 1$$

Hence $g_n(y) = n(y - \theta)^{n-1}$ for $\theta \leq y \leq \theta + 1$. For ease of calculation, we find

$$E(Y_{(n)} - \theta) = \int_{\theta}^{\theta+1} (y - \theta)n(y - \theta)^{n-1} \, dy = \int_{0}^{1} nz^{n-1} \, dz = \frac{n}{n+1}$$

and

$$E(Y_{(n)} - \theta)^2 = \int_{\theta}^{\theta+1} n(y - \theta)^{n+1} \, dy = \frac{n}{n+2}$$

We can then deduce that $E(Y_{(n)}) = \theta + \dfrac{n}{n+1}$ and that

$$E\left(Y_{(n)}^2\right) = \frac{n}{n+2} + 2\theta E(Y_{(n)}) - \theta^2 = \frac{n}{n+2} + 2\theta\left(\theta + \frac{n}{n+1}\right) - \theta^2$$

$$= \frac{n}{n+2} + \theta^2 + 2\theta \times \frac{n}{n+1}$$

Finally, $E(\hat{\theta}_2) = E(Y_{(n)}) - \frac{n}{n+1} = \theta$

and

$$V(\hat{\theta}_2) = V(Y_{(n)}) = \frac{n}{n+2} + \theta^2 + 2\theta \times \frac{n}{n+1} - \left(\theta + \frac{n}{n+1}\right)^2 = \frac{n}{n+2} - \frac{n^2}{(n+1)^2}$$

$$= \frac{n}{(n+2)(n+1)^2}$$

Both $\hat{\theta}_1$ and $\hat{\theta}_2$ are unbiased, and the efficiency of $\hat{\theta}_1$ relative to $\hat{\theta}_2$ is

$$\frac{V(\hat{\theta}_2)}{V(\hat{\theta}_1)} = \frac{\frac{n}{(n+2)(n+1)^2}}{\frac{1}{12n}} = \frac{12n^2}{(n+2)(n+1)^2}$$

9.5 From Exercise 7.8(b), we know that S^2 is unbiased and that $V(S^2) = \frac{2\sigma^4}{n-1}$.
Consider $\hat{\sigma}_2^2 = \left(\frac{1}{2}\right)(Y_1 - Y_2)^2$. Since Y_1 and Y_2 are both normally distributed, the quantity $(Y_1 - Y_2)$ is normal with mean 0 and variance $2\sigma^2$. Finally,

$$Z^2 = \left(\frac{Y_1 - Y_2}{\sqrt{2\sigma^2}}\right)^2 = \frac{(Y_1 - Y_2)^2}{2\sigma^2}$$

has by definition a χ^2 distribution with 1 degree of freedom. Thus,

$$E(Z^2) = 1 \qquad \text{so that} \qquad E\left[\frac{(Y_1 - Y_2)^2}{2}\right] = \sigma^2$$

and

$$V(Z^2) = 2 \qquad \text{so that} \qquad V\left[\frac{(Y_1 - Y_2)^2}{2}\right] = 2\sigma^4$$

The efficiency of $\hat{\sigma}_1^2$ relative to $\hat{\sigma}_2^2$ is

$$\frac{V\left(\hat{\sigma}_2^2\right)}{V\left(\hat{\sigma}_1^2\right)} = \frac{2\sigma^4}{\left(\frac{2\sigma^4}{n-1}\right)} = n-1$$

9.7 Since Y is exponential, $E(Y) = \theta$ and $V(Y) = \theta^2$.

$V(\hat{\theta}_1) = \theta^2$, since it was given that $\hat{\theta}_1$ is an unbiased estimator and $\text{MSE}(\hat{\theta}_1) = \theta^2$.

$$V(\hat{\theta}_2) = V(\overline{Y}) = \frac{\theta^2}{n}$$

$$\text{eff}(\hat{\theta}_1, \hat{\theta}_2) = \frac{V(\hat{\theta}_2)}{V(\hat{\theta}_1)} = \frac{\left(\frac{\theta^2}{n}\right)}{\theta^2} = \frac{1}{n}$$

9.9   Using Theorem 9.1, we have

$$\lim_{n\to\infty} V(\hat{\theta}_1) = \lim_{n\to\infty} \frac{1}{12n} = 0$$

$$\lim_{n\to\infty} V(\hat{\theta}_2) = \lim_{n\to\infty} \frac{n}{(n+2)(n+1)} = \lim_{n\to\infty} \frac{1}{\left[1+\left(\frac{2}{n}\right)\right](n+1)^2} = 0$$

Hence $\hat{\theta}_1$ and $\hat{\theta}_2$ are consistent for θ.

9.11   From Example 9.2, $\overline{X}$ and $\overline{Y}$ are consistent estimators of μ_1 and μ_2, respectively. Hence, by Theorem 9.2, $\overline{X} - \overline{Y}$ converges in probability to $\mu_1 - \mu_2$.

9.13   Given $f(y)$, calculate

$$E(Y) = \frac{\theta}{\theta+1} \qquad \text{and} \qquad E(Y^2) = \frac{\theta}{\theta+2}$$

Thus

$$V(Y) = \frac{\theta}{\theta+2} - \frac{\theta^2}{(\theta+1)^2} = \frac{\theta}{(\theta+2)(\theta+1)^2}$$

Hence, $E(\overline{Y}) = \frac{\theta}{\theta+1}$. Since $\sigma^2 = V(Y)$ is finite, the law of large numbers (Example 9.2) holds, and $\overline{Y}$ is consistent for $\frac{\theta}{\theta+1}$.

9.15   a. $E(Y_i) = \mu$

      b. $P\left(|Y_i - \mu| \le 1\right) = P(-1 < Y_i - \mu < 1) = P(-1 < z < 1) = 2(.5 - .1587) = .6826$

      c.   No, since $\lim_{n\to\infty} .6826 \ne 1$

9.17 $P\left(|Y_1 - \theta| \le \epsilon\right) = F(\theta+\epsilon) - F(\theta-\epsilon) = 1 - \left[\left(1 - \frac{\theta-\epsilon}{\theta}\right)^n\right] = \left(\frac{\theta-\theta+\epsilon}{\theta}\right)^n = \left(\frac{\epsilon}{\theta}\right)^n$

Now,

$$\lim_{n\to\infty} \left(\frac{\epsilon}{\theta}\right)^n = 0, \text{ not } 1.$$

Thus, $Y_{(1)}$ is not a consistent estimator of θ.

9.19 $P\left(\left|Y_{(n)} - \theta\right| \le \epsilon\right) = F(\theta + \epsilon) - F(\theta - \epsilon) = 1 - \left(\frac{\theta - \epsilon}{\theta}\right)^{\alpha n}$

Now,

$$\lim_{n \to \infty} 1 - \left(\frac{\theta - \epsilon}{\theta}\right)^{\alpha n} = 1 - 0 = 1$$

Thus, $Y_{(1)}$ is a consistent estimator of θ.

9.21 $\overline{Y}$ converges in probability to μ_y, thus $\overline{Y}$ converges in probability to $\alpha\beta$.

9.23 By the law of large numbers, $\overline{X}$ is consistent for λ_1 and $\overline{Y}$ is consistent for λ_2.

Using Theorem 9.2, we see that the estimator $\dfrac{\overline{X}}{\overline{X} + \overline{Y}}$ is consistent for $\dfrac{\lambda_1}{\lambda_1 + \lambda_2}$.

9.25 The likelihood of the sample, L, is

$$L = P(X_1 = x_1, X_2 = x_2, \ldots, X_n = x_n) = p^{\Sigma x_i}(1 - p)^{n - \Sigma x_i}$$

If we consider $g(\Sigma x_i, p) = p^{\Sigma x_i}(1 - p)^{n - \Sigma x_i}$ and $h(X_1, X_2, \ldots, X_n) = 1$, then Theorem 9.4 states that ΣX_i is sufficient for p.

9.27 Refer to Definition 9.3. Each Y_i has a Poisson distribution with mean λ_i; hence ΣY_i has a Poisson distribution with mean $n\lambda$. The conditional distribution of $Y_1, Y_2, \ldots, Y_n$ given ΣY_i is

$$P(Y_1 = y_1, Y_2 = y_2, \ldots, Y_n = y_n | \Sigma Y_i = x) = \frac{p(Y_1 = y_1, Y_2 = y_2, \ldots, Y_n = y_n)}{P(\Sigma Y_i = x)}$$

$$= \frac{\prod_{i=1}^{n} \frac{e^{-\lambda}\lambda^{y_i}}{y_i!}}{P(\Sigma Y_i = x)} = \frac{\left(\frac{e^{-n\lambda}\lambda^{\Sigma y_i}}{\prod_{i=1}^{n} y_i!}\right)}{\left(\frac{e^{-n\lambda}\lambda^{\Sigma y_i}}{(\Sigma y_i)!}\right)} = \begin{cases} \dfrac{(\Sigma y_i)!}{\prod_{i=1}^{n} y_i!}, & \text{if } \displaystyle\sum_{i=1}^{n} y_i = x \\[4mm] 0, & \text{otherwise} \end{cases}$$

which is independent of λ. Hence ΣY_i is sufficient for λ.

9.29 The likelihood is $L = \left(\dfrac{m^n}{\alpha^n}\right)\left(\displaystyle\prod_{i=1}^{n} y\right)^{m-1} e^{-\Sigma y_i^m/\alpha}$. Let $g(\Sigma y_i^m, \alpha) = \left(\dfrac{1}{\alpha^n}\right) e^{-\Sigma y_i^m/\alpha}$

and $h(y_1, \ldots, y_n) = m^n \left(\displaystyle\prod_{i=1}^{n} y_i\right)^{m-1}$, so that Theorem 9.4 is satisfied. Thus, ΣY_i^m is sufficient for α.

9.31 $L(y_1, \ldots, y_n | \alpha, \theta) = \frac{\alpha^n \Pi y_i^{\alpha-1}}{\theta^{n\alpha}} = g\left(\Pi Y_i, \alpha\right)$

By Theorem 9.4, where $h(y_1, \ldots, y_n) = 1$, $\prod_{i=1}^{n} Y_i$ is sufficient for α.

9.33 $L(y_1, \ldots, y_n | \theta) = a(\theta)^n \, \Pi \, b(y_i) e^{-c(\theta)\Sigma d(y_i)}$

Letting $g(\Sigma \, d(Y_i), \theta) = a(\theta)^n e^{-c(\theta)\Sigma d(y_i)}$ and $h(y_1, \ldots, y_n) = \Pi b(y_i)$, by Theorem 9.4,

$\sum_{i=1}^{n} d(Y_i)$ is sufficient for θ.

9.35 We can write $(y | \alpha, \theta) = \left(\frac{\alpha}{\theta^\alpha}\right) e^{(\alpha-1)\ln y}$. Letting

$$a(\alpha) = \frac{\alpha}{\theta^\alpha}, \qquad b(y) = 1, \qquad c(\alpha) = \alpha - 1 \qquad \text{and} \qquad d(y) = \ln y,$$

we see that $f(y | \alpha, \theta)$ is in the exponential family. It was shown in Exercise 9.34 that $\Sigma \, d(Y_i)$ was a sufficient statistic. Since $d(y) = \ln y$, $\sum_{i=1}^{n} \ln Y_i$ is sufficient for α.

It was shown in Exercise 9.31 that $\prod_{i=1}^{n} Y_i$ was sufficient for α. Since $\sum_{i=1}^{n} \ln Y_i$

$= \ln \prod_{i=1}^{n} Y_i$, we have no contradiction.

9.37 For the uniform distribution, $f(y_1, y_2, \ldots, y_n | \theta) = \frac{1}{\theta^n}$ for $0 \leq y_i \leq \theta$. Further, if $U = Y_{(n)}$, then $g_n(u | \theta) = \frac{nu^{n-1}}{\theta^n}$ from Section 6.6 in the text. Hence the conditional distribution of $Y_1, Y_2, \ldots, Y_n$ given U is

$$f((y_1, y_2, \ldots, y_n | u) = \frac{f(y_1, y_2, \ldots, y_n | \theta)}{g_n(u | \theta)} = \frac{1}{nu^{n-1}}$$

which is not dependent on θ. Hence $Y_{(n)}$ is sufficient for θ.

Alternate solution:

Define the <u>indicator</u> <u>function</u> $I(a < y < b)$ by

$$I(a < y < b) = \begin{cases} 1, & \text{if } a < y < b \\ 0, & \text{otherwise} \end{cases}$$

Then we can write the uniform density as $f(y | \theta) = \frac{1}{\theta} I(0 < y < \theta)$. The likelihood is

$$L(y_1, \ldots, y_n|\theta) = \frac{1}{\theta^n} \prod_{i=1}^{n} I(0 < y_i < \theta) = \frac{1}{\theta^n} I(0 < y_{(n)} < \theta)$$

Theorem 9.4 is satisfied with

$$g(y_{(n)}, \theta) = \frac{1}{\theta^n} I(0 < y_{(n)} < \theta) \text{ and } h(y_1, \ldots, y_n) = 1.$$

Thus, $Y_{(n)}$ is sufficient.

9.39 Since $g_1(y|\theta) = ne^{-n(y-\theta)}$ from Exercise 6.56, and since $f(y_1, y_2, \ldots, y_n|\theta)$

$= e^{-\Sigma(y_i - \theta)}$, the conditional distribution of $Y_1, Y_2, \ldots, Y_n$ given $Y_{(1)}$ is

$$f(y_1, y_2, \ldots, y_n|y_{(1)}) = \frac{e^{-\Sigma y_i}}{ne^{-ny_{(1)}}}$$

which is independent of θ. Hence $Y_{(1)}$ is sufficient for θ.

Alternate solution:

The density is $f(y|\theta) = e^{-(y-\theta)} I(\theta < y < \infty)$ (see the solutions to Exercises 9.37 and 9.38). The likelihood is

$$L(y_1, \ldots, y_n|\theta) = e^{-\Sigma(y_i - \theta)} \prod_{1}^{n} I(\theta < y_i < \infty) = e^{-\Sigma y_i} e^{n\theta} I(\theta < y_{(1)} < \infty).$$

Let $g(y_{(1)}, \theta) = e^{n\theta} I(\theta < y_{(1)} < \infty)$ and let $h(y_1, \ldots, y_n) = e^{-\Sigma y_i}$. Thus, $Y_{(1)}$ is sufficient for θ.

9.41 $f(y|\alpha, \beta) = (\alpha\beta^\alpha)^n y^{-(\alpha+1)} I(\beta < y)$

$$L(y_1, \ldots, y_n|\alpha, \beta) = (\alpha\beta^\alpha)^n \Pi y_i^{-(\alpha+1)} I(\beta < y_{(1)})$$

By Theorem 9.4, where $h(y_1, \ldots, y_n) = 1$, we see that $\prod_{i=1}^{n} Y_i$ and $Y_{(1)}$ are jointly sufficient for α and β.

9.43 From Example 9.8, $\hat{\sigma}_1^2 = \left(\frac{1}{n-1}\right) \Sigma (X_i - \overline{X})^2$ is the MVUE for σ^2 from the sample

$X_1, \ldots, X_n$. Similarly, $\hat{\sigma}_2^2 = \left(\frac{1}{n-1}\right) \Sigma (Y - \overline{Y})^2$ is the MVUE for σ^2 from the sample $Y_1, \ldots, Y_n$. Now,

$$\hat{\sigma}_2^2 = \frac{\Sigma (X_i - \overline{X})^2 + \Sigma (Y_i - \overline{Y})^2}{2n - 2} = \frac{\hat{\sigma}_1^2}{2} + \frac{\hat{\sigma}_2^2}{2}$$

is unbiased. Also, $V\left(\hat{\sigma}_1^2\right) = V\left(\hat{\sigma}_2^2\right)$. Using the result in Exercise 8.2, we see that $\hat{\sigma}_2$ has minimum variance and is thus an MVUE.

9.45 With Y a Poisson random variable with parameter λ, it is necessary to find the MVUE for

$$E(c) = 3E(Y^2) = 3\left(V(Y) + [E(Y^2)]\right) = 3(\lambda + \lambda^2)$$

In Exercise 9.27 it was determined that $\sum_{i=1}^{n} Y_i$ is sufficient for λ and thus for λ^2 and $3(\lambda + \lambda^2)$. If a function of $\sum_{i=1}^{n} Y_i$ that is unbiased for $3(\lambda + \lambda^2)$ can be found, then this function will be MVUE. Note that

$$E(\overline{Y}^2) = V(\overline{Y}) + [E(\overline{Y})]^2 = \tfrac{\lambda}{n} + \lambda^2$$

and

$$E\left(\frac{\overline{Y}}{n}\right) = \tfrac{1}{n}E(\overline{Y}) = \tfrac{\lambda}{n}$$

so that

$$\lambda^2 = E(\overline{Y}^2) - E\left(\frac{\overline{Y}}{n}\right)$$

and

$$\lambda = E(\overline{Y}).$$

$$E(c) = 3E\left[\overline{Y}^2 - \left(\frac{\overline{Y}}{n}\right) + \overline{Y}\right] \text{ and the MVUE is } 3\left[\overline{Y}^2 + \overline{Y}\left(1 - \tfrac{1}{n}\right)\right].$$

9.47 In Exercise 9.37, $Y_{(n)}$ was shown to be sufficient for θ. Then, it was shown in Example 9.1 that

$$E(Y_{(n)}) = \int_0^\theta \frac{ny^n}{\theta^n}\, dy = \left(\frac{n}{n+1}\right)\theta$$

Thus, $\left(\frac{n+1}{n}\right)Y_{(n)}$ is unbiased and is the MVUE for θ.

9.49 (1) Let $T = 1$ if $Y_1 = 1$, $Y_2 = 0$, and $T = 0$ otherwise. Then

$$E(T) = (1)P(Y_1 = 1, Y_2 = 0 = P(Y_1 = 1)P(Y_2 = 0) = p(1 - p)$$

(2) $P(T = 1 | W = w) = \dfrac{P(T = 1, Y = y)}{P(W = w)} = \dfrac{P(Y_1 = 1, Y_2 = 0, W = w)}{\binom{n}{w}p^w q^{n-w}}$

Since a success must occur on trial 1, but not on trial 2, the remaining $w - 1$ successes must be distributed in $n - 2$ trials, which can be done in $\binom{n-2}{w-1}$ ways.

Hence

$$P(T = 1 | W = w) = \frac{\binom{n-2}{w-1} p^w q^{n-w}}{\binom{n}{y} p^w q^{n-w}} = \frac{w(n-w)}{n(n-1)}$$

(3) $E(T|W) = (1)P(T = 1|W) = \left(\frac{W}{n}\right)\frac{(n-W)}{(n-1)} = \left(\frac{W}{n}\right)\left(\frac{n}{n-1}\right)\left(\frac{n-W}{n}\right)$

$$= \frac{n}{n-1}\left[\left(\frac{W}{n}\right)\left(1 - \frac{W}{n}\right)\right]$$

Since T is unbiased by (1) and $W = \Sigma Y_i$ is sufficient for p [and hence for $p(1 - p)$], then $\left(\frac{n}{n-1}\right)\overline{Y}(1 - \overline{Y})$ is the MVUE for $p(1 - p)$.

9.51 $L\left(y_1, y_2, \ldots, y_n | \mu_1 \sigma^2\right) = \prod_{i=1}^{n} \frac{1}{\sqrt{2\pi}\sigma} e^{(1/2\sigma^2)(y_i - \mu)^2}$

$$= \frac{1}{(2\pi)^{n/2}\sigma^n} \exp\left[-\frac{1}{2\sigma^2}\sum_{i=1}^{n}(y_i - \mu)^2\right]$$

Then

$$\frac{L(x_1, x_2, \ldots, x_n|\theta)}{L(y_1, y_2, \ldots, y_n|\theta)} = \frac{\dfrac{1}{(2\pi)^{n/2}\sigma^n}\exp\left[-\dfrac{1}{2\sigma^2}\sum\limits_{i=1}^{n}(x_i - \mu)^2\right]}{\dfrac{1}{(2\pi)^{n/2}\sigma^n}\exp\left[-\dfrac{1}{2\sigma^2}\sum\limits_{i=1}^{n}(y_i - \mu)^2\right]}$$

$$= \exp\left[-\frac{1}{2\sigma^2}\left(\sum_{i=1}^{n}(x_i - \mu)^2 - \sum_{i=1}^{n}(y_i - \mu)^2\right)\right]$$

$$= \exp\left[-\frac{1}{2\sigma^2}\left(\sum_{i=1}^{n}x_i^2 - \sum_{i=1}^{n}y_i^2 - 2\mu\left[\sum_{i=1}^{n}x_i - \Sigma y_i\right]\right)\right].$$

Let $g(Y_1, Y_2, Y_3, \ldots, Y_n) = \left(\sum\limits_{i=1}^{n}Y_i, \sum\limits_{i=1}^{n}Y_i^2\right)$. Then, the above ratio is free of (μ, σ^2) if and only if $\sum\limits_{i=1}^{n}x_i^2 = \sum\limits_{i=1}^{n}y_i^2$ and $\sum\limits_{i=1}^{n}x_i = \sum\limits_{i=1}^{n}y_i$, i.e., $g(x_1, x_2, x_3, \ldots, x_n)$ $= g(y_1, y_2, y_3, \ldots, y_n)$. Then, by the Lehmann-Scheffe′ method discussed in Exercise 9.50, $\left(\sum\limits_{i=1}^{n}Y_i, \sum\limits_{i=1}^{n}Y_i^2\right)$ jointly form a minimal sufficient statistic for μ and σ^2.

9.53 To use the method of moments, calculate

$$\mu = \int_{-\infty}^{\infty} yf(y)\, dy = \int_{0}^{1} (\theta + 1)y^{\theta+1}\, dy = \frac{(\theta+1)y^{\theta+2}}{(\theta+2)}\Bigg]_{0}^{1} = \frac{\theta+1}{\theta+2}$$

Equating sample and population moments, we obtain the estimator of θ.

$$\overline{Y} = \frac{\hat{\theta}+1}{\hat{\theta}+2} \qquad \text{or} \qquad (\hat{\theta}+2)\overline{Y} = \hat{\theta}+1 \qquad \text{or} \qquad \hat{\theta} = \frac{2\overline{Y}-1}{1-\overline{Y}}$$

Because of the law of large numbers, $\overline{Y}$ is consistent for $\mu = \frac{\theta+1}{\theta+2}$. Then, using Theorem 9.2, we can show that $\hat{\theta}$ is consistent for θ, since $\hat{\theta}$ converges to

$$\frac{2\left(\frac{\theta+1}{\theta+2}\right)-1}{1-\left(\frac{\theta+1}{\theta+2}\right)} = \frac{2(\theta+1)-(\theta+2)}{(\theta+2)-(\theta+1)} = \theta$$

$\hat{\theta}$ is not a function of $-\sum_{i=1}^{n} \ln(Y_i)$ or of $\exp\left[-\left(-\sum_{i=1}^{n} \ln(Y_i)\right)\right] = \prod_{i=1}^{n} Y_i$. This implies that $\hat{\theta}$ is not the MVUE.

9.55 Since the first population moment is $\mu_1' = \mu$, which does not involve σ^2, we equate the second sample and population moments. That is,

$$m_2' = \frac{\sum Y_i^2}{n} \qquad\qquad \mu_2' = \sigma^2 + \mu^2 = \sigma^2 + 0 = \sigma^2$$

and the moment estimator of σ^2 is $\hat{\sigma}^2 = \sum \frac{Y_i^2}{n}$.

9.57 Let $Y_i = 1$ if the ith ball is black, and 0 otherwise. Note that for Y_i to equal 1 we need to fill position i with a black ball, which can be done in θ ways, and the other $n-1$ positions may be filled with any combination of the other $N-1$ balls, which can be done in P_{n-1}^{N-1} ways (see Definition 2.7). All together, there are P_n^N ways to fill the n positions. Thus,

$$P(Y_i = 1) = \frac{\theta P_{n-1}^{N-1}}{P_n^N} = \frac{\theta}{N}$$

(See Example 5.29.) Hence $E(Y_i) = \mu_1' = \frac{\theta}{N}$. Letting μ_1' equal $\overline{Y}$, we obtain

$$\overline{Y} = \frac{\hat{\theta}}{N} \qquad\qquad \text{so that} \qquad\qquad \hat{\theta} = N\overline{Y} = N\frac{\sum Y_i}{n} = \frac{NY}{n}$$

where $Y = \sum_{i=1}^{n} Y_i$.

9.59 The density shown in this exercise is that of a beta random variable with $\alpha = \theta$ and $\beta = \theta$. Hence

$$\mu_1' = E(Y) = \frac{\theta}{\theta + \theta} = \frac{\theta}{2\theta} = \frac{1}{2}$$

which does not involve θ and hence is of no use in finding a moment estimator. Thus we equate second sample and population moments. From Exercise 4.78,

$$E(Y^2) = \frac{(\theta + 1)\theta}{2\theta(2\theta + 1)} = \frac{\theta + 1}{2(2\theta + 1)}.$$

Hence with $m_2' = \frac{\Sigma y_i^2}{n}$, we solve for $\hat{\theta}$ in

$$m_2' = \frac{\theta + 1}{2(2\theta + 1)} \quad \text{or} \quad 2m_2'(2\theta + 1) = \theta + 1 \quad \text{or} \quad \theta\left(4m_2' - 1\right) = 1 - 2m_2'$$

$$\text{or} \quad \hat{\theta} = \frac{1 - 2m_2'}{4m_2' - 1}$$

9.61 $\mu_1' = E(Y) = \mu = \frac{3\theta}{2} = \overline{Y}$

Now,

$$m_1' = \hat{\theta} = \frac{2\overline{Y}}{3}.$$

9.63 $E(Y) = \int\limits_{\beta}^{\infty} \alpha\beta^\alpha y^{-\alpha} \, dy = \left(\alpha\beta^\alpha\right)\left(\frac{y^{-\alpha+1}}{-\alpha + 1}\right)\Big]_{\beta}^{\infty} = \left(\frac{\alpha\beta^\alpha}{-\alpha + 1}\right)(y^{-\alpha+1})\Big]_{\beta}^{\infty}$

$$= 0 - \left[\left(\frac{\alpha\beta^\alpha}{-\alpha + 1}\right)(\beta^{-\alpha+1})\right] \quad \text{if } \alpha > 1$$

$$= \frac{\alpha\beta}{\alpha - 1} \quad \text{if } \alpha > 1$$

is undefined if $\alpha < 1$.

9.65 The likelihood is

$$L(\theta) = f(y_1, y_2, \ldots, y_n | \theta) = \frac{1}{\theta^n} e^{-(\Sigma Y_i)/\theta}.$$

Then

$$\ln\left[L(\theta)\right] = -n \ln \theta - \frac{\Sigma Y_i}{\theta}$$

$$\frac{d \ln\left[L(\theta)\right]}{d\theta} = -\frac{n}{\theta} + \frac{\Sigma Y_i}{\theta^2} = 0$$

or

$$-n + \frac{\Sigma Y_i}{\theta} = 0$$

$$\hat{\theta} = \frac{\Sigma Y_i}{n} = \overline{Y}$$

which is the MVUE of θ as shown in Example 9.9.

Assuming that $\theta > 0$, $t(\theta) = \theta^2$ is a one-to-one function of θ. Therefore, $\overline{Y}^2$ is the MLE of θ^2, which is not the MVUE of θ^2 as shown in Example 9.9.

9.67 **a.** The likelihood function is

$$L = \prod_{i=1}^{n} \frac{1}{(2\theta + 1)} I(0 < y_i < 2\theta + 1) = \frac{1}{(2\theta + 1)^n} I(0 < y_{(n)} < 2\theta + 1)$$

where I is the indicator function as defined in the solution to Exercise 9.37. To maximize L, we minimize $(2\theta + 1)^n$ subject to $0 > \left(\frac{1}{2}\right)(y_{(n)} - 1)$. Thus,

$$\hat{\theta} = \left(\frac{1}{2}\right)(Y_{(n)} - 1).$$

b. The variance of the distribution is $\dfrac{(2\theta + 1)^2}{12}$. Since $\theta > 1$, this is a one-to-one

function. Thus, the MLE for $\dfrac{(2\theta + 1)^2}{12}$ is

$$\frac{(2\hat{\theta} + 1)^2}{12} = \frac{\left[2\left(\frac{1}{2}\right)(Y_{(n)} - 1) + 1\right]^2}{12} = \frac{Y_{(n)}^2}{12}.$$

9.69 **a.** The likelihood function, defined as the joint density of Y_1, Y_2, ..., Y_n evaluated at y_1, y_2, ..., y_n, is given by

$$L = \prod_{i=1}^{n} \frac{1}{\Gamma(\alpha)\theta^{\alpha}} y_i^{\alpha-1} e^{-y_i/\theta} = \frac{1}{[\Gamma(\alpha)]^n \theta^{n\alpha}} e^{-\Sigma y_i/\theta} \prod_{i=1}^{n} y_i^{\alpha-1} = K\left(\frac{1}{\theta^{n\alpha}}\right) e^{-\Sigma y_i/\theta}$$

where K is a constant, independent of θ. Then $\ln L = \ln K - n\alpha \ln \theta - \left(\dfrac{\Sigma y_i}{\theta}\right)$, and if α is known,

$$\frac{d}{d\theta} \ln L = \frac{\Sigma y_i}{\theta^2} - \frac{n\alpha}{\theta}$$

Equating the derivative to 0, we obtain $\hat{\theta}$.

$$\frac{\Sigma y_i}{\hat{\theta}^2} - \frac{n\alpha}{\hat{\theta}} = 0 \qquad \text{or} \qquad \hat{\theta} = \frac{\Sigma Y_i}{n\alpha} = \frac{\overline{Y}}{\alpha}$$

b.   Taking expectations and recalling that $E(Y_i) = \alpha\theta$ and $V(Y_i) = \alpha\theta^2$, we have

$$E(\hat{\theta}) = \frac{\sum_{i=1}^{n} E(Y_i)}{n\alpha} = \frac{n\alpha\theta}{n\alpha} = \theta \qquad \text{and} \qquad V(\hat{\theta}) = \frac{\sum_{i=1}^{n} V(Y_i)}{n^2\alpha^2} = \frac{n\alpha\theta^2}{n^2\alpha^2} = \frac{\theta^2}{n\alpha}$$

c.   By the law of large numbers, we know that $\overline{Y}$ is a consistent estimator of $\mu = \alpha\theta$. That is, $\overline{Y}$ converges in probability to $\alpha\theta$. Then, by Theorem 9.2, the quantity $\frac{\overline{Y}}{\alpha} = \hat{\theta}$ converges in probability to $\frac{\mu}{\alpha} = \theta$, so that $\hat{\theta}$ must be a consistent estimator of θ.

d.   Using Lehmann and Scheffe''s method, we have

$$\frac{L(x_1, x_2, \ldots, x_n | \alpha, \theta)}{L(y_1, y_2, \ldots, y_n | \alpha, \theta)} = \frac{(\Pi\, x_i)^{\alpha-1} e^{-\Sigma x_i/\theta}}{(\Pi\, y_i)^{\alpha-1} e^{-\Sigma y_i/\theta}}$$

In order for this ratio to be free of θ, we need $\Sigma x_i = \Sigma y_i$ so that ΣY_i is the minimal sufficient statistic.

e.   Let $U = \sum_{i=1}^{n} Y_i$. The moment-generating function of U is

$$m_U(t) = \prod_{i=1}^{n} m_{Y_i}(t) = \frac{1}{(1 - \theta t)^{n\alpha}} = \frac{1}{(1 - \theta t)^{10}}.$$

A random variable that possesses a χ^2 distribution is one whose moment-generating function is $\dfrac{1}{(1 - 2t)^k}$, where $2k$ are the degrees of freedom. It is necessary to transform U to obtain a random variable with such a moment-generating function.

Consider $X = \frac{2U}{\theta}$, with

$$m_X(t) = m_U\left(\frac{2t}{\theta}\right) = \frac{1}{(1 - 2t)^{10}}$$

Hence $X = \frac{2U}{\theta}$ has a χ^2 distribution with $2(10) = 20$ degrees of freedom. Using X as a pivotal statistic, write

$$P\left(\chi^2_{.05,\,20} < \frac{2U}{\theta} < \chi^2_{.95,\,.20}\right) = .90 \qquad \text{or} \qquad P\left(\frac{2U}{\chi^2_{.95,\,20}} < \theta < \frac{2U}{\chi^2_{.05,\,20}}\right) = .90$$

and the 90% confidence interval is

$$\frac{2\sum_{i=1}^{n} Y_i}{31.41} < \theta < \frac{2\sum_{i=1}^{n} Y_i}{10.85}$$

9.71 Let p_1, p_2, p_3 be the proportions of voters in the population favoring candidates A, B, and C, respectively. Further, define the random variables n_1, n_2, and n_3 as the number of voters in a random sample of size n who favor candidates A, B, and C, respectively. Note that

$$\sum_{i=1}^{3} p_i = 1 \qquad\text{and}\qquad \sum_{i=1}^{3} n_i = n$$

so that we may write $p_3 = 1 - p_1 - p_2$ and $n_3 = n - n_1 - n_2$. The random variables n_1, n_2, and n_3 follow a multinomial probability distribution (see Section 5.9 of the text), and the likelihood function is

$$L = \frac{n!}{n_1!\,n_2!\,n_3!}\, p_1^{n_1}\, p_2^{n_2}\,(1 - p_1 - p_2)^{n_3}$$

so that

$$\ln L = \ln K + n_1 \ln p_1 + n_2 \ln p_2 + n_3 \ln (1 - p_1 - p_2)$$

Differentiating with respect to p_1 and p_2, we have

$$\frac{d \ln L}{dp_1} = \frac{n_1}{p_1} - \frac{n_3}{1 - p_1 - p_2} \qquad\text{and}\qquad \frac{d \ln L}{dp_2} = \frac{n_2}{p_2} - \frac{n_3}{1 - p_1 - p_2}$$

Set these two equations equal to 0 and solve simultaneously for $\hat{p}_1$ and $\hat{p}_2$.

$$(*) \quad n_1(1 - \hat{p}_1 - \hat{p}_2) - n_3 \hat{p}_1 = 0 \qquad\text{and}\qquad n_2(1 - \hat{p}_1 - \hat{p}_2) - n_3 \hat{p}_2 = 0$$

Adding the two equations, we have

$$(n_1 + n_2)(1 - \hat{p}_1 - \hat{p}_2) = (\hat{p}_1 + \hat{p}_2)n_3 \qquad\text{or}\qquad \hat{p}_1 + \hat{p}_2 = \frac{n_1 + n_2}{n}$$

Thus, in $(*)$,

$$n_1\left[1 - \left(\frac{n_1 + n_2}{n}\right)\right] = n_3 \hat{p}_1 \qquad\text{or}\qquad \hat{p}_1 = \frac{(n - n_1 - n_2)\theta_1}{n}\left(\frac{1}{n_3}\right) = \frac{n_1}{n}$$

Similarly,

$$\hat{p}_2 = \frac{n_2}{n} \qquad\text{and}\qquad \hat{p}_3 = 1 - \hat{p}_1 - \hat{p}_2 = \frac{n_3}{n}$$

For the data given in this exercise, $\hat{p}_1 = .30$, $\hat{p}_2 = .38$, and $\hat{p}_3 = .32$. To estimate $p_1 - p_2$, we use $\hat{p}_1 - \hat{p}_2 = \frac{n_1}{n} - \frac{n_2}{n} = -.08$. From Theorem 5.13 of the text, we have $V(n_i) = np_i q_i$ and $\text{Cov}(n_i, n_j) = -np_i p_j$, so that the variance of $\hat{p}_1 - \hat{p}_2$ is

$$V(\hat{p}_1) + V(\hat{p}_2) - 2\,\text{Cov}(\hat{p}_1, \hat{p}_2) = \frac{1}{n^2}V(n_1) + \frac{1}{n^2}V(n_2) - \frac{2}{n^2}\text{Cov}(n_1, n_2)$$

$$= \frac{p_1 q_1}{n} + \frac{p_2 q_2}{n} + 2\,\frac{p_1 p_2}{n}$$

which may be estimated as

$$V(\hat{p}_1 - \hat{p}_2) = \frac{(.30)(.70)}{100} + \frac{(.38)(.62)}{100} + 2\frac{(.30)(.38)}{100} = .006736$$

and the approximate bound on the error of estimation is

$$2\sqrt{V(\hat{p} - \hat{p})} = 2\sqrt{.006736} = .1641.$$

9.73 $P(Y = y) = \binom{2}{y}p^y(1-p)^{2-y}$

Letting $\hat{p} = $ maximum likelihood estimator,

$$P(Y = 0) = (1-p)^2 \qquad\qquad \hat{p} = \frac{1}{4}$$

$$P(Y = 1) = 2p(1-p) \qquad\qquad \hat{p} = \frac{1}{4} \text{ or } \frac{3}{4}$$

$$P(Y = 2) = p^2 \qquad\qquad \hat{p} = \frac{3}{4}$$

9.75 $L = \left(\frac{1}{2\theta}\right)^n \qquad 0 \le y_i \le 2\theta,\ i = 1, 2, \dots, n$

Note that L increases as θ decreases, and 2θ must be $\ge Y_{(n)}$.

Thus, $\theta = \dfrac{Y_{(n)}}{2}$.

9.77 The parameter to be estimated is $R = \dfrac{p}{1-p}$. Since $\hat{p} = \frac{Y}{n}$ is the maximum-likelihood estimate of p (see Example 9.14), the result proved in Exercise 9.73 implies that $\hat{R} = \dfrac{\hat{p}}{1-\hat{p}}$ is the maximum-likelihood estimator of R.

9.79 From Example 9.18, the MLE for $t(p) = p$ is $\hat{t}(n) = \hat{p} = \frac{Y}{n}$ and $\dfrac{\partial\,t(p)}{\partial p} = 1$. Also, from Example 9.18,

$$E\left[\frac{-\partial^2 \ln[p(Y|p)]}{\partial p^2}\right] = \frac{1}{p(1-p)}.$$

Then, a $100(1-\alpha)\%$ confidence interval is

$$\hat{p} \pm z_{\alpha/2}\sqrt{\frac{1}{n\left[\dfrac{1}{p(1-p)}\right]\Big|_{p=\hat{p}}}} = \hat{p} \pm z_{\alpha/2}\sqrt{\frac{\hat{p}(1-\hat{\beta})}{n}}.$$

This is the confidence interval derived in Section 8.6.

9.81 The MLE for $e^{-\lambda}$ was shown to be $e^{-\overline{Y}}$ in Exercise 9.64. Also,

$$t(\lambda) = e^{-\lambda}$$

$$\frac{\partial\, t(\lambda)}{\partial \lambda} = -e^{-\lambda}$$

$$p(y|\lambda) = \frac{\lambda^y e^{-\lambda}}{y!}$$

$$\ln\left[p(y|\lambda)\right] = y \ln \lambda - \lambda - \ln(y!)$$

$$\frac{\partial\, \ln\left[p(y|\lambda)\right]}{\partial \lambda} = \frac{y}{\lambda} - 1$$

$$\frac{\partial^2\, \ln\left[p(y|\lambda)\right]}{\partial \lambda^2} = -\frac{y}{\lambda^2}$$

$$E\left[\frac{-\partial^2\, \ln\left[p(Y|\lambda)\right]}{\partial \lambda^2}\right] = \frac{E(Y)}{\lambda^2} = \frac{1}{\lambda}.$$

Then, an approximate large-sample $100(1-\alpha)\%$ confidence interval for $e^{-\lambda}$ is

$$t(\hat{\lambda}) \pm z_{\alpha/2} \sqrt{\left.\left(\frac{\left(\frac{\partial\, t(\lambda)}{\partial \lambda}\right)^2}{nE\left[\frac{-\partial^2\, \ln f(Y|\lambda)}{\partial \lambda^2}\right]}\right)\right|_{\lambda = \hat{\lambda}}} = e^{-\overline{Y}} \pm z_{\alpha/2} \sqrt{\left.\left(\frac{e^{-2\lambda}}{n\left(\frac{1}{\lambda}\right)}\right)\right|_{\lambda = \overline{Y}}}$$

$$= e^{-\overline{Y}} \pm z_{\alpha/2} \sqrt{\frac{\overline{Y} e^{-2\overline{Y}}}{n}}$$

9.83 From Example 9.15, since μ is known, we need to solve

$$\frac{d \ln L}{d\sigma^2} = -\frac{n}{2\sigma^2} + \frac{\Sigma (y_i - \mu)^2}{2\sigma^4} = 0 \qquad \text{or} \qquad \hat{\sigma}^2 = \frac{\Sigma (Y - \mu)^2}{n}$$

9.85 From Exercise 9.69, the MLE for θ is $\overline{Y}$. Hence we use the result of Exercise 9.76 and deduce that the MLE of $e^{-t/\theta} = \overline{F}(t)$ is $e^{-t/\overline{Y}}$.

9.87 Let Y_i be the number drawn on the ith draw for $i = 1, 2, \ldots, n$. Then, by definition, $P(Y_i = k) = \frac{1}{N}$ for $k = 1, 2, \ldots, N$, and the Y_i are independent.

a.   In order to use the method of moments, calculate

$$\mu_1' = E(Y) = \sum_{k=1}^{N} kP(Y = k) = \sum_{k=1}^{N} k\left(\frac{1}{N}\right) = \frac{N(N+1)}{2N} = \frac{N+1}{2}$$

Then the moment estimator will be

$$\frac{\hat{N}_1 + 1}{2} = \overline{Y} \qquad \text{or} \qquad \hat{N}_1 = 2\overline{Y} - 1$$

b.   First calculate

$$E(Y^2) = \sum_{k=1}^{N} k^2 P(Y = k) = \frac{1}{N} \sum_{k=1}^{N} k^2 = \frac{N(N+1)(2N+1)}{6N} = \frac{(N+1)(2N+1)}{6}$$

so that

$$V(Y) = \frac{(N+1)(2N+1)}{6} - \frac{(N+1)^2}{4} = \frac{(N+1)(N-1)}{12}$$

Now

$$E(\hat{N}_1) = 2E(\overline{Y}) - 1 = 2E(Y) - 1 = (N+1) - 1 = N$$

$$V(\hat{N}_1) = 4V(\overline{Y}) = 4\frac{V(Y)}{n} = \frac{4(N^2 - 1)}{12n} = \frac{1}{3n}(N^2 - 1)$$

9.89   Refer to Exercise 9.88.   Using the results of parts (b) and (c), we have

$$\hat{N}_3 = \frac{n+1}{n}[\max(Y_1, Y_2, \ldots, Y_5)] = \frac{6}{5}(210) = 252$$

and an approximate bound on the error of estimation is

$$2\sqrt{V(\hat{N}_3)} = 2\sqrt{\frac{N^2}{n(n+2)}} = 2\sqrt{\frac{(252)^2}{5(7)}} = 2\sqrt{1814.4} = 85.192$$

CHAPTER 10 HYPOTHESIS TESTING

10.1 See Definition 10.1.

10.3 **a.** With $n = 20$ and $p = .8$, it is necessary to find c such that $\alpha = P(Y \leq c | p = .8)$ $= .01$. From Table 1, Appendix III, this value is $c = 11$.

b. With the rejection region given as $Y \leq 11$,
$$\beta = P(Y > 11 | p = .6) = 1 - P(Y \leq 11 | p = .6) = 1 - .404 = .596$$

c. $\beta = P(Y > 11 | p = .4) = 1 - P(Y \leq 11 | p = .4) = 1 - .943 = .057$

10.5 **a.** Since it is necessary to test a claim that the average amount saved is \$900, the hypothesis to be tested is two-tailed:
$$H_0:\ \mu = 900 \qquad \text{vs.} \qquad H_a:\ \mu \neq 900$$

b. The rejection region with $\alpha = .01$ is determined by a critical value of z such that
$$P\big[|z| < z_0\big] = .01$$

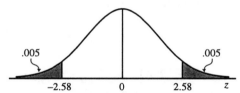

Figure 10.1

This value is $z_0 = 2.58$ (see Figure 10.1) and the rejection region is $|z| > 2.58$.

c. The test statistic is
$$z = \frac{\overline{y} - \mu}{\frac{\sigma}{\sqrt{n}}} \approx \frac{\overline{y} - \mu}{\frac{s}{\sqrt{n}}} = \frac{885 - 900}{\frac{50}{\sqrt{35}}} = -1.77$$

The observed value, $z = -1.77$, does not fall in the rejection region, and H_0 is not rejected. We cannot conclude that the average savings is different than claimed.

10.7 We are to test H_0: $\mu = 130$ vs. H_a: $\mu < 130$. The test statistic is

$$Z = \frac{\bar{y} - \mu_0}{\frac{\sigma}{\sqrt{n}}} = \frac{128.6 - 130}{\frac{2.1}{\sqrt{40}}} = -4.22 \text{ using } S \text{ to estimate } \sigma$$

RR: Reject H_0 if $Z < -Z_{.05} = -1.645$.

Conclusion: Reject H_0 at $\alpha = .05$.

10.9 We are to test H_0: $\mu_1 - \mu_2 = 0$ vs. H_a: $\mu_2 - \mu_2 \neq 0$. The test statistic and rejection region are

$$Z = \frac{1.65 - 1.43}{\sqrt{\frac{(.26)^2}{30} + \frac{(.22)^2}{35}}} = 3.65$$

RR: Reject H_0 if $|Z| > 2.575$.

Conclusion: Reject H_0 at $\alpha = .01$. The soils do appear to differ with respect to average shear strength, at the 1% significance level.

10.11 **a.–b.** Since there is no prior knowledge as to which mean should be larger, the hypothesis of interest is two-tailed:

$$H_0: \ \mu_1 - \mu_2 = 0, \qquad \text{vs.} \qquad H_a: \ \mu_1 - \mu_2 \neq 0.$$

c. The test statistic is

$$z \approx \frac{\bar{y}_1 - \bar{y}_2}{\sqrt{\frac{s_1^2}{n_1} + \frac{s_2^2}{n_2}}} = \frac{2980 - 3205}{\sqrt{\frac{(1140)^2 + (963)^2}{40}}} = -.954.$$

The rejection region, with $\alpha = .10$, is two-tailed or $|z| > 1.645$. The null hypothesis is not rejected. There is insufficient evidence to indicate a difference in the two means.

10.13 Let $p =$ proportion of adults unable to name an elected official that they admire. Then we want to test

$$H_0: \ p = .5 \qquad \text{vs.} \qquad H_a: \ p > .5$$

The test statistic is

$$z = \frac{\hat{p} - p_0}{\sqrt{\frac{p_0(1 - p_0)}{n}}} = \frac{.6 - .5}{\sqrt{\frac{(.5)(.5)}{1429}}} = 7.56.$$

With $\alpha = .01$, $z_{.01} = 2.326$ and we reject H_0 if $z > 2.326$. H_0 is rejected and we conclude that, of these adults, a majority ($> .5$) is unable to name an elected official that they admire.

10.15 The hypothesis of interest is

$$H_0: \quad p_1 - p_2 = 0 \qquad \text{vs.} \qquad H_a: \quad p_1 - p_2 \neq 0$$

where p_1 = the proportion of all colleges with an increase in applications in 1992 and p_2 = the proportion of all colleges with an increase in applications in 1991.

The test statistic, based on the sample data, will be

$$z = \frac{(\widehat{p}_1 - \widehat{p}_2) - 0}{\sqrt{\frac{p_1 q_1}{n_1} + \frac{p_2 q_2}{n_2}}}.$$

In order to evaluate the denominator, estimates for p_1 and p_2 must be obtained. Because we are assuming that $p_1 = p_2$, the best estimate for this common value will be (with $\widehat{p}_1 = .71$ and $\widehat{p}_2 = .50$)

$$\widehat{p} = \frac{n_1 \widehat{p}_1 + n_2 \widehat{p}_2}{n_1 + n_2} = \frac{875 + 613}{1232 + 1225} = .606.$$

The test statistic is, then,

$$z = \frac{\widehat{p}_1 - \widehat{p}_2}{\sqrt{\widehat{p}\widehat{q}\left(\frac{1}{n_1} + \frac{1}{n_2}\right)}} = \frac{.71 - .50}{\sqrt{(.606)(.394)\left(\frac{1}{1232} + \frac{1}{1225}\right)}} = 10.65.$$

The rejection region, with $\alpha = .01$, is $|z| > 2.5$ and H_0 is rejected. There is evidence of a difference in the proportions between 1991 and 1992.

10.17 The object of this experiment is to make a decision about the binomial parameter p, which is the probability that a customer prefers color A. Hence the null hypothesis will be that a customer has no preference for A, and the alternative will be that he has a preference. The null hypothesis is

$$H_0: \quad p = P(\text{customer prefers } A) = \tfrac{1}{3},$$

while the alternative hypothesis is

$$H_a: \quad p > \tfrac{1}{3}$$

Note that a one-tailed test of hypothesis is implied. The test statistic will be constructed by using $\widehat{p}$, the sample fraction favoring color A. This variable has mean $\mu_{\widehat{p}} = p$ and standard deviation $\sigma_{\widehat{p}} = \sqrt{\frac{pq}{n}}$. Thus the test statistic becomes

$$z = \frac{\hat{\theta} - \theta_0}{\sigma_{\hat{\theta}}} = \frac{\hat{p} - p_0}{\sqrt{\frac{p_0 q_0}{n}}} = \frac{\left(\frac{400}{1000}\right) - \left(\frac{1}{3}\right)}{\sqrt{\frac{\left(\frac{1}{3}\right)\left(\frac{2}{3}\right)}{1000}}} = 4.47$$

Notice that the value for $\sigma_{\hat{p}}$ was determined by assuming the null hypothesis to be true, that is, $p = \frac{1}{3}$. The rejection region for this one-tailed test (using $\alpha = .05$) is $z > 1.645$. The test statistic, $z = 4.47$, falls in the rejection region, and hence the null hypothesis is rejected. We conclude that customers have a preference for color A.

10.19 The following conditions must hold in order that the z statistic be appropriate:

(1) $\hat{\theta}$ is a normally distributed random variable.

(2) $\sigma_{\hat{\theta}}$ is a known quantity, or n is large, so that a good approximation for $\sigma_{\hat{\theta}}$ can be obtained from the sample observations.

10.21 Two binomial populations are involved. The hypothesis to be tested is

$$H_0:\ p_1 - p_2 = 0 \text{ (i.e., } p_1 = p_2) \qquad \text{vs.} \qquad H_a:\ p_1 - p_2 > 0 \text{ (i.e., } p_1 > p_2)$$

The test statistic, based on the sample data, will be

$$Z = \frac{(\hat{p}_1 - \hat{p}_2) - (p_1 - p_2)}{\sqrt{\frac{p_1 q_1}{n_1} + \frac{p_2 q_2}{n_2}}}$$

In order to evaluate the denominator, estimates for p_1 and p_2 must be obtained, using the assumption that $p_1 - p_2 = 0$. Because we are assuming that $p_1 = p_2$, the best estimate for this common value will be

$$\hat{p} = \frac{y_1 + y_2}{n_1 + n_2} = \frac{46 + 34}{200 + 200} = .2$$

The test statistic then becomes $z = \dfrac{\hat{p}_1 - \hat{p}_2 - 0}{\sqrt{\hat{p}\hat{q}\left[\left(\frac{1}{n}\right) + \left(\frac{1}{n}\right)\right]}} = \dfrac{.23 - .17}{\sqrt{(.2)(.8)\left(\frac{1}{100}\right)}} = 1.5.$

Rejection region: With $\alpha = .05$, the null hypothesis will be rejected if $z > 1.645$.

Conclusion: The test statistic does not fall in the rejection region. Hence the null hypothesis is not rejected. The researcher's belief is not supported.

10.23 Let p_1 = proportion of homeless men currently working and p_2 = proportion of domiciled men currently working. The hypothesis of interest is

$$H_0:\ p_1 - p_2 = 0 \qquad \text{vs.} \qquad H_a:\ p_1 - p_2 < 0.$$

Calculate

$$\widehat{p}_1 = \frac{34}{112} = .30, \ \widehat{p}_2 = .38, \text{ and } \widehat{p} = \frac{x_1 + x_2}{n_1 + n_2} = \frac{34 + 98}{112 + 260} = .355$$

The test statistic is then

$$z = \frac{\widehat{p}_1 - \widehat{p}_2}{\sqrt{\widehat{p}\widehat{q}\left(\frac{1}{n_1} + \frac{1}{n_2}\right)}} = \frac{.30 - .38}{\sqrt{(.355)(.645)\left(\frac{1}{112} + \frac{1}{260}\right)}} = -1.48.$$

The rejection region with $\alpha = .01$ is $z < -2.326$ and H_0 is not rejected. There is no evidence that the proportion of homeless men working is less than the proportion of domiciled men.

10.25 For testing

$$H_0: \ \mu = 130 \qquad \text{vs.} \qquad H_a: \ \mu = 128$$

we reject H_0 if $\frac{\overline{y} - \mu_0}{\frac{\sigma}{\sqrt{n}}} < -1.645$ or, if $\overline{y} < \mu_0 - \frac{1.645\sigma}{\sqrt{n}} = 130 - \frac{1.645(2.1)}{\sqrt{40}} = 129.45.$

If the mean is really 128, then

$$\beta = P\left[\overline{y} > 129.45\right] = P\left[\frac{\overline{y} - \mu_a}{\frac{\sigma}{\sqrt{n}}} > \frac{129.45 - 128}{\frac{2.1}{\sqrt{40}}}\right] = P[Z > 4.37] = 0$$

10.27 In Exercise 10.18 we found that the rejection region for this test could be written in terms of $\widehat{p}$ as $\widehat{p} \le .1342$. Refer to Figure 10.4. In order to calculate β we must find the area over the acceptance region (that is, the area to the right of $\widehat{p}_c$), where we assume the true value of p is .15. Thus

Figure 10.4

$$\beta = P\left(\widehat{p} > \widehat{p}_c\right) = P\left(\widehat{p} > .1342\right).$$

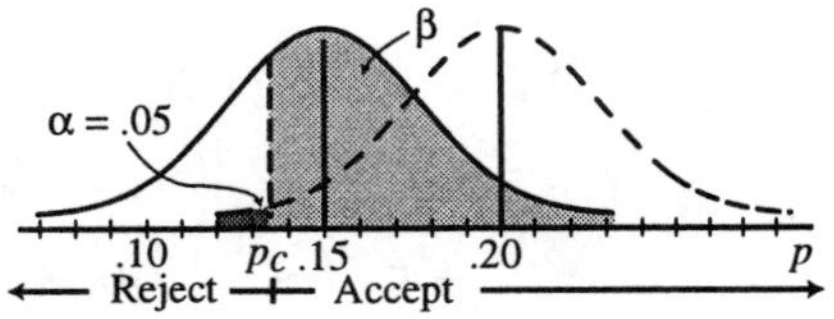

The corresponding z value is

$$z = \frac{\widehat{p}_c - p}{\sqrt{\frac{pq}{n}}} = \frac{.1342 - .15}{\sqrt{\frac{(.15)(.85)}{100}}} = -.44$$

Hence $\beta = P(Z > -.44) = P(Z < .44) = 1 - .3300 - .6700.$

10.29 Refer to Exercise 10.22. The rejection region, written in terms of $\overline{y}$, is

$$z > 1.645 \qquad \text{or} \qquad \frac{\overline{y} - 5}{\left(\frac{3.1}{\sqrt{500}}\right)} > 1.645 \qquad \text{or} \qquad \overline{y} \ge 5.228$$

Then

$$\beta = P(\text{accept } H_0 | \mu = 5.5) = P(\bar{y} < 5.228 | \mu = 5.5) = P\left[Z < \frac{5.228 - 5.5}{\frac{3.1}{\sqrt{500}}}\right]$$

$$= P(Z < -1.96) = .025$$

10.31 a. The hypothesis of interest is H_0: $\mu_1 - \mu_2 = 0$ vs. H_a: $\mu_1 - \mu_2 > 0$ and the test statistic is approximated as

$$z = \frac{(\bar{y}_1 - \bar{y}_2) - 0}{\sqrt{\left(\frac{s_1^2}{n_1}\right) + \left(\frac{s_2^2}{n_2}\right)}} = \frac{32.19 - 31.68}{\sqrt{\frac{(4.34)^2 + (4.56)^2}{37}}} = .49$$

The rejection region with $\alpha = .05$ is $z > 1.645$. Hence H_0 is not rejected. There is insufficient evidence to indicate that $\mu_1 > \mu_2$.

b. The rejection region, written in terms of $\bar{Y}_1 - \bar{Y}_2$, is

$$Z > 1.645 \qquad \text{or} \qquad \frac{\bar{Y}_1 - \bar{Y}_2}{\hat{\sigma}_{\bar{Y}_1 - \bar{Y}_2}} > 1.645$$

or

$$\bar{Y}_1 - \bar{Y}_2 > 1.645 \sqrt{\frac{(4.34)^2 + (4.56)^2}{37}} = 1.702$$

Then

$$\beta = P(\text{accept } H_0 | \mu_1 - \mu_2 = 3) = P(\bar{Y}_1 - \bar{Y}_2 < 1.702) = P\left[Z < \frac{1.702 - 3}{\hat{\sigma}_{\bar{Y}_1 - \bar{Y}_2}}\right]$$

$$= P(Z < -1.25) = P(Z > 1.25) = .1056$$

10.33 $\bar{y} \pm z_{.005} \dfrac{s}{\sqrt{n}} = 885 \pm 2.58 \left(\dfrac{50}{\sqrt{35}}\right) = (863.195,\ 906.805)$

The value $\mu_0 = 900$ is inside the interval. The null hypothesis should not be rejected because 900 is a believable value for the true mean. This is consistent with the conclusion in Exercise 10.5.

10.35 $\hat{p} - z_{.01} \sqrt{\dfrac{\hat{p}(1 - \hat{p})}{n}} = .6 - 2.326 \sqrt{\dfrac{(.6)(.4)}{1429}} = .6 - .03 = .57$

Because the interval .57 to 1.0 does not contain .5, the alternative hypothesis of Exercise 10.13 should be accepted at the 99% confidence level. This does not conflict with the answer to Exercise 10.13.

10.37 Using the result of Exercise 10.36, the bound is

$$\bar{y} + z_{.05}\frac{s}{\sqrt{n}} = 128.6 + 1.645\left(\frac{2.1}{\sqrt{40}}\right) = 128.6 + .546 = 129.146.$$

This bound is less than the hypothesized value, i.e., 130. Therefore, the alternative hypothesis should be accepted, which does not conflict with the answer to Exercise 10.7.

10.39 We are to test

$$H_0:\ \mu_1 - \mu_2 = 0 \qquad \text{vs.} \qquad H_a:\ \mu_1 - \mu_2 \neq 0.$$

The test statistic is

$$z = \frac{74 - 71}{\frac{9^2}{50} + \frac{10^2}{50}} = 1.58$$

The p-value is

$$p\text{-value} = P\big(|Z| > 1.58\big) = 2(.0571) = .1142$$

Since $.1142 > .05$, we would not reject H_0 in a test at level $\alpha = .05$.

10.41 a. The hypothesis of interest is

$$H_0:\ \mu_1 = 3.8 \qquad \text{vs.} \qquad H_a:\ \mu_1 < 3.8,$$

where μ_1 is the average drop in FVC for men on the physical fitness program. The test statistic is

$$z = \frac{\bar{y}_1 - \mu_1}{\frac{s_1}{\sqrt{n_1}}} = \frac{3.6 - 3.8}{\frac{1.1}{\sqrt{30}}} = -.996,$$

and the p-value is

$$p\text{-value} = P[z < -1.00] = .5 - .3413 = .1587.$$

b. Since $\alpha = .05$ is smaller than the p-value $= .1587$, H_0 cannot be rejected. We cannot support the contention that the mean decrease in FVC for men is less than 3.8.

c. The hypothesis of interest is

$$H_0:\ \mu_2 = 3.1 \qquad \text{vs.} \qquad H_a:\ \mu_2 < 3.1,$$

where μ_2 is the average drop in FVC for men on the physical fitness program. The test statistic is

$$z = \frac{\overline{y}_2 - \mu_2}{\frac{s_2}{\sqrt{n_2}}} = \frac{2.7 - 3.1}{\frac{1.2}{\sqrt{30}}} = -1.826,$$

and the p-value is

$$p\text{-value} = P[z < -1.83] = .5 - .4664 = .0336.$$

d. Since $\alpha = .05$ is larger than the p-value, $.0336$, H_0 can be rejected (at $\alpha = .0336$ or any larger value). The data support the contention that the mean decrease in FVC is less than 3.1 for women on the physical fitness program.

10.43 We are to test

$$H_0\!: \; p = .05 \qquad\qquad \text{vs.} \qquad\qquad H_a\!: \; p < .05$$

with $\widehat{p} = \frac{45}{1124} = .04$.

The test statistic is

$$z = \frac{\widehat{p} - .05}{\sqrt{\frac{(.05)(.95)}{n}}} = \frac{.04 - .05}{\sqrt{\frac{(.05)(.95)}{1124}}} = -1.538$$

The p-value is

$$p\text{-value} = P(Z) < -1.538 = .0618$$

At level $\alpha = .01$, we would not reject H_0, since $.0618 > .01$.

10.45 We wish to test

$$H_0\!: \; p = .60 \qquad\qquad \text{vs.} \qquad\qquad H_a\!: \; p \neq .60.$$

The sample proportion is $\widehat{p} = \frac{108}{200} = .54$. The test statistic is

$$z = \frac{.54 - .60}{\sqrt{\frac{(.6)(.4)}{200}}} = -1.732$$

The p-value is

$$p\text{-value} = P\big(|Z| > |-1.732|\big) = 2P(Z > 1.7321) = 2(.0418) = .0836$$

10.47 The z test is inappropriate as a test statistic when the sample size is small and α is unknown. The estimator S can be used as an approximation to α when n is large (say $n \geq 30$), but the approximation affects the distribution of the test statistic when n is small. Conditions will often be satisfied to permit the use of Student's t when n is small.

10.49 The hypothesis to be tested is

$$H_0: \ \mu = 800 \qquad \text{vs.} \qquad H_a: \ \mu < 800.$$

However, there are only five measurements on which to base the test, and a t statistic must be employed. The test statistic is

$$T = \frac{\overline{Y} - \mu}{\frac{s}{\sqrt{n}}}$$

where

$$\overline{y} = \frac{\sum_i y_i}{n} = \frac{3975}{5} = 795$$

and

$$s^2 = \frac{\sum_i - \frac{\left(\sum_i y_i\right)^2}{n}}{n-1} = \frac{3{,}160{,}403 - \frac{15{,}800{,}625}{5}}{4} = 69.5$$

Then

$$t = \frac{\overline{y} - \mu}{\frac{s}{\sqrt{n}}} = \frac{795 - 800}{\frac{\sqrt{69.5}}{\sqrt{5}}} = \frac{-5}{3.728} = -1.341$$

The rejection region is determined by a t value based on $(n-1) = 4$ degrees of freedom. Indexing $t_{.05}$ in Table 5, the rejection region is $t < -2.132$. Since the observed value of the test statistic does not fall in the rejection region, we do not reject H_0. Refer to Figure 10.6 at the right and notice that the t distribution is similar to the z distribution. The shaded area constitutes the size of the rejection region.

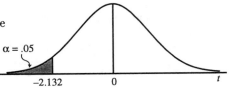

Figure 10.6

To find bounds on the p-value, note in Table 5 that a value of $t = -1.553$ would yield a p-value of .10. The observed value of t is $-1.341 > -1.533$. Thus, p-value $> .10$.

10.51 a. The hypothesis to be tested is

$$H_0: \ \mu = 45 \qquad \text{vs.} \qquad H_a: \ \mu < 45$$

Calculate

$$\overline{y} = \frac{\sum\limits_{i=1}^{n} y_i}{n} = \frac{712.01}{18} = 39.556$$

$$s^2 = \frac{\sum y_i^2 - \frac{\left(\sum y_i\right)^2}{n}}{n-1} = \frac{29,040.4275 - \frac{(712.01)^2}{18}}{17} = 50.94593$$

and $s = 7.138$. The test statistic is

$$t = \frac{\overline{y} - \mu_0}{\frac{s}{\sqrt{n}}} = \frac{39.556 - 45}{\frac{7.138}{\sqrt{18}}} = -3.24$$

With 17 degrees of freedom, the p-value is less than .005. Because this is so small, we reject H_0 and conclude that the average is significantly less than 45 cents.

b.   The 95% confidence interval, based on $n - 1 = 17$ degrees of freedom, is

$$\overline{y} \pm t_{.025} \frac{s}{\sqrt{n}} = 39.556 \pm 2.110 \left(\frac{7.138}{\sqrt{18}} \right) = 39.556 \pm 3.550$$

or $36.006 < \mu < 43.106$.

10.53 a.   The hypothesis of interest is

$$H_0: \ \mu = 400 \qquad \text{vs.} \qquad H_a: \ \mu \neq 400$$

The test statistic is

$$t = \frac{\overline{y} - \mu_0}{\frac{s}{\sqrt{n}}} = \frac{365 - 400}{\frac{46}{\sqrt{20}}} = -3.4$$

With 19 degrees of freedom, the p-value is less than $2(.005) = .01$. Because the p-value is so small, we reject H_0 and conclude that the mean number of units of vitamin D is not 400.

b.   The 95% confidence interval is

$$\overline{y} \pm t_{.025} \frac{s}{\sqrt{n}} = 365 \pm 2.093 \left(\frac{46}{\sqrt{20}} \right) = 365 \pm 21.53 = (343.47, \ 386.53).$$

c.   Because the interval in Part (b) does not contain the value 400, we conclude that the true mean is not 400. This conclusion is consistent with the formal test in Part (a).

10.55   This is similar to Examples 10.14 and 10.15. The hypothesis to be tested is

$$H_0: \ \mu_1 - \mu_2 = 0 \qquad \text{vs.} \qquad H_a: \ \mu_1 - \mu_2 \doteq 0.$$

We must assume that the data come from two normal populations with a common variance. We obtain an estimate for the common variance σ^2 by calculating

$$s^2 = \frac{(n_1 - 1)s_1^2 + (n_2 - 1)s_2^2}{n_1 + n_2 - 2} = \frac{10(52) + 13(71)}{11 + 14 - 2} = \frac{1443}{23} = 62.74$$

The test statistic is

$$t = \frac{\bar{y}_1 - \bar{y}_2 - D_0}{\sqrt{s^2\left[\left(\frac{1}{n_1}\right) + \left(\frac{1}{n_2}\right)\right]}} = \frac{64 - 69}{\sqrt{62.74\left[\left(\frac{1}{11}\right) + \left(\frac{1}{14}\right)\right]}} = -1.57$$

The rejection region is $|t| > t_{.025,\,23} = 2.069$. Since the observed value of the test statistic does not exceed $t = 2.069$ in absolute value, the null hypothesis is not rejected.

Note in Table 5 that $1.319 < 1.57 < 1.714$. The p-values associated with 1.319 and 1.714 are 2(.10) and 2(.05). Thus, $.10 < p\text{-value} < .20$.

10.57 Refer to Exercise 10.56. We are to test

$$H_0\colon \ \mu_1 - \mu_2 \le .01 \qquad\qquad \text{vs.} \qquad\qquad H_a\colon \ \mu_1 - \mu_2 > .01.$$

The test statistic is

$$t = \frac{(.041 - .026) - .01}{\sqrt{s^2\left[\left(\frac{1}{10}\right) + \left(\frac{1}{13}\right)\right]}} = .989$$

From Table 5, the p-value associated with $t = 1.323$ is .10. Hence, since $.989 < 1.323$, the p-value is greater than .10.

10.59 a. The hypothesis to be tested is

$$H_0\colon \ \mu_1 - \mu_2 = 0 \qquad\qquad \text{vs.} \qquad\qquad H_a\colon \ \mu_1 - \mu_2 \ne 0.$$

where μ_1 is the average compartment pressure for runners, and μ_2 is the average compartment pressure for cyclists. The pooled estimator of σ^2 is calculated as

$$s_p^2 = \frac{(n_1 - 1)s_1^2 + (n_2 - 1)s_2^2}{n_1 + n_2 - 2} = \frac{9(3.92)^2 + 9(3.98)^2}{18} = 15.6034$$

and the test statistic is

$$t = \frac{(\bar{y}_1 - \bar{y}_2) - 0}{\sqrt{s_p^2\left(\frac{1}{n_1} + \frac{1}{n_2}\right)}} = \frac{14.5 - 11.1}{\sqrt{15.6034\left(\frac{1}{10} + \frac{1}{10}\right)}} = 1.92$$

The rejection region is two-tailed, based on $n_1 + n_2 - 2 = 18$ degrees of freedom. With $\alpha = .05$, from Table 5 the rejection region is $|t| > t_{.025} = 2.101$. We do not reject H_0; there is insufficient evidence to indicate a difference in the means.

With 18 degrees of freedom, we can say that the p-value is between $2(.025)$ and $2(.05)$; i.e., $.05 < p$-value $< .10$.

b.   The hypothesis to be tested is

$$H_0: \; \mu_1 - \mu_2 = 0 \qquad\qquad \text{vs.} \qquad\qquad H_a: \; \mu_1 - \mu_2 \neq 0.$$

where $\mu_1 = $ mean compartment pressure for runners at 80% maximal O_2 consumption and $\mu_2 = $ mean compartment pressure for cyclists at 80% maximal O_2 consumption.

First,

$$s_p^2 = \frac{(n_1 - 1)s_1^2 + (n_2 - 1)s_2^2}{n_1 + n_2 - 2} = \frac{9(3.49)^2 + 9(4.95)^2}{18} = 15.3413.$$

then the test statistic is

$$t = \frac{(\bar{y}_1 - \bar{y}_2) - 0}{\sqrt{s_p^2\left(\frac{1}{n_1} + \frac{1}{n_2}\right)}} = \frac{12.2 - 11.5}{\sqrt{18.3413\left(\frac{1}{10} + \frac{1}{10}\right)}} = .365$$

The rejection region is $|t| > t_{.025} = 2.101$. Do not reject H_0. We conclude that there is not sufficient evidence to indicate a difference in the mean compartment pressure between runners and cyclists at 80% maximal O_2 consumption.

The associated p-value is, with 18 degrees of freedom, greater than $2(.10)$ or the p-value $> .20$.

10.61 As in Section 10.3, the hypothesis to be tested is

$$H_0: \; \mu = 30.31 \qquad\qquad \text{vs.} \qquad\qquad H_a: \; \mu < 30.31.$$

However, there are only 20 measurements on which to base the test, and a t statistic must be employed. The test statistic is

$$t = \frac{\bar{y} - \mu}{\frac{s}{\sqrt{n}}}$$

where

$$\bar{y} = \frac{\sum y_i}{n} = \frac{578.7}{20} = 28.935,$$

$$s^2 = \frac{\sum y_i^2 - \frac{\left(\sum y_i\right)^2}{n}}{n - 1} = \frac{18,462.09 - 16,744.6845}{19} = 90.3898,$$

$$s = \sqrt{90.3898} = 9.507.$$

Then

$$t = \frac{\bar{y} - \mu}{\frac{s}{\sqrt{n}}} = \frac{28.935 - 30.31}{\frac{9.507}{\sqrt{20}}} = -.647.$$

The rejection region, with $n - 1 = 19$ degrees of freedom and $\alpha = .05$, is

$$t < -t_{.05,\,19} = -1.729,$$

as shown in Figure 10.7. Since the observed value of $t = -.647$ does not fall in the rejection region, H_0 is not rejected. There is no reason to believe that the mean has decreased.

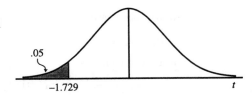

Figure 10.7

10.63 a. The hypothesis to be tested is

$$H_0: \mu_1 - \mu_2 = 0 \qquad \text{vs.} \qquad H_a: \mu_1 - \mu_2 \neq 0$$

Calculate

$$s_p^2 = \frac{(n_1 - 1)s_1^2 + (n_2 - 1)s_2^2}{n_1 + n_2 - 2} = \frac{14(42)^2 + 14(45)^2}{28} = 1894.5$$

and the test statistic is

$$t = \frac{(\bar{y}_1 - \bar{y}_2) - 0}{\sqrt{s_p^2\left(\frac{1}{n_1} + \frac{1}{n_2}\right)}} = \frac{446 - 534}{\sqrt{1894.5\left(\frac{1}{15} + \frac{1}{15}\right)}} = -5.54$$

The rejection region is two-tailed, based on $n_1 + n_2 - 2 = 28$ degrees of freedom. From Table 5, the approximate p-value is p-value $< 2(.005) = .01$. Thus, at the $\alpha = .05$ level there is sufficient evidence to indicate that a difference exists in the mean verbal scores for the two groups.

b. Yes.

c. The hypothesis to be tested is

$$H_0: \mu_1 - \mu_2 = 0 \qquad \text{vs.} \qquad H_a: \mu_1 - \mu_2 \neq 0$$

Calculate

$$s^2 = \frac{(n_1 - 1)s_1^2 + (n_2 - 1)s_2^2}{n_1 + n_2 - 2} = \frac{14(57)^2 + 14(52)^2}{28} = 2976.5$$

and the test statistic is

$$t = \frac{\bar{y}_1 - \bar{y}_2}{\sqrt{s^2\left(\frac{1}{n_1} + \frac{1}{n_2}\right)}} = \frac{548 - 517}{\sqrt{2976.5\left(\frac{1}{15} + \frac{1}{15}\right)}} = 1.56$$

The rejection region is two-tailed, based on $n_1 + n_2 - 2 = 28$ degrees of freedom. From Table 5, the approximate p-value is $2(.1) > p\text{-value} > 2(.05)$ or $.2 > p\text{-value} > .1$. There is no evidence of a significant difference in the mean math scores for the two groups.

d. Yes.

10.65 The hypothesis to be tested is

$$H_0: \ \sigma^2 = .01 \qquad \text{vs.} \qquad H_a: \ \sigma^2 > .01.$$

The test statistic is

$$\chi^2 = \frac{(n-1)s^2}{\sigma_0^2} = \frac{7(.018)}{.01} = 12.6$$

A one-tailed test is required. Hence, a critical value of χ^2 (denoted by χ_c^2) must be found such that $P\left(\chi^2 > \chi_c^2\right) = .05$. Indexing $\chi_{.05}^2$ with $(n-1) = 7$ degrees of freedom (see Table 6), the critical value is found to be $\chi_{.05}^2 = 14.07$ (see Figure 10.8).

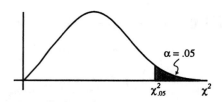

Figure 10.8

The value of the test statistic is not in the rejection region. Consequently, we cannot reject the hypothesis that $\sigma^2 = .01$.

Note in Table 6 that $12.02 < 12.6 \leq 14.07$. The p-values associated with 12.02 and 14.07 are .10 and .05. Thus, $.05 < p\text{-value} < .10$.

We must assume that the data (carton weights) are from a normal population.

10.67 a. The rejection region given is

$$\left\{ \frac{S_1^2}{S_2^2} > F_{\nu_2, \alpha/2}^{\nu_1} \ \text{ or } \ \frac{S_1^2}{S_2^2} < \left(F_{\nu_1, \alpha/2}^{\nu_2} \right)^{-1} \right\}$$

which, by reversing the second inequality and inverting fractions, becomes

$$\left\{ \frac{S_1^2}{S_2^2} > F_{\nu_2, \alpha/2}^{\nu_1} \ \text{ or } \ \frac{S_2^2}{S_1^2} > F_{\nu_1, \alpha/2}^{\nu_2 \ -1} \right\}$$

b.
$$P\left(\frac{S_L^2}{S_S^2} > F_{\nu_S, \alpha/2}^{\nu_L}\right) = \left\{\frac{S_1^2}{S_2^2} > F_{\nu_2, \alpha/2}^{\nu_1} \text{ or } \frac{S_2^2}{S_1^2} > F_{\nu_1, \alpha/2}^{\nu_2}\right\}$$

$$= \left\{\frac{S_1^2}{S_2^2} > F_{\nu_2, \alpha/2}^{\nu_1} \text{ or } \frac{S_1^2}{S_2^2} < \left(F_{\nu_2, \alpha/2}^{\nu_2}\right)^{-1}\right\} = \alpha$$

10.69 It is possible to test the null hypothesis H_0: $\sigma_1^2 = \sigma_2^2$ against any one of the three alternative hypotheses:

(1) H_a: $\sigma_1^2 \neq \sigma_2^2$ (2) H_a: $\sigma_1^2 < \sigma_2^2$ (3) H_a: $\sigma_1^2 > \sigma_2^2$

The first alternative would be preferred by the manager of the dairy. He does not know anything about the variability of the two machines and would wish to detect departures from equality of the type $\sigma_1^2 > \sigma_2^2$ or $\sigma_2^2 > \sigma_1^2$. These alternatives are implied in (1). The salesman for company A would prefer that the experimenter select the second alternative. Rejection of the null hypothesis would imply that his machine had smaller variability. Moreover, rejection of the null hypothesis in favor of (2) is more likely than rejection of H_0 in favor of (1) if, in fact, $\sigma_1^2 < \sigma_2^2$. The salesman for company B would prefer the third alternative for a similar reason.

10.71 The hypothesis to be tested is

$$H_0: \sigma = .7 \qquad \text{vs.} \qquad H_a: \sigma > .7.$$

This is equivalent to testing the hypothesis

$$H_0: \sigma^2 = .49 \qquad \text{vs.} \qquad H_a: \sigma^2 > .49.$$

Once the value for S^2 is calculated, the test statistic $\chi^2 = \dfrac{(n-1)S^2}{\sigma_0^2}$ will be used to test the above hypothesis. Then

$$s^2 = \frac{\sum_i y_i^2 - \dfrac{\left(\sum_i y_i\right)^2}{n}}{n-1} = \frac{497,036 - \dfrac{(1410)^2}{4}}{3} = 3.667$$

and the test statistic is

$$\chi^2 = \frac{(n-1)s^2}{\sigma_0^2} = \frac{3(3.6667)}{.49} = \frac{11}{.49} = 22.45$$

Since $22.45 > 12.8381 = \chi_{.005, 3}^2$, the p-value is less than .005.

10.73 Refer to Exercise 10.56. The hypothesis of interest is

$$H_0: \sigma_1^2 = \sigma_2^2 \qquad \text{vs.} \qquad H_a: \sigma_1^2 > \sigma_2^2$$

and the test statistic is

$$F = \frac{S_1^2}{S_2^2} = \frac{(.017)^2}{(.006)^2} = 8.03$$

The rejection region with $\alpha = .05$ is $F > F^9_{12, .05} = 2.80$ and H_0 is rejected. We conclude that $\sigma_1^2 > \sigma_2^2$.

10.75 a.   Refer to Example 10.23 in the text. The uniformly most powerful test is found to be the z test of Section 10.3. That is, reject H_0: $\mu = 7$ if

$$Z = \frac{\overline{Y} - 7}{\sqrt{\frac{\sigma^2}{20}}} = \frac{\overline{Y} - 7}{\sqrt{\frac{5}{20}}} \geq 1.645 \qquad \text{or} \qquad \overline{Y} \geq 1.645\sqrt{.25} + 7 = 7.82$$

b.   The power of the test is $1 - \beta = P(\overline{Y} > 7.82 | \mu)$.

For $\mu = 7.5$, $1 - \beta = P\left(Z > \frac{7.82 - 7.5}{.5}\right) = P(Z > .64) = .2611$

For $\mu = 8.0$, $1 - \beta = P\left(Z > \frac{7.82 - 8}{.5}\right) = P(Z > -.36) = .6406$

For $\mu = 8.5$, $1 - \beta = P\left(Z > \frac{7.82 - 8.5}{.5}\right) = P(Z > -1.36) = .9131$

For $\mu = 9.0$, $1 - \beta = P\left(Z > \frac{7.82 - 9}{.5}\right) = P(Z > -2.36) = .9909$

c.   The graph is omitted here.

10.77 Using the sample size formula derived at the end of Section 10.4 in the text, we have

$$n = \frac{(z_\alpha + z_\beta)^2 \sigma^2}{(\mu_a - \mu_0)^2} = \frac{(1.96 + 1.96)^2 25}{(10 - 5)^2} = 15.3664$$

or $n = 16$ is the desired sample size.

10.79 a.   Under H_0 the likelihood function is

$$L(\theta_0) = \frac{1}{\left(2\theta_0^3\right)^4} \left(\prod_{i=1}^{4} y_i^2\right) e^{-\Sigma y_i / \theta_0}$$

Under H_a it is

$$L(\theta_a) = \frac{1}{\left(2\theta_a^3\right)^4} \left(\prod_{i=1}^{4} y_i^2\right) e^{-\Sigma y_i / \theta_a}$$

Using Theorem 10.1, we obtain the most powerful critical region as

$$\frac{L(\theta_0)}{L(\theta_a)} = \frac{\theta_a^{12}}{\theta_0^{12}} \exp\left[-\Sigma y_i \left(\frac{1}{\theta_0} - \frac{1}{\theta_a}\right)\right] \leq k \quad \text{or} \quad \exp\left[\Sigma y_i \left(\frac{1}{\theta_0} - \frac{1}{\theta_a}\right)\right] \leq k\left(\frac{\theta_0}{\theta_a}\right)^{12}$$

or

$$-\sum y_i\left(\frac{1}{\theta_0}-\frac{1}{\theta_a}\right)\le \ln k\left(\frac{\theta_0}{\theta_a}\right)^{12} \quad\text{or}\quad -\sum y_i \le \frac{\ln k\left(\frac{\theta_0}{\theta_a}\right)^{12}}{\left(\frac{1}{\theta_0}\right)-\left(\frac{1}{\theta_a}\right)} \quad\text{or}\quad \sum y_i \ge -k'$$

If H_0 is true, Y_i has a gamma distribution with $\alpha = 3$ and $\beta = \theta_0$, and $\frac{2Y_i}{\theta_0}$ has a χ^2 distribution with 6 degrees of freedom. Hence $2\left(\sum Y_i\right)\theta_0$ has a χ^2 distribution with 24 degrees of freedom. (Refer to Exercise 7.47 and the method of moment-generating functions.) The critical region can be written as

$$\frac{2\sum Y_i}{\theta_0} \ge \frac{-2k'}{\theta_0} = k''$$

where k'' is chosen so that the test will have size α.

b. The choice of critical region did not depend on the particular value of θ_a but only upon the fact that $\theta_a > \theta_0$. Hence, for any $\theta > \theta_0$, the above critical region is most powerful and the test given in part (a) is uniformly most powerful for the alternative $\theta > \theta_0$.

10.81 a. $L(\lambda) = f\left(\frac{y_1}{\lambda}\right)f\left(\frac{y_2}{\lambda}\right)\cdots f\left(\frac{y_n}{\lambda}\right) = \prod_{i=1}^{n} f\left(\frac{y_i}{\lambda}\right) = \frac{\lambda^{\sum y_i}e^{-n\lambda}}{\prod_{i=1}^{n} y_i!}.$

Then, by the Neymann–Pearson Lemma, the test that maximizes the power at θ_a has a rejection region determined by

$$\frac{L(\lambda_0)}{L(\lambda_1)} < k$$

or

$$\frac{\left(\frac{\lambda_0^{\sum y_i}e^{-n\lambda_0}}{\prod_{i=1}^{n} y_i!}\right)}{\left(\frac{\lambda_a^{\sum y_i}e^{-n\lambda_a}}{\prod_{i=1}^{n} y_i!}\right)} < k$$

or

$$\left(\frac{\lambda_0}{\lambda_a}\right)^{\Sigma y_i} e^{n(\lambda_a - \lambda_0)} < k$$

or

$$\sum_{i=1}^{n} y_i \ln\left(\frac{\lambda_0}{\lambda_a}\right) + n(\lambda_a - \lambda_0) < \ln k$$

or

$$\sum_{i=1}^{n} y_i \ln\left(\frac{\lambda_0}{\lambda_a}\right) < \ln k - n(\lambda_a - \lambda_0)$$

or, since $\lambda_0 < \lambda_a$

$$\sum_{i=1}^{n} y_i > \frac{\ln k - n(\lambda_a - \lambda_0)}{\ln\left(\frac{\lambda_0}{\lambda_a}\right)}$$

or with $k' = \dfrac{\ln k - n(\lambda_a - \lambda_0)}{\ln\left(\dfrac{\lambda_0}{\lambda_a}\right)}$

we have $\displaystyle\sum_{i=1}^{n} Y_i > k'$.

b. $\displaystyle\sum_{i=1}^{n} Y_i \sim$ Poisson with mean $n\lambda$. Then, for a given α, the constant k' is the value such that $P\left(\displaystyle\sum_{i=1}^{n} Y_i > k' \text{ when } \lambda = \lambda_0\right) = \alpha$.

c. The form of the rejection region does not depend upon the particular value assigned to λ_a. Therefore, the test derived in part (a) is the uniformly most powerful.

d. The form is similar to that in part (a). We start with

$$\frac{L(\lambda_0)}{L(\lambda_a)} < k$$

or, since $\lambda_a < \lambda_0$,

$$\sum y_i \ln\left(\frac{\lambda_0}{\lambda_a}\right) < \ln k - n(\lambda_0 - \lambda_a)$$

$$\sum Y_i < \frac{\ln k - n(\lambda_0 - \lambda_a)}{\ln\left(\frac{\lambda_0}{\lambda_a}\right)}$$

or $\sum_{i=1}^{n} Y_i < k'$

with $k' = \dfrac{\ln k - n(\lambda_0 - \lambda_a)}{\ln\left(\dfrac{\lambda_0}{\lambda_a}\right)}.$

10.83 a. First,

$$L(\theta) = \frac{1}{\theta^n} e^{-\Sigma y_i/\theta}.$$

Then, the rejection region for the most powerful test is

$$\frac{L(\theta_0)}{L(\theta_a)} < k$$

or

$$\frac{\frac{1}{\theta_0^n} e^{-\Sigma y_i/\theta_0}}{\frac{1}{\theta_a^n} e^{-\Sigma y_i/\theta_a}} < k$$

or

$$n \ln\left(\frac{\theta_a}{\theta_0}\right) + \sum_{i=1}^{n} y_i \left(\frac{1}{\theta_a} - \frac{1}{\theta_0}\right) < \ln k$$

or

$$\sum_{i=1}^{n} \frac{y_i}{n} < \left[\frac{\ln k}{n} - \ln\left(\frac{\theta_a}{\theta_0}\right)\right]\left(\frac{\theta_a \theta_0}{\theta_0 - \theta_a}\right)$$

or, finally,

$$\sum y_i < k'$$

for

$$k' = \left[\frac{\ln k}{n} - \ln\left(\frac{\theta_a}{\theta_0}\right)\right]\left(\frac{\theta_a \theta_0}{\theta_0 - \theta_a}\right).$$

b. The test derived in part (a) is the uniformly most powerful test, since the form of the rejection region does not depend on θ_a.

10.85 a. The density function for y, given θ, is

$$f(y) = \left(\frac{1}{\theta}\right) I(0 < y < \theta)$$

where $I(0 < y < \theta)$ equals 1 if $0 < y < \theta$ and 0 otherwise. The likelihood is

$$L(\theta) = \left(\frac{1}{\theta^n}\right) \prod_{i=1}^{n} I(0 < y_i < \theta) = \left(\frac{1}{\theta^n}\right) I(0 < y_{(n)} < \theta)$$

Thus, the most powerful test rejects H_0 if

$$\frac{L(\theta_0)}{L(\theta_a)} = \left(\frac{\theta_a}{\theta_0}\right)^n \frac{I(0 < y_{(n)} < \theta_0)}{I(0 < y_{(n)} < \theta_a)} < k$$

The rejection rule depends on the data only through $y_{(n)}$. (If $\theta_a < y_{(n)} < \theta_0$, the left-hand side of the above inequality is undefined and we know H_0 is true.) We will reject H_0 if $y_{(n)}$ is small.

We will consider the rejection region given by $y_{(n)} < k$, choosing k so that $P(Y_{(n)} < k | \theta = \theta_0) = \alpha$. The density for $Y_{(n)}$ is given as $\frac{ny^{n-1}}{\theta_0^n}$ for $0 \le y \le \theta_0$. Hence

$$\alpha = P(Y_{(n)} < k) = \int_0^k \frac{ny^{n-1}}{\theta_0^n}\, dy = \frac{k^n}{\theta_0^n}$$

so that $k = \theta_0 \alpha^{1/n}$. The most powerful critical region is thus $y_{(n)} < \theta_0 \alpha^{1/n}$.

b. Notice that the choice of critical region depended only upon the fact that $\theta_a < \theta_0$, since this implied rejection of H_0 for small values of $Y_{(n)}$. Therefore, the test in part (a) is uniformly most powerful.

10.87 The null hypothesis specifies $\Omega_0 = \{\sigma^2: \sigma^2 - \sigma_0^2\}$, while $\Omega = \Omega_0 \cup \Omega_a = \{\sigma^2: \sigma^2 \ge \sigma_0^2\}$. In the restricted space Ω_0, the likelihood function is

$$L(\Omega_0) = \prod_{i=1}^{n} \frac{1}{(2\pi)^{1/2}\sigma_0} \exp\left[\frac{(y_i - \mu)^2}{2\sigma_0^2}\right]$$

The maximum likelihood estimate of μ is $\overline{Y}$, so that

$$L(\widehat{\Omega}) = \frac{1}{(2\pi)^{n/2}\sigma_0^n} \exp\left[\frac{\sum_{i=1}^{n}(y_i - \overline{y})^2}{2\sigma_0^2}\right]$$

In Ω

$$L(\Omega) = \frac{1}{(2\pi)^{n/2}\sigma^n} \exp\left[\frac{\sum(y_i - \mu)^2}{2\sigma^2}\right]$$

The maximum likelihood estimate of μ is $\widehat{\mu} = \overline{Y}$, while

$$\widehat{\sigma}^2 = \max\left[\sigma_0^2, \ \widehat{\sigma}^2 = \frac{\sum (Y_i - \overline{Y})^2}{n}\right] \quad \text{and} \quad L(\widehat{\Omega}) = \frac{1}{(2\pi)^{n/2}\widehat{\sigma}^2} \exp\left[-\frac{\sum (y_i - \overline{y})^2}{2\widehat{\sigma}^2}\right]$$

The likelihood ratio statistic is

$$\lambda = \frac{L(\widehat{\Omega}_0)}{L(\widehat{\Omega})} = \left(\frac{\widehat{\sigma}^2}{\sigma_0^2}\right)^{n/2} \exp\left[-\frac{\sum (y_i - \overline{y})^2}{2\sigma_0^2} + \frac{\sum (y_i - \overline{y})^2}{2\widehat{\sigma}^2}\right] = 1 \quad \text{if} \quad \widehat{\sigma} \leq \sigma_0$$

$$= \left[\frac{\sum (y_i - \overline{y})^2}{n\sigma_0^2}\right]^{n/2} \exp\left[-\frac{\sum (y_i - \overline{y})^2}{2\sigma_0^2}\right] e^{n/2} \quad \text{if} \quad \widehat{\sigma} > \sigma_0$$

Hence the rejection region $\lambda \leq k$ is equivalent to

$$g(\chi^2) = (\chi^2)^{n/2} e^{-\chi^2/2} n^{-n/2} e^{n/2} \leq k$$

where χ^2 is $\dfrac{(n-1)s^2}{\sigma_0^2}$, the χ^2 statistic given in

Section 10.9.

Note that if $\widehat{\sigma} \leq \sigma_0$, $g(\chi^2) = 1$. Further, if

$\widehat{\sigma} > \sigma_0$, $g(\chi^2)$ is a monotonically decreasing

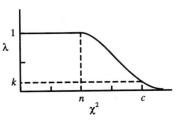

Figure 10.9

function of χ^2. Hence the region $\lambda \leq k$ is equivalent to $\chi^2 \geq c$, where c is

determined so that the test has size α. A rough sketch of $g(\chi^2) = \lambda$ against χ^2 is

shown in Figure 10.9.

10.89 Let $X_1, X_2, \ldots, X_n$ be the random sample drawn from population 1, and let $Y_1, Y_2, \ldots, Y_m$ be the random sample from population 2. Assuming that H_0 is true,

$$\frac{\sum (X_i - \overline{X})^2 + \sum (Y_i - \overline{Y})^2}{\sigma_0^2} = \frac{(n-1)S_1^2 + (m-1)S_2^2}{\sigma_0^2} = \chi^2$$

has a χ^2 distribution with $m + n - 2$ degrees of freedom. If H_a is true, then S_1^2 and S_2^2 will tend to be larger than σ_0^2 (since they will be estimates of $\sigma^2 > \sigma_0^2$).

Under H_0 the likelihood is

$$L(\widehat{\Omega}_0) = \frac{1}{(2\pi)^{n/2}} \frac{1}{\sigma_0^n} \exp\left\{-\tfrac{1}{2}\chi^2\right\}$$

In the space Ω, the likelihood is maximized either at $\hat{\sigma} = \sigma_0$ or at $\hat{\sigma} = \sigma_a$. If $\hat{\sigma} = \sigma_0$, then $\dfrac{L(\widehat{\Omega}_0)}{L(\widehat{\Omega})} = 1$. Thus, for $k < 1$, $\dfrac{L(\widehat{\Omega}_0)}{L(\widehat{\Omega})} \le k$ only if $\hat{\sigma} = \sigma_a$. In this case,

$$\frac{L(\widehat{\Omega}_0)}{L(\widehat{\Omega})} = \left(\frac{\sigma_a}{\sigma_0}\right)^n \exp\left\{-\frac{1}{2}\chi^2 + \frac{1}{2}\frac{(n-1)S_1^2 + (m-1)S_2^2}{\sigma_a^2}\right\}$$

which is a decreasing function of χ^2. Thus, we reject H_0 if χ^2 is too large. The rejection region is $\chi_2 > \chi_a^2$.

10.91 a.   In Ω the likelihood function is

$$L(\Omega) = \left(\frac{1}{\theta_1^m}\right)e^{-\Sigma x_i/\theta_1}\left(\frac{1}{\theta_2^n}\right)e^{-\Sigma y_i/\theta_2}$$

From Exercise 9.69(a), the maximum likelihood estimators of θ_1 and θ_2 are $\widehat{\theta}_1 = \overline{X}$ and $\widehat{\theta}_2 = \overline{Y}$, and

$$L(\widehat{\Omega}) = \frac{1}{\overline{X}^m \overline{Y}^n}e^{-(n+m)}$$

In the restricted space Ω_0,

$$L(\Omega_0) = \frac{1}{\theta^{m+n}}e^{-(\Sigma x_i + \Sigma y_i)/\theta},$$

and, as in Exercise 9.69, the maximum likelihood estimate of θ is

$$\widehat{\theta} = \frac{\Sigma X_i + \Sigma Y_i}{m+n} = \frac{m\overline{X} + n\overline{Y}}{m+n} \quad \text{so that} \quad L(\widehat{\Omega}_0) = \left(\frac{m\overline{X} + n\overline{Y}}{m+n}\right)^{-(m+n)}e^{-(m+n)}$$

Then

$$\lambda = \frac{L(\widehat{\Omega}_0)}{L(\widehat{\Omega})} = \frac{\overline{X}^m \overline{Y}^n}{\left(\dfrac{m\overline{X} + n\overline{Y}}{m+n}\right)^{m+n}}$$

b.   Notice that X_i has a gamma distribution with $\alpha = 1$ and $\beta = \theta_1$. Hence if H_0 is true, $\dfrac{2X_i}{\theta_1}$ has a gamma distribution with $\alpha = 1$, $\beta = 2$, which is equivalent to a χ^2 distribution with $\nu = 2$ degrees of freedom. Also, $\dfrac{2\left(\sum\limits_{i=1}^{m} Y_i\right)}{\theta}$ is the sum of the m independent χ^2 variates and has a χ^2 distribution with $2m$ degrees of freedom. Similarly, $\dfrac{2\left(\sum\limits_{i=1}^{n} Y_i\right)}{\theta}$ has a χ^2 distribution with $2n$ degrees of freedom. Since X and Y are independent, an F statistic can be formed.

$$\frac{\left(\dfrac{2\sum X_i}{2m\theta}\right)}{\left(\dfrac{2\sum Y_i}{2n\theta}\right)} = \frac{\overline{X}}{\overline{Y}}$$

has an F distribution with $2m$ and $2n$ degrees of freedom. Write

$$\lambda = \frac{\overline{X}^m\,\overline{Y}^n}{\left(\dfrac{m\overline{X} + n\overline{Y}}{m+n}\right)^{m+n}} = \frac{1}{\left[\dfrac{m\overline{X} + n\overline{Y}}{\overline{X}(m+n)}\right]^m \left[\dfrac{m\overline{X} + n\overline{Y}}{\overline{Y}(m+n)}\right]^n}$$

$$= \frac{1}{\left[\dfrac{m}{m+n} + \dfrac{n}{F(m+n)}\right]^m \left[\left(\dfrac{m}{m+n}\right)F + \dfrac{n}{m+n}\right]^n}$$

Note that λ is small if F is either too large or too small.

Then the rejection region, $\lambda \le k$, is equivalent to $F \ge c_1$ and $F \le c_2$, where c_1 and c_2 are chosen so that the test has size α.

10.93 a. Under the null hypothesis, with $\Omega_0 = \{\theta_0\}$, the likelihood is maximized at θ_0. Then, under the alternative hypothesis, with $\Omega = \{\theta_0, \theta_a\}$, the likelihood is maximized at either θ_0 or θ_a. Thus, $L(\widehat{\Omega}_0) = L(\theta_0)$ and $L(\widehat{\Omega}) = \max\{L(\theta_0, L(\theta_a)\}$, so that

$$\lambda = \frac{L(\widehat{\Omega}_0)}{L(\widehat{\Omega})} = \frac{L(\theta_0)}{\max\{L(\theta_0), L(\theta_a)\}} = \frac{1}{\max\left\{1, \dfrac{L(\theta_a)}{L(\theta_0)}\right\}}$$

b. First, recognize that

$$\lambda = \frac{1}{\max\left\{1, \dfrac{L(\theta_a)}{L(\theta_0)}\right\}} = \min\left\{1, \frac{L(\theta_0)}{L(\theta_a)}\right\}.$$

Now, as mentioned in Example 10.24, we restrict the attention to $k < 1$. Then

$$\lambda < k$$

if and only if

$$\min\left\{1, \frac{L(\theta_0)}{L(\theta_a)}\right\} < k < 1$$

if and only if

$$\frac{L(\theta_0)}{L(\theta_a)} < k.$$

c. These results imply that in the case of both simple null and alternative hypotheses, the likelihood ratio test is equivalent to the most powerful test as given by the Neymann–Pearson Lemma.

10.95 Refer to Exercise 10.94, where

$$\lambda^{2/(n_1+n_2)} = \frac{1}{1 + \dfrac{t^2}{n_1 + n_2 - 2}}$$

In the restricted space Ω_0, $\mu_1 = \mu_2$ so that $t = \sqrt{t^2}$ could be either positive or negative. Hence, small values of λ imply large positive or negative values of t, and a two-tailed t test is implied.

10.97 a. Test

$$H_0: \ \mu_1 - \mu_2 = 0 \qquad \text{vs.} \qquad H_a: \ \mu_1 - \mu_2 \neq 0$$

where μ_1 = mean nitrogen density of chemical compounds and μ_2 = mean nitrogen density of atmosphere. Then

$$s_p^2 = \frac{(n_1 - 1)s_1^2 + (n_2 - 1)s_2^2}{n_1 + n_2 - 2} = \frac{9(.002310)^2 + 8(.000574)^2}{17} = .000001064.$$

The test statistic is

$$t = \frac{\overline{y}_1 - \overline{y}_2}{\sqrt{s_p^2\left(\frac{1}{n_1} + \frac{1}{n_2}\right)}} = \frac{2.29971 - 2.310217}{\sqrt{.000001064\left(\frac{1}{10} + \frac{1}{9}\right)}} = -22.17.$$

The p-value is less than $2(.005) = .010$. Thus, there is sufficient evidence to indicate that a difference exists in the mean mass of nitrogen per flask for chemical compounds and air.

b. The 95% confidence interval is

$$\overline{y}_1 - \overline{y}_2 \pm t_{.025,\,17} \sqrt{s_p^2\left(\frac{1}{n_1} + \frac{1}{n_2}\right)}$$

$$2.29971 - 2.310217 \pm 2.110\sqrt{.000001064\left(\frac{1}{10} + \frac{1}{9}\right)} = (-.01151, -.00951).$$

c. Yes, there is sufficient evidence, since the interval does not contain 0.

d. No.

10.99 **a.** Let $p = P(\text{customer prefers brand } A)$. Then the hypothesis of interest is

$$H_0: \ p = .2 \qquad \text{vs.} \qquad H_a: \ p > .2.$$

b. It is decided to reject H_0 if $y \geq 92$. Hence

$$\alpha = P(\text{reject } H_0 | H_0 \text{ true}) = P(Y \geq 92 | p = .2)$$

If $p = .2$, then $E(Y) = np = 400(.2) = 80$ and $\sigma = \sqrt{npq} = \sqrt{64} = 8$. Using the normal approximation to the binomial distribution, we have the approximation

$$\alpha = P(Y > 91.5) = P\left(Z > \frac{91.5 - 80}{8}\right) = P(Z > 1.44) = .0749$$

10.101 The test is performed as follows:

(1) $H_0: \ \mu_1 - \mu_2 = 0 \qquad \text{vs.} \qquad H_a: \ \mu_1 - \mu_2 \doteq 0$

(2) Test statistic: $z = \dfrac{\bar{y}_1 - \bar{y}_2}{\sqrt{\left(\frac{s_1^2}{n_1}\right) + \left(\frac{s_2^2}{n_2}\right)}} = \dfrac{118 - 109}{\sqrt{\left(\frac{102}{64}\right) + \left(\frac{87}{64}\right)}} = 5.24$

(3) Rejection region: With $\alpha = .05$, the null hypothesis is rejected if $|z| > 1.96$.

(4) The null hypothesis is rejected, and we conclude that there is a difference in mean stopping time. The p-value is $P(|z| > 5.24) = 0$.

10.103 **a.** The hypothesis of interest is

$$H_0: \ \sigma_1^2 = \sigma_2^2 \qquad \text{vs.} \qquad H_a: \ \sigma_1^2 \neq \sigma_2^2$$

and the test statistic is

$$F = \frac{s_1^2}{s_2^2} = \frac{.273}{.094} = 2.904.$$

The null hypothesis will be rejected if $F > F_{9,9} = 3.18$, with $\alpha = 2(.05) = .10$. Hence, H_0 is not rejected.

b. The 90% confidence interval for σ_B^2 is

$$\frac{(n_2 - 1)s_2^2}{\chi_{.05}^2} < \sigma_B^2 < \frac{(n_2 - 1)s_2^2}{\chi_{.95}^2}$$

$$\frac{9(.094)}{16.919} < \sigma_B^2 < \frac{9(.094)}{3.32511}$$

$$.050 < \sigma_B^2 < .254.$$

Intervals constructed in this manner enclose σ_B^2 90% of the time. Hence, we are fairly certain that σ_B^2 is between .050 and .254.

10.105 **a.** The procedure is as follows:

(1) H_0: $\mu_1 - \mu_2 = 0$ vs. H_a: $\mu_1 - \mu_2 \neq 0$

(2) Calculate

$$\bar{x}_1 = \frac{585}{8} = 73.125, \qquad\qquad \bar{x}_2 = \frac{466}{6} = 77.667,$$

$$\sum (x_i - \bar{x}_1)^2 = 42{,}845 - \frac{(585)^2}{8} = 66.875,$$

$$\sum (x_i - \bar{x}_2)^2 = 36{,}246 - \frac{(466)^2}{6} = 53.3333,$$

$$s^2 = \frac{66.875 + 53.333}{8 + 6 - 2} = 10.017.$$

(3) Test statistic:

$$t = \frac{\bar{x}_1 - \bar{x}_2}{\sqrt{s^2\left(\frac{1}{n_1} + \frac{1}{n_2}\right)}} = \frac{-4.542}{\sqrt{2.9217}} = -2.657.$$

(4) The level of significance is $p = 2P[t > 2.657]$ with $8 + 6 - 2 = 12$ degrees of freedom. From Table 5, $t = 2.657$ is between $t_{.025}$ and $t_{.01}$ so that $.02 < p < .05$. Hence, H_0 will be rejected for $\alpha > .02$. If $\alpha = .01$, H_0 would not be rejected. There is evidence of a difference in mean efficiencies.

b. The 90% confidence interval for $\mu_1 - \mu_2$ is

$$(73.125 - 77.667) \pm 1.782\sqrt{10.017\left(\tfrac{1}{8} + \tfrac{1}{6}\right)} = -4.542 \pm 3.046$$

or

$$-7.588 < (\mu_1 - \mu_2) < -1.496.$$

10.107 Let $P = X + Y - W$. Then P has a normal distribution with mean $\mu_1 + \mu_2 - \mu_3$ and variance $(1 + a + b)\sigma^2$. Also, $\bar{P} = \bar{X} + \bar{Y} - \bar{W}$ has a normal distribution with

$$\mu_{\bar{P}} = \mu_1 + \mu_2 - \mu_3 \qquad \text{and} \qquad \sigma_{\bar{P}}^2 = (1 + a + b)\frac{\sigma^2}{n}$$

Hence

$$Z = \frac{(\bar{X} + \bar{Y} - \bar{W}) - (\mu_1 + \mu_2 - \mu_3)}{\sqrt{(1 + a + b)\left(\frac{\sigma^2}{n}\right)}} \text{ is a standard normal variate.}$$

Secondly, the quantities

$$\frac{\sum (X_i - \overline{X})^2}{\sigma^2}, \qquad \frac{\sum (Y_i - \overline{Y})^2}{a\sigma^2}, \qquad \frac{\sum (W_i - \overline{W})^2}{b\sigma^2}$$

are independently distributed as χ^2 variables, each with $(n-1)$ degrees of freedom, so that

$$\frac{1}{\sigma^2}\left[\sum (X_i - \overline{X})^2 + \frac{\sum (Y_i - \overline{Y})^2}{a} + \frac{\sum (W_i - \overline{W})^2}{b} \right]$$

has a χ^2 distribution with $(3n-3)$ degrees of freedom and is independent of $\overline{X}$, $\overline{Y}$, and $\overline{W}$. A t statistic can now be formed.

$$T = \frac{Z}{\sqrt{\frac{V}{\nu}}}$$

$$= \frac{(\overline{X} + \overline{Y} - \overline{W}) - (\mu_1 + \mu_2 - \mu_3)}{\left\{ \frac{1 + a + b}{n(3n-3)\left[\sum (X_i - \overline{X})^2 + \left(\frac{1}{a}\right) \sum (Y_i - \overline{Y})^2 + \left(\frac{1}{b}\right) \sum (W_i - \overline{W})^2 \right]} \right\}^{1/2}}$$

has a Student's t distribution with $(3n-3)$ degrees of freedom. The rejection region for the test is $|t| > t_{\alpha/2, 3n-3}$

10.109 In Ω we have

$$L(\Omega) = \frac{1}{\theta_1^n} e^{-\Sigma (y_i - \theta_2)/\theta_1}$$

From Exercise 9.82, the maximum likelihood estimator (MLE) for θ_2 is $\widehat{\theta}_2 = Y_{(1)}$. To find the MLE of θ_1, consider

$$\ln L = -n \ln \theta_1 - \frac{1}{\theta_1} \sum (y_i - \theta_2)$$

and

$$\frac{d \ln L}{d\theta_1} = -\frac{n}{\theta_1} + \frac{1}{\theta_1^2} \sum (y_i - \theta_2) = 0$$

or

$$\widehat{\theta}_1 = \frac{\sum (y_i - \widehat{\theta}_2)}{n}$$

In Ω_0, $\theta_1 = \theta_{1,0}$ and $\widehat{\theta}_2 = Y_{(1)}$ as before. Hence

$$\lambda = \frac{L(\widehat{\Omega}_0)}{L(\widetilde{\Omega})} = \left(\frac{\widehat{\theta}_1}{\theta_{1,0}}\right)^n \exp\left[-\frac{\sum (Y_i - Y_{(1)})}{\theta_{1,0}} + \frac{\sum (Y_i - Y_{(1)})}{\widetilde{\theta}_1} \right]$$

$$= \left[\frac{\sum (Y_i - Y_{(1)})}{n\theta_{1,0}} \right]^n \exp \left[-\frac{\sum (Y_i - Y_{(1)})}{\theta_{1,0}} + n \right]$$

Values of $\lambda \leq k$ will cause rejection of the null hypothesis.

CHAPTER 11 LINEAR MODELS AND ESTIMATION BY LEAST SQUARES

11.1 Calculate the following:

$$\sum x_i = 0 \qquad\qquad \sum y_i = 7.5 \qquad\qquad \sum x_i y_i = -6$$

$$\sum x_i^2 = 10 \qquad\qquad \sum y_i^2 = 15.25 \qquad\qquad n = 5$$

$$S_{xy} = \sum x_i y_i - \frac{1}{n}\left(\sum x_i\right)\left(\sum y_i\right) = -6$$

$$S_{xx} = \sum x_i^2 - \frac{1}{n}\left(\sum x_i\right)^2 = 10$$

Then

$$\widehat{\beta}_1 = \frac{S_{xy}}{S_{xx}} = -\frac{6}{10} = -.6 \qquad \text{and} \qquad \widehat{\beta}_0 = \overline{y} - \widehat{\beta}_1 \overline{x} = \frac{7.5}{5} - 0 = 1.5$$

The least squares straight line is $\widehat{y} = 1.5 - .6x$

The observed points and the fitted lines are shown in Figure 11.1.

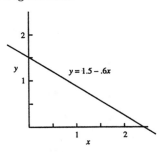

Figure 11.1

11.3 Calculate

$$\sum x_i = 36 \qquad\qquad \sum y_i = 346.9 \qquad\qquad \sum x_i y_i = 1764.4$$

$$\sum x_i^2 = 204 \qquad\qquad \sum y_i^2 = 16{,}045.29 \qquad\qquad n = 8$$

$$S_{xy} = 203.35 \qquad\qquad S_{xx} = 42$$

Then

$$\widehat{\beta}_1 = \frac{S_{xy}}{S_{xx}} = \frac{54{,}243}{54{,}714} = 4.84167$$

and

$$\widehat{\beta}_0 = \overline{y} - \widehat{\beta}_1\overline{x} = 43.3625 - 21.7875 = 21.575$$

The least squares straight line is $\widehat{y} = 21.575 + 4.842x$.

11.5   **a.**   The data are plotted in Figure 11.2.

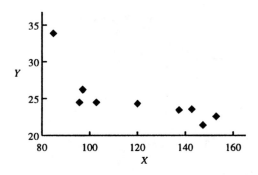

Figure 11.2

   b.   Calculate

$$\sum x_i = 1076 \qquad\qquad \sum y_i = 216 \qquad\qquad \sum x_i y_i = 25{,}431$$

$$\sum x_i^2 = 133{,}336 \qquad\qquad \sum y_i^2 = 5228 \qquad\qquad n = 9$$

$$S_{xy} = -393.0 \qquad\qquad S_{xx} = 4694.22$$

Then

$$\widehat{\beta}_1 = \frac{S_{xy}}{S_{xx}} = -.0837 \qquad \text{and} \qquad \widehat{\beta}_0 = \overline{y} - \widehat{\beta}_1\overline{x} = 24 + 10.0092 = 34.0092$$

c. The least squares line is graphed in Figure 11.3.

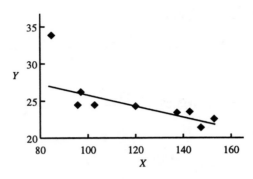

Figure 11.3

d. The estimate of y when $x = 125$ is $\hat{y} = 34.0092 - .0837(125) = 23.5467$.

11.7 $\sum\limits_{i=1}^{n} x_i y_i = 134{,}542$ $\qquad\qquad\qquad\qquad$ $\sum\limits_{i=1}^{n} x_i^2 = 53{,}514$

$\hat{\beta}_1 = \dfrac{134{,}542}{53{,}514} = 2.514$ $\qquad\qquad\qquad\qquad$ $\hat{y}_i = 2.514 x_i$

11.9 **a.** Using the model $y = \beta_0 + \beta_1 x + \epsilon$, calculate

$\sum\limits_{i=1}^{14} x_i = 86.48$ $\qquad$ $\sum\limits_{i=1}^{14} y_i = 3787$ $\qquad$ $\sum\limits_{i=1}^{14} y_i^2 = 1{,}257{,}465$

$\sum\limits_{i=1}^{14} x_i^2 = 732.4876$ $\qquad$ $\sum\limits_{i=1}^{5} x_i y_i = 17{,}562.8$ $\qquad$ $S_{xy} = -5830.04$

$S_{xx} = 198.29$ $\qquad\qquad$ $\hat{\beta}_1 = \dfrac{S_{xy}}{S_{xx}} = -29.402$

$\hat{\beta}_0 = \bar{y} - \hat{\beta}_1 \bar{x} = \dfrac{3787}{14} - (-29.402)\left(\dfrac{86.48}{14}\right) = 452.119$

The least squares line is $\hat{y} = 452.119 - 29.402x$.

b. The graph is omitted.

11.11 Since $\hat{\beta}_0 = \bar{y} - \hat{\beta}_1 \bar{x}$,

$$\text{SSE} = \sum \left[y_i - \left(\bar{y} - \hat{\beta}_1 \bar{x}\right) - \hat{\beta}_1 x_i \right]^2$$

$$= \sum (y_i - \bar{y})^2 + \hat{\beta}_1^2 \sum (x_i - \bar{x})^2 - 2\hat{\beta}_1 \sum (x_i - \bar{x})(y_i - \bar{y})$$

$$= \sum (y_i - \overline{y})^2 + \widehat{\beta}_1 \times \frac{\sum (x_i - \overline{x})(y_i - \overline{y})}{\sum (x_i - \overline{x})^2} \times \sum (x_i - \overline{x})^2$$

$$- 2\widehat{\beta}_1 \sum (x_i - \overline{x})(y_i - \overline{y})$$

$$= \sum (y_i - \overline{y})^2 - \widehat{\beta}_1 \sum (x_i - \overline{x})(y_i - \overline{y}) = S_{yy} - \widehat{\beta}_1 S_{xy}$$

11.13 **a.** Calculate

$$S_{yy} = \sum (y_i - \overline{y})^2 = 16{,}045.29 - \frac{(346.9)^2}{8} = 1002.8388$$

and

$$S_{xy} = \sum (x_i - \overline{x})(y_i - \overline{y}) = \frac{1626.8}{8} = 203.35$$

Then

$$\text{SSE} = 1002.8388 - 4.84167(203.35) = 18.2858 \quad \text{and} \quad s^2 = \frac{\text{SSE}}{6} = 3.0476$$

b. Using the coding formula given here, we obtain the y values and the corresponding x^* values shown below.

y	27.6	32.5	35.9	39.3	44.2	48.8	55.7	62.9
x^*	-7	-5	-3	-1	1	3	5	7

Then

$$\sum x^* = 0 \qquad\qquad \sum y = 346.9 \qquad\qquad \sum x^* y = 406.7$$

$$\sum x^{*2} = 168 \qquad\qquad \sum y^2 = 16{,}045.29 \qquad\qquad n = 8$$

$$S_{x^* y} = 406.7 \qquad\qquad S_{x^* x^*} = 168$$

so that

$$\widehat{\beta}_1^* = \frac{S_{xy}}{S_{xx}} = 2.42 \qquad \text{and} \qquad \widehat{\beta}_0^* = \overline{y} = 43.3625$$

The fitted model is $\widehat{y} = 43.35 + 2.42x^*$. Calculate $\sum (y_i - \overline{y})^2 = 1002.8388$, as in part (a), and

$$\text{SSE} = S_{xy} - \widehat{\beta}_1 S_{xx} = 1002.8388 - 2.42(406.7) = 18.2858$$

and

$$s^2 = \frac{\text{SSE}}{6} = 3.0476$$

as in part (a).

11.15 a. Similar to previous exercises. Calculate

$$\sum x_i = 160 \qquad \sum y_i = 106 \qquad \sum x_i y_i = 1848$$

$$\sum x_i^2 = 2880 \qquad \sum y_i^2 = 1236 \qquad n = 10$$

$$S_{xy} = 152.0 \qquad S_{xx} = 320 \qquad S_{yy} = 112.4$$

Then

$$\widehat{\beta}_1 = \frac{S_{xy}}{S_{xx}} = .475 \qquad \text{and} \qquad \widehat{\beta}_0 = \overline{y} - \widehat{\beta}\overline{x} = 10.6 - .475(1.6) = 3.000$$

b. The least squares line, $\widehat{y} = 3.000 + 4.75x$, is shown in Figure 11.6.

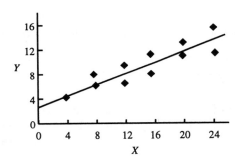

Figure 11.6

c. Calculate $S_{yy} - \widehat{\beta}_1 S_{xy} = 112.4 - .475(152) = 40.2$

and

$$s^2 = \frac{\text{SSE}}{n-2} = \frac{40.2}{8} = 5.025.$$

11.17 Calculate

$$\text{Cov}\left(\widehat{\beta}_0, \widehat{\beta}_1\right) = \text{Cov}\left(\overline{Y} - \widehat{\beta}_1\overline{x}, \widehat{\beta}_1\right) = \text{Cov}\left(\overline{Y}, \widehat{\beta}_1\right) - \overline{x}\,\text{Cov}\left(\widehat{\beta}_1, \widehat{\beta}_1\right)$$

$$= 0 - \overline{x}\,V(\widehat{\beta}_1) = \frac{-\overline{x}\sigma^2}{\sum (x_i - \overline{x})^2}$$

This will equal zero if and only if $\overline{x} = 0$ or $\sum x_i = 0$. Since $\widehat{\beta}_1$ and $\widehat{\beta}_0$ are normally distributed, $\text{Cov}\left(\widehat{\beta}_0, \widehat{\beta}_1\right) = 0$ if and only if $\widehat{\beta}_1$ and $\widehat{\beta}_0$ are independent.

11.19 **a.** Refer to Exercise 11.1. Using the information calculated there, we have

$$S_{yy} = \sum_{i=1}^{n} y_i^2 - \frac{1}{n}\left(\sum_{i=1}^{n} y_i\right)^2 = 15.25 - \frac{1}{5}(7.5)^2 = 4.0$$

$$S_{xy} = -6$$

Then

$$\text{SSE} = S_{yy} - \widehat{\beta}_1 S_{xy} = 4 - (-.6)(-6) = .4$$

and

$$s^2 = \frac{\text{SSE}}{n-2} = \frac{.4}{3} = .1333$$

The test of the hypothesis

$$H_0: \ \beta_1 = 0 \qquad \text{vs.} \qquad H_a: \ \beta_1 \neq 0$$

is the familiar t test given in Chapter 10. The test statistic is

$$t = \frac{\widehat{\beta}_1 - \beta_{10}}{s\sqrt{c_{11}}} = \frac{-.6}{\sqrt{.1333}\ \sqrt{.1}} - 5.20$$

The rejection region for $\alpha = .05$ and 3 degrees of freedom will be all values of t such that $|t| > t_{.025,3} = 3.182$. Since the calculated value of t falls in the rejection region, there is sufficient evidence to conclude that the slope β_1 differs from zero.

b. Using the methods given in Chapter 8, we can obtain a 95% confidence interval for β_1:

$$\widehat{\beta}_1 \pm t_{.025,3}\ s\sqrt{c_{11}} \qquad \text{or} \qquad -.6 \pm 3.182\sqrt{.1333}\ \sqrt{.1} \qquad \text{or} \qquad -.6 \pm .367$$

or

$$-.967 < \beta_1 \leq -.233.$$

11.21 **a.** Refer to Exercise 11.15. We wish to test

$$H_0: \ \beta_1 = 0 \qquad \text{vs.} \qquad H_a: \ \beta_1 \neq 0.$$

To carry out the test we need to estimate the value of $\sigma^2 c_{11} = V(\widehat{\beta}_1)$. From Exercise 11.15, $s^2 = \widehat{\sigma}^2 = 5.025$. From Section 11.4, $V(\widehat{\beta}_1) = \dfrac{\sigma^2}{\sum (x_i - \overline{x})^2}$.

Now

$$\sum (x_i - \overline{x}) = 2880 - \frac{(160)^2}{10} = 320 = \frac{1}{c_{11}}$$

Thus,

$$V(\widehat{\beta}_1) = \frac{\sigma^2}{320} = \frac{5.025}{30}$$

The test statistic is

$$t = \frac{\widehat{\beta}_1}{\sqrt{\dfrac{s^2}{\sum (x_i - \overline{x})^2}}} = \frac{.475}{\sqrt{\dfrac{5.025}{320}}} = 3.791$$

The *p*-value is $P(|t_8| > 3.791) < P(|t_8| > 3.355) = 2(.005) = .01$ (see Table 5).

b. We would reject H_0 when $\alpha = .05$, since the *p*-value $< .05$.

c. No. For large values of x we would expect y (the number of mistakes) to increase more and more slowly as x increases.

d. A 95% confidence interval for β_i is

$$\widehat{\beta}_1 \pm t_{.025}\, s\, \sqrt{c_{11}} \qquad \text{or} \qquad .475 \pm 2.306\sqrt{\frac{5.025}{320}} \qquad \text{or} \qquad .475 \pm .289$$

We are highly confident that β_1, the slope between x and y, is in this interval.

11.23 a. Section 11.4 of the text states the following (assuming that ϵ_i is normally distributed):

(1) $Z = \dfrac{\widehat{\beta}_i - \beta_i}{\sqrt{\sigma^2 c_{ii}}}$ has a standard normal distribution.

(2) $\dfrac{[n - (k+1)]S^2}{\sigma^2}$ has a χ^2 distribution with $n - (k+1)$ degrees of freedom.

(3) S^2 and $\widehat{\beta}_i$ are independent.

Thus, constructing the *t* statistic, we have, from Chapter 7, that

$$T = \frac{2}{\sqrt{\dfrac{\chi^2}{\nu}}} = \frac{\widehat{\beta}_i - \beta_i}{\sqrt{\left(\dfrac{S^2 \sigma^2}{\sigma^2}\right) c_{ii}}} = \frac{\widehat{\beta}_i - \beta_i}{S\sqrt{c_{ii}}}$$

has a *t*-distribution with $n - 2$ degrees of freedom.

b. Using the above T as a pivotal statistic, write

$$P(-t_{\alpha/2} < T < t_{\alpha/2}) = 1 - \alpha \qquad \text{or} \qquad P\left(-t_{\alpha/2} < \frac{\widehat{\beta}_i - \beta_i}{S\sqrt{c_{ii}}} < t_{\alpha/2}\right) = 1 - \alpha$$

or

$$P\left(\widehat{\beta}_i - t_{\alpha/2} S \sqrt{c_{ii}} < \beta_i < \widehat{\beta}_i + t_{\alpha/2} S \sqrt{c_{ii}}\right) = 1 - \alpha$$

and a $(1 - \alpha)100\%$ confidence interval for β_i is $\widehat{\beta}_i \pm t_{\alpha/2} S \sqrt{c_{ii}}$.

11.25 We know the following:

(1) $E(\widehat{\beta}_1 - \widehat{Y}_1) = \beta_1 - Y_1$.

(2) $V(\widehat{\beta}_1 - \widehat{Y}_1) = V(\widehat{\beta}_1) + V(\widehat{Y}_1) = \sigma^2\left[\dfrac{1}{S_{xx}} + \dfrac{1}{\sum (c_i - \bar{c})^2}\right]$

(3) $\widehat{\beta}_1$ and $\widehat{Y}_1$ are normally distributed.

Hence under H_0,

$$Z = \frac{((\widehat{\beta}_1 - \widehat{Y}_1) - 0}{\sqrt{\sigma^2\left[\dfrac{1}{S_{xx}} + \dfrac{1}{\sum (c_i - \bar{c})^2}\right]}}$$

is a standard normal variate. Further, from Section 11.5 we have the following:

(1) $\dfrac{(n - 2)S_1^2}{\sigma^2}$ has a χ^2 distribution with $(n - 2)$ degrees of freedom.

(2) $\dfrac{(m - 2)S_2^2}{\sigma^2}$ has a χ^2 distribution with $(m - 2)$ degrees of freedom.

(3) $\dfrac{(n - 2)S_1^2 + (m - 2)S_2^2}{\sigma^2}$ has a χ^2 distribution with $(m + n - 4)$ degrees of freedom.

Since S_1^2 is independent of $\widehat{\beta}_1$ and S_2^2 is independent of $\widehat{Y}_1$, the quantity

$$\frac{(n - 2)S_1^2 + (m - 2)S_2^2}{\sigma^2} = \frac{\text{SSE}_1 + \text{SSE}_2}{\sigma^2}$$

must be independent of $\widehat{\beta}_1 - \widehat{Y}_1$ and a t statistic can be constructed.

$$T = \frac{Z}{\sqrt{\dfrac{\chi^2}{\nu}}} = \frac{\widehat{\beta}_1 - \widehat{Y}_1}{\sqrt{\dfrac{\text{SSE}_1 + \text{SSE}_2}{m + n - 4}\left[\dfrac{1}{S_{xx}} + \dfrac{1}{\sum (c_i - \bar{c})^2}\right]}}$$

has a t distribution with $(m + n - 4)$ degrees of freedom. H_0 will be rejected for very large (negative or positive) values of T.

11.27 Using the coding $x = \dfrac{\text{year} - 1971.5}{.5}$, we obtain the following calculations:

$$\sum x = 0 \qquad\qquad \sum y = 215.9 \qquad\qquad \sum xy = -174.9$$

$$\sum x^2 = 330 \qquad\qquad \sum y^2 = 4760.43 \qquad\qquad n = 10$$

$$S_{xy} = -174.9 \qquad\qquad S_{xx} = 330 \qquad\qquad S_{yy} = 99.149$$

Then

$$\widehat{\beta}_1 = \frac{S_{xy}}{S_{xx}} = \frac{-174.9}{330} = -.53$$

and

$$\text{SSE} = 99.149 - (-.53)(-174.9) = 6.452 \qquad \text{and} \qquad s^2 = \frac{\text{SSE}}{8} = .8065$$

to test the hypothesis

$$H_0: \ \beta_1 = 0 \qquad\qquad \text{vs.} \qquad\qquad H_a: \ \beta_1 < 0$$

we use the test statistic

$$t = \frac{\widehat{\beta}_1}{\sqrt{\dfrac{s^2}{\sum (x_i - \overline{x})^2}}} = \frac{-.53}{\sqrt{\dfrac{.8065}{330}}} = -10.72$$

The rejection region with $\alpha = .05$ is $t < -1.86$ and H_0 is rejected. We conclude that the rate of tuberculosis is decreasing with time.

11.29 Using the coded x^*'s given in Exercise 11.14, we obtain

$$\widehat{\beta}_1^* = .655 \qquad\qquad\qquad s^2 = 10.97$$

$$\sum (x_i^* - \overline{x^*})^2 = 2360.2388$$

To test $H_0: \ \beta_1^* = 0$, we use the test statistic

$$t = \frac{.655}{\sqrt{\dfrac{10.97}{2360.2388}}} = 9.61$$

The rejection region is $|t| > t_{.025,\,8} = 2.306$, and H_0 is rejected. There is evidence of a linear relationship between X^* and Y (and hence between X and Y).

11.31 $V\left(a_0\widehat{\beta}_0 + a_1\widehat{\beta}_1\right) = \left(\dfrac{a_0^2 \dfrac{\sum x_i^2}{n} + a_1^2 - 2a_0a_1\overline{x}}{S_{xx}}\right)$

Letting $a_0 = 1$ and $a_1 = x^*$, we have

$$V\left(\widehat{\beta}_0 + \widehat{\beta}_1 x^*\right) = \frac{\dfrac{\sum x_i^2}{n} + (x^*)^2 - 2x^*\overline{x}}{S_{xx}} = \frac{\dfrac{\sum x_i^2 - \frac{1}{n}\left(\sum x_i\right)^2}{n} + (x^*)^2 - 2x^*\overline{x} + \overline{x}^2}{S_{xy}}$$

$$= \frac{\dfrac{S_{xx}}{n} + (x^* - \overline{x})^2}{S_{xx}} = \frac{1}{n} + \frac{(x^* - \overline{x})^2}{S_{xx}}.$$

Since $(x^* - \overline{x})^2 \geq 0$ for all x^*, $V\left(\widehat{\beta}_0 + \widehat{\beta}_1 x^*\right)$ is minimized for $(x^* - \overline{x})^2 = 0$ or $x^* = \overline{x}$.

11.33 From Exercises 11.4 and 11.14, $\widehat{y} = 7.15$ when $x = 12$; $s^2 = 10.97$; $S_{xx} = 2369.929$. With the result of Exercise 11.31, the 95% confidence interval is

$$\widehat{y} \pm t_{.025,8}\sqrt{s^2\left[\frac{1}{n} + \frac{(x^* - \overline{x})^2}{S_{xx}}\right]} \quad \text{or} \quad 7.15 \pm 2.306\sqrt{10.97\left[\frac{1}{10} + \frac{(12 - 15.504)^2}{2359.929}\right]}$$

or 7.15 ± 2.48 or 4.67 to 9.63.

11.35 Refer to Exercise 11.12. When $x = 65$, $\widehat{y} = 25.395$, and a 95% confidence interval for $E(Y)$ is

$$\widehat{y} \pm t_{.025,10}\sqrt{s^2\left[\frac{1}{12} + \frac{(65 - \overline{x})^2}{S_{xx}}\right]} \quad \text{or} \quad 25.395 \pm 2.228\sqrt{19.0333\left[\frac{1}{12} + \frac{(65 - 60)^2}{6000}\right]}$$

or 25.395 ± 2.875

11.37 **a.** Using $\widehat{\beta}_0 = \overline{Y} - \widehat{\beta}_1\overline{x}$ and $\widehat{\beta}_1$ as estimators, we have

$$\widehat{\mu}_y = \overline{Y} - \widehat{\beta}_1\overline{x} + \widehat{\beta}_1\mu_x = \overline{Y} + \widehat{\beta}_1(\mu_x - \overline{x})$$

b. Calculate

$$V(\widehat{\mu}_y) = V(\overline{Y}) + (\mu_x - \overline{x})^2 V(\widehat{\beta}_1) = \frac{\sigma^2}{n} + (\mu_x - \overline{x})^2 \times \frac{\sigma^2}{S_{xx}} = \sigma^2\left[\frac{1}{n} + \frac{(\mu_x - \overline{x})^2}{S_{xx}}\right]$$

From Exercise 11.2,

$S_{yy} = 53{,}932.9$ $\qquad\qquad\qquad$ $\text{SSE} = S_{yx} = \widehat{\beta}_1$

$$S_{xy} = 53,832.9 - (.99139)(54,243) = 56.845444$$

Thus,

$$\sigma^2 = \frac{56.845444}{8} = 7.1056805$$

and

$$\widehat{\mu}_y = 72.1 + .9913916(74 - 72) = 74.09$$

and

$$V(\widehat{\mu}_y) = 7.1056805\left[\frac{1}{10} + \frac{(74-72)^2}{54,714}\right] = .711$$

11.39 Refer to Exercises 11.3 and 11.13. When $x = 9$, $\widehat{y} = 65.15$, and the 95% prediction interval is

$$65.15 \pm 2.447\sqrt{3.05\left[1 + \frac{1}{8} + \frac{(9-4.5)^2}{42}\right]} \quad \text{or} \quad 65.15 \pm 5.42 \quad \text{or} \quad 59.73 \text{ to } 70.57$$

11.41 When $x = 12$, $\widehat{y} = 7.15$, and the 95% prediction interval is

$$7.15 \pm 2.306\sqrt{s^2\left[1 + \frac{1}{10} + \frac{(12-15.504)^2}{2359.929}\right]} \quad \text{or} \quad 7.15 \pm 8.03 \quad \text{or} \quad -.86 \text{ to } 15.18$$

11.43 Refer to Exercise 11.10. When $x = .6$, $\widehat{y} = .3933$. To calculate SSE and s^2, refer to Exercise 11.11. The 95% prediction interval is

$$.3933 \pm 2.306\sqrt{.0019394\left[1 + \frac{1}{10} + \frac{(.6-.325)^2}{.20625}\right]} \quad \text{or} \quad .3933 \pm .12 \quad \text{or} \quad .27 \text{ to } .51$$

11.45 From Exercise 11.14, $S_{xy} = 1546.459$ and $S_{yy} = 1101.1686$. Also,

$$S_{xx} = 4763.979 - \left[\frac{(155.04)^2}{10}\right] = 2359.929.$$

Then

$$r = \frac{1546.553}{\sqrt{2360.2388(1101.1686)}} = .9593$$

The hypothesis of interest is

$$H_0: \ \rho = 0 \qquad \text{vs.} \qquad H_a: \ \rho \neq 0.$$

As in Exercise 11.44, since $\rho_0 = 0$, we use

$$t = \frac{r\sqrt{n-2}}{\sqrt{1-r^2}} = \frac{.9593\sqrt{8}}{\sqrt{1-(.9593)^2}} = 9.608.$$

From Table 7, with 8 degrees of freedom, the p-value $< 2(.005) = .01$. With $\alpha = .01$. we reject H_0 and conclude that the correlation is significantly different from zero.

11.47 We know $\widehat{\beta}_1 = \dfrac{S_{xy}}{S_{xx}}$, $S = \sqrt{\left(\dfrac{1}{n-2}\right)\left(S_{yy} - \widehat{\beta}_1 S_{xy}\right)}$, and $r = \dfrac{S_{xy}}{\sqrt{S_{xx}S_{yy}}} = \widehat{\beta}_1\sqrt{\dfrac{S_{xx}}{S_{yy}}}$.

With this information,

$$T = \frac{\widehat{\beta}_1 - 0}{\frac{s}{\sqrt{S_{xx}}}} = \frac{\sqrt{S_{xx}}\,\widehat{\beta}_1\left(\sqrt{n-2}\right)}{\sqrt{S_{yy} - \widehat{\beta}_1 S_{xy}}} = \frac{\sqrt{\dfrac{S_{xx}}{S_{yy}}}\left(\widehat{\beta}_1\right)\sqrt{n-2}}{\sqrt{1 - \left(\widehat{\beta}_1\dfrac{S_{xy}}{S_{yy}}\right)}} = \frac{r\sqrt{n-2}}{\sqrt{1-r^2}}.$$

11.49 Refer to Example 11.9 and use the results given there. Then the 90% prediction interval is

$$.979 \pm 2.132(.045)\sqrt{1 + \tfrac{1}{6} + \frac{(1.5 - 1.457)^2}{.234}} \quad \text{or} \quad .979 \pm .104 \quad \text{or} \quad .875 \text{ to } 1.083.$$

11.51 a.   Let $W = \ln Y$ so that $E(W) = \ln \alpha_0 - \alpha_1 x$. After calculating individual values of W for each of the $n = 10$ observations, we have

$\sum x = 55$	$\sum w = 35.505412$	$\sum xw = 194.49729$
$\sum x^2 = 385$	$\sum w^2 = 126.07188$	$n = 10$
$S_{xw} = -.782481$	$S_{xx} = 82.5$	$S_{ww} = .008448$

Hence

$$\widehat{\beta}_1 = \frac{S_{xw}}{S_{xx}} = \frac{-.782481}{82.5} = -.00948 \qquad \text{and} \qquad \widehat{\beta}_0 = \overline{w} - \widehat{\beta}_1\overline{x} = 3.6027$$

Transforming back to the original variables, we have

$$\widehat{\alpha}_1 = -\widehat{\beta}_1 = +.0095 \qquad \text{and} \qquad \widehat{\alpha}_0 = e^{\widehat{\beta}_0} = 36.70$$

and the prediction equation is $\widehat{y} = 36.70e^{-.0095x}$.

b. In order to find a confidence interval for α_0, we first find a confidence interval for β_0 and then transform the endpoints of the interval. The least squares estimator of β_0 is $\widehat{\beta}_0$, which has variance (given in Section 11.5)

$$V(\widehat{\beta}_0) = \frac{\sigma^2 \sum x_i^2}{n \sum (x_i - \bar{x})^2}$$

It is necessary to calculate

$$\text{SSE} = S_{ww} - \widehat{\beta}_1 S_{xw} = .008448 - (-.00948)(-.782481) = .0010265$$

and

$$s^2 = \frac{\text{SSE}}{8} = .0001283.$$

Then the 90% confidence interval for β_0 is

$$\widehat{\beta}_0 \pm t_{.05,8}\sqrt{V(\widehat{\beta}_0)} \qquad \text{or} \qquad 3.6027 \pm 1.86\sqrt{.0001283\left[\frac{385}{10(82.5)}\right]}$$

or

$$3.6027 \pm .0144 \qquad \text{or} \qquad 3.5883 \text{ to } 3.6171.$$

Transforming, we have

$$3.5883 \le \beta_0 \le 3.6171 \qquad \text{or} \qquad 36.17 \le \left(e^{\beta_0} = \alpha_0\right) \le 37.23$$

11.53 Suppose $E(Y) = 1 - e^{-\beta t}$. Then $E(1 - Y) = e^{-\beta t}$ and $E[\ln(1 - Y)] = -\beta t$. Let $Y^* = \ln(1 - Y)$ and $\beta^* = -\beta$. The new equation is then $E(Y^*) = \beta^* t$ or $Y^* = \beta^* t + \epsilon$, where ϵ is normally distributed with mean 0 and variance σ^2. Using the moment of least squares, we must minimize $\text{SSE} = \sum (y^* - \widehat{\beta}^* t)^2$. Differentiating with respect to $\widehat{\beta}^*$, we have

$$\frac{d\,\text{SSE}}{d\widehat{\beta}^*} = 2 \sum t_i(y_i - \widehat{\beta}^* t_i) = 0 \quad \text{or} \quad \sum t_i y_i - \widehat{\beta}^* \sum t_i^2 = 0 \quad \text{or} \quad \widehat{\beta}^* = \frac{\sum t_i y_i}{\sum t_i^2}$$

Using this estimate of β^* with $\text{SSE} = \sum \left(y_i - \widehat{\beta}^* t_i\right)^2 = \sum y_i^2 - \widehat{\beta}^* \sum t_i y_i$, we obtain the confidence interval for $-\beta^* = \beta$:

$$-\widehat{\beta}^* \pm t_{n-2}\sqrt{\frac{s^2}{\sum t_i^2}}$$

Note that

$$V(\hat{\beta}^*) = \frac{1}{\left(\sum t_i^2\right)^2} \sum t_i^2(\sigma^2) = \frac{\sigma^2}{\sum t_i^2}$$

11.55 $X = \begin{bmatrix} 1 & -1 \\ 1 & 0 \\ 1 & 1 \\ 1 & 2 \\ 1 & 3 \end{bmatrix}$ $\quad Y = \begin{bmatrix} 3 \\ 2 \\ 1 \\ 1 \\ .5 \end{bmatrix}$ $\quad X'Y = \begin{bmatrix} 7.5 \\ 1.5 \end{bmatrix}$ $\quad X'X = \begin{bmatrix} 5 & 5 \\ 5 & 15 \end{bmatrix}$

The student may verify that

$$(X'X)^{-1} = \begin{bmatrix} .3 & -.1 \\ -.1 & .1 \end{bmatrix} \quad \text{and} \quad \hat{\beta} = (X'X)^{-1}(X'Y) = \begin{bmatrix} 2.1 \\ -.6 \end{bmatrix}$$

so that the least squares line is

$$\hat{y} = 2.1 - .6x$$

The line is shown in Figure 11.8. Notice that for this exercise $X'X$ was no longer diagonal and hence the calculation of $(X'X)^{-1}$ was a bit more tedious.

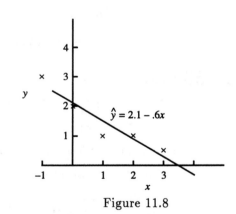

Figure 11.8

11.57 a. Using the model $y = \beta_0 + \beta_1 x + \epsilon$, calculate

$$(X'X) = \begin{bmatrix} 10 & 0 \\ 0 & 330 \end{bmatrix} \quad X'Y = \begin{bmatrix} 299.9 \\ 458.3 \end{bmatrix} \quad \hat{\beta} = (X'X)^{-1}(Y'Y) = \begin{bmatrix} 29.99000 \\ 1.3887879 \end{bmatrix}$$

and the least squares line is

$$\hat{y} = 29.99 + 1.39x.$$

b. Using the model $y = \beta_0 + \beta_1 x + \beta_2 c^2 + \epsilon$, calculate

$$X'X = \begin{bmatrix} 10 & 0 & 330 \\ 0 & 330 & 0 \\ 330 & 0 & 19{,}338 \end{bmatrix} \qquad X'Y = \begin{bmatrix} 299.9 \\ 458.3 \\ 8220.7 \end{bmatrix}$$

11.59 Since the vector a_i is a vector of k 0's and one 1 (in the jth position), we can write

$$\beta_i = a_i' \beta \qquad \text{and} \qquad \widehat{\beta}_i = a_i' \widehat{\beta}$$

in vector notation. Then with $U = a_i' \widehat{\beta}$,

$$E(\widehat{\beta}_i) = E\left(a_i' \widehat{\beta}\right) = a_i' E(\widehat{\beta}) = a_i' \widehat{\beta} = \beta_i$$

$$V(\widehat{\beta}_i) = V\left(a_i' \widehat{\beta}\right) = \left[a_i'(X'X)^{-1} a_i\right] \sigma^2$$

But

$$a_i'(X'X)^{-1} a_i = \begin{bmatrix} c_{0i} & c_{1i} & c_{2i} & \cdots & c_{ki} \end{bmatrix} \begin{bmatrix} 0 \\ 0 \\ . \\ . \\ 1 \\ . \\ . \\ 0 \end{bmatrix} = c_{ii}$$

so that

$$V(\widehat{\beta}_i) = c_{ii} \sigma^2.$$

11.61 If the minimum value is to occur at $x_0 = 1$, then this implies $\beta_1 + 2\beta_2 = 0$. To test this claim, let $a' = \begin{bmatrix} 0 & 1 & 2 \end{bmatrix}$, so that $U = a' \widehat{\beta} = \widehat{\beta}_1 + 2\widehat{\beta}_2$ and the hypothesis to be tested is

$$H_0: \ E(U) = 0 \qquad \text{vs.} \qquad H_a: \ E(U) \neq 0.$$

From Exercise 11.56, we have $\widehat{\beta}$ and $(X'X)^{-1}$, with $s^2 = .14285$. Then

$$a'(X'X)^{-1}a = \begin{bmatrix} -.095238 & .035714 & .02381 \end{bmatrix} \begin{bmatrix} 0 \\ 1 \\ 2 \end{bmatrix} = .083334$$

and $U = \hat{\beta}_1 + 2\hat{\beta}_2 = .142861$. The test statistic is

$$t = \frac{U - E(U)}{\sqrt{s^2[a'(X'X)^{-1}a]}} = \frac{.142861}{\sqrt{(.14285)(.08333)}} = 1.31$$

The rejection region with $\alpha = .05$ is $|t| > 2.776$, and H_0 is not rejected.

11.63 When $T_1 = 50$, $P = 20$, $C = 1$, and $T_2 = 200$, we have $x_1 = -1$, $x_2 = 1$, $x_3 = -1$, and $x_4 = 1$, so that

$$a' = \begin{bmatrix} 1 & -1 & 1 & -1 & 1 \end{bmatrix} \quad \text{and the} \quad a'(X'X)^{-1}a = \frac{5}{16} = .3125$$

The estimate of $E(Y)$ at this particular setting is

$$\hat{y} = 21.125 + 3.1375 - 1.2125 + .1625 - 1.275 = 21.9375$$

and the confidence interval is

$$\hat{y} \pm t_{\alpha/2}s\sqrt{a'(X'X)^{-1}a} \quad \text{or} \quad 21.94 \pm 1.796\sqrt{8.98}\sqrt{.3125} \quad \text{or} \quad 21.94 \pm 3.01$$

or

18.93 to 24.95.

11.65 Refer to Exercises 11.62 and 11.63. For the given levels

$$\hat{y} = 21.9375 \quad \text{and} \quad a'(X'X)^{-1}a = .3135.$$

Hence the 90% prediction interval is

$$21.94 \pm 1.796\sqrt{8.98(1.3135)} \quad \text{or} \quad 21.94 \pm 6.17 \quad \text{or} \quad 15.77 \text{ to } 28.11$$

11.67 First, note that with $k = 1$, $\text{SSE}_R = S_{yy}$. Then, $g = 0$, and $\text{SSE}_c = S_{yy} - \hat{\beta}_1 S_{xy}$.

$$F = \frac{(\text{SSE}_R - \text{SSE}_c)(k - g)}{\dfrac{\text{SSE}_c}{n - (k + 1)}} = \frac{S_{yy} - S_{yy} + \hat{\beta}_1 S_{xy}}{\dfrac{\text{SSE}_c}{n - 2}} = \frac{\hat{\beta}_1 S_{xy}}{\dfrac{\text{SSE}_c}{n - 2}} = \frac{\dfrac{\hat{\beta}_1 S_{xy}}{S_{xy}}}{\dfrac{s^2}{S_{xx}}} = \frac{\hat{\beta}_1^2}{\dfrac{s^2}{S_{xx}}}$$

$$= t^2$$

In Exercise 11.66, $F = 114.94$, and in Exercise 11.27, $t = -10.72$. Thus, $t^2 = 1$ which equals F (except for roundoff error).

11.69 The hypothesis of interest is H_0: $\beta_1 - \beta_4 = 0$. For the complete model,

$$Y = \beta_0 + \beta_1 x_1 + \beta_2 x_2 + \beta_3 x_3 + \beta_4 x_4 + \epsilon, \qquad SSE_c = 98.8125 \text{ with 11 d.f.}$$

For the reduced model, the 2nd and 5th columns of the X matrix are deleted, so that

$$(X'X)^{-1} = \begin{bmatrix} \frac{1}{16} & & \\ & \frac{1}{16} & \\ & & \frac{1}{16} \end{bmatrix} \qquad X'Y = \begin{bmatrix} 338 \\ -19.4 \\ -2.6 \end{bmatrix}$$

Then

$$SSE = Y'Y - \widehat{\beta}'X'Y = 7446.52 - 7164.195 = 282.325$$

with 13 d.f. The test statistic is

$$F = \frac{\dfrac{SSE_R - SSE_c}{2}}{\dfrac{SSE_c}{11}} = 10.21$$

The rejection region with $\alpha = .05$ is $F > F_{2,11} = 3.98$, and H_0 is rejected. There is reason to believe that either T_1 or T_2 or both affect the yield.

11.71 Refer to Example 11.13. For the reduced model, $S^2 = \dfrac{326.623}{8} = 40.83$,

$$(X'X)^{-1} = \begin{bmatrix} \frac{1}{11} & & \\ & \frac{1}{11} & \\ & & \frac{1}{11} \end{bmatrix} \qquad a' = \begin{bmatrix} 1 & 1 & -1 \end{bmatrix}$$

Then $\widehat{y} = a'\widehat{\beta} = 93.73 + 4.00 - 7.35 = 90.38$, $a'(X'X)^{-1}a = .31620$, and the 95% confidence interval for $E(Y)$ is

$$\widehat{y} \pm t_{.025}\sqrt{s^2\, a'(X'X)^{-1}a} \qquad \text{or} \qquad 90.38 \pm 2.306\sqrt{40.83(.32620)}$$

or

$$90.38 \pm 8.42 \qquad \text{or} \qquad 81.96 \text{ to } 98.80.$$

11.73 **a.** Calculate

$$\sum x = -2682.8 \qquad \sum y = 6.826 \qquad \sum xy = -1847.0073$$

$$\sum x^2 = 720{,}039.3 \qquad \sum y^2 = 5.6326 \qquad n = 10$$

$$S_{xy} = 15.728 \qquad S_{xx} = 297.716 \qquad S_{yy} = .9732$$

Then

$$\widehat{\beta}_1 = \frac{S_{xy}}{S_{xx}} = -.053 \quad \text{and} \quad \widehat{\beta}_0 = \frac{6.826}{10} - \widehat{\beta}_1\left(\frac{-2682.8}{10}\right) = -13.54$$

The least squares line is $\widehat{y} = -13.54 - .053x$.

b. Note that

$$\text{SSE} = S_{yy} - \widehat{\beta}_1 S_{xy} = .9732 - (-.053)(-15.72802) = .14225$$

and

$$s^2 = \frac{.14225}{8} = .01778$$

Also,

$$V(\widehat{\beta}_1) = \frac{\sigma^2}{\sum (x_i - \overline{x})^2} = \frac{s^2}{297.716}. \quad \text{Thus,}$$

$$t = \frac{\widehat{\beta}_1}{\sqrt{V(\widehat{\beta}_1)}} = \frac{-.053}{\sqrt{\frac{.01778}{297.716}}} = -6.84$$

With $\alpha = .01$, we reject H_0 if $t < -2.896$. Thus, we reject H_0.

c. With $x = -273$, $\widehat{y} = -13.54 + .053(.273) = .93$. A 95% prediction interval is

$$\widehat{y} \pm t_{.025}\, s\sqrt{1 + \frac{1}{n} + \frac{(x_0 - \overline{x})^2}{\sum (x_i - \overline{x})^2}}$$

or

$$.93 \pm 2.306\sqrt{.01778}\sqrt{1 + \frac{1}{10} + \frac{(273 - 268.28)^2}{297.716}}$$

or

$$.93 \pm .33$$

11.75 a.

$$X = \begin{bmatrix} 1 & -3 & 5 & -1 \\ 1 & -2 & 0 & 1 \\ 1 & -1 & -3 & 1 \\ 1 & 0 & -4 & 0 \\ 1 & 1 & -3 & -1 \\ 1 & 2 & 0 & -1 \\ 1 & 3 & 5 & 1 \end{bmatrix} \qquad Y = \begin{bmatrix} 1 \\ 0 \\ 0 \\ 1 \\ 2 \\ 3 \\ 3 \end{bmatrix} \qquad X'Y = \begin{bmatrix} 10 \\ 14 \\ 10 \\ -3 \end{bmatrix}$$

$$X'X = \begin{bmatrix} 7 & 0 & 0 & 0 \\ 0 & 28 & 0 & 0 \\ 0 & 0 & 84 & 0 \\ 0 & 0 & 0 & 6 \end{bmatrix} \qquad (X'X)^{-1} = \begin{bmatrix} \frac{1}{7} & 0 & 0 & 0 \\ 0 & \frac{1}{28} & 0 & 0 \\ 0 & 0 & \frac{1}{84} & 0 \\ 0 & 0 & 0 & \frac{1}{6} \end{bmatrix}$$

$$\hat{\beta} = (X'X)^{-1}X'Y = \begin{bmatrix} 1.4285 \\ .5000 \\ .1190 \\ -.5000 \end{bmatrix}$$

and the fitted model is $\hat{y} = 1.4825 + .5000x_1 + .1190x_2 - .5000x_3$.

b. When $x_1 = 1$, $x_2 = -3$, and $x_3 = -1$, the predicted value of y is

$$\hat{y} = 1.4285 + .5000 - .3570 + .5000 = 2.0715$$

whereas the observed response at this setting was $y = 2$. The difference appears because the former is a predicted value based on $n = 7$ points, while the latter is the actual observed response.

c. Calculate

$$\text{SSE} = Y'Y - \hat{\beta}'X'Y = 24 - 23.9757 = .0243$$

and

$$s^2 = \frac{\text{SSE}}{n-4} = \frac{.0243}{3} = .008$$

In order to test the hypothesis

$$H_0: \ \beta_3 = 0 \qquad \text{vs.} \qquad H_a: \ \beta_3 \neq 0,$$

we use the test statistic

$$t = \frac{\hat{\beta}_3 - \beta_3}{s\sqrt{c_{33}}} = \frac{-.5000}{\sqrt{.008\left(\frac{1}{6}\right)}} = \frac{-.5000}{.0365} = -13.7$$

The rejection region, with $\alpha = .05$ and 3 degrees of freedom, will be $|t| > t_{.025,3}$ = 3.182, and the null hypothesis is rejected.

d. Refer to Section 11.8 of the text and write

$$\hat{Y} = U = \sum_{i=0}^{k} a_i \hat{\beta}_i$$

In this exercise, $\hat{Y} = U = \hat{\beta}_0 + \hat{\beta}_1 - 3\hat{\beta}_2 - \hat{\beta}_3$ and $a' = \begin{bmatrix} 1 & 1 & -3 & -1 \end{bmatrix}$. A 95% confidence interval for $E(Y)$ is given by

$$\hat{Y} \pm t_{\alpha/2} S \sqrt{a'(X'X)^{-1}a}$$

where

$$a'(X'X)^{-1}a = \begin{bmatrix} \frac{1}{7} & \frac{1}{28} & -\frac{3}{84} & -\frac{1}{6} \end{bmatrix} \begin{bmatrix} 1 \\ 1 \\ 1 \\ -3 \\ -1 \end{bmatrix} = .45238$$

Hence the 95% confidence interval is

$$2.0715 \pm 3.182\sqrt{.008}\ \sqrt{.45238} \qquad \text{or} \qquad 2.07 \pm .19$$

e. Refer to Section 11.9. The 95% prediction interval for Y will be

$$y \pm t_{\alpha/2} s\sqrt{1 + a'(X'X)^{-1}a} \qquad \text{or} \qquad 2.07 \pm 3.182\sqrt{.008}\ \sqrt{1.45238}$$

or

$$2.07 \pm .34$$

11.77 It is known that

$$V(\hat{\beta}_1) = \frac{\sigma^2}{\sum_{i=1}^{n}(X_i - \overline{X})^2}$$

This will be minimized when $\sum_{i=1}^{n} (X_i - \overline{X})^2$ is maximized. However, $\sum_{i=1}^{n} (X_i - \overline{X})^2$ will be maximized when all the X_i are as far away from $\overline{X}$ as possible. That is, take $\frac{n}{2}$ at $X = -9$ and $\frac{n}{2}$ at $X = 9$.

11.79 a. $Y = \begin{bmatrix} 8.0 \\ 9.0 \\ 9.1 \\ 10.2 \\ 10.4 \\ 10.0 \\ 10.3 \\ 12.2 \\ 12.6 \\ 13.9 \end{bmatrix}$ $X = \begin{bmatrix} 1 & 0 & -2 & 0 \\ 1 & 0 & -1 & 0 \\ 1 & 0 & 0 & 0 \\ 1 & 0 & 1 & 0 \\ 1 & 0 & 2 & 0 \\ 1 & 1 & -2 & -2 \\ 1 & 1 & -1 & -1 \\ 1 & 1 & 0 & 0 \\ 1 & 1 & 1 & 1 \\ 1 & 1 & 2 & 2 \end{bmatrix}$ $X'Y = \begin{bmatrix} 105.7 \\ 59.0 \\ 16.1 \\ 10.1 \end{bmatrix}$

$$X'X = \begin{bmatrix} 10 & 5 & 0 & 0 \\ 5 & 5 & 0 & 0 \\ 0 & 0 & 20 & 10 \\ 0 & 0 & 10 & 10 \end{bmatrix}$$

The student may verify that

$$(X'X)^{-1} = \begin{bmatrix} .2 & -.2 & 0 & 0 \\ -.2 & .4 & 0 & 0 \\ 0 & 0 & .1 & -.1 \\ 0 & 0 & -.1 & .2 \end{bmatrix} \quad \text{and} \quad \widehat{\beta} = \begin{bmatrix} 9.34 \\ 2.46 \\ .60 \\ .46 \end{bmatrix}$$

b. The observed points and the two growth lines are shown in Figure 11.10. They are

$$\widehat{y}_A = \widehat{\beta}_0 + \widehat{\beta}_2 x_2 = 9.34 = .60x_2$$

and

$$\widehat{y}_B = (\widehat{\beta}_0 + \widehat{\beta}_1) + (\widehat{\beta}_2 + \widehat{\beta}_3)x_3$$
$$= 11.80 + 1.01x_2$$

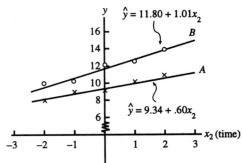

c. If we are interested in the growth of bacteria A, then $x_1 = 0$, $x_2 = 0$, $x_1 x_2 = 0$, and the prediction is

$$\widehat{y} = \widehat{\beta}_0 = 9.34$$

For bacteria B, $x_1 = 1$, $x_2 = 0$, $x_1 x_2 = 0$ and

$$\widehat{y} = \widehat{\beta}_0 + \widehat{\beta}_1 = 9.34 + 2.46 = 11.80$$

The observed growths for bacteria A and

B were 9.1 and 12.2, respectively.

Figure 11.10

d. Calculate

$$\text{SSE} = \boldsymbol{Y'Y} - \widehat{\boldsymbol{\beta}}'\boldsymbol{X'Y} = 1146.91 - 1146.179 = .731$$

and

$$s^2 = \frac{\text{SSE}}{n - (k+1)} = \frac{.731}{6} = .121833 \cdot$$

Refer to Figure 11.10. The difference in the rates of growth for the two types of bacteria is represented by the difference in the slopes of the two lines. Since for bacteria A the model is $y = \beta_0 + \beta_2 x_2 + \epsilon$ and for bacteria B the model is $y = (\beta_0 + \beta_1) + (\beta_2 + \beta_3)x_2 + \epsilon$, the growth rates will be different only if $\beta_3 \neq 0$. Hence it is necessary to test

$$H_0: \ \beta_3 = 0 \qquad\qquad \text{vs.} \qquad\qquad H_a: \ \beta_3 \neq 0$$

using the test statistic

$$t = \frac{\widehat{\beta}_3 - \beta_3}{s\sqrt{c_{33}}} = \frac{.41}{\sqrt{.121833}\ \sqrt{175}} = 2.63$$

With $\alpha = .05$ and 6 degrees of freedom, the rejection region will be $|t| > t_{.025,6} = 2.447$. The null hypothesis is rejected, and there is sufficient evidence to detect a difference in the growth rates.

e. It is given that $x_1 = 1$ and $x_2 = 1$, so that $\widehat{y} = 12.81$,

$$\boldsymbol{a}' = \begin{bmatrix} 1 & 1 & 1 & 1 \end{bmatrix} \quad \text{and} \quad \boldsymbol{a}'(\boldsymbol{X'X})^{-1}\boldsymbol{a} = \begin{bmatrix} 0 & \frac{1}{5} & 0 & \frac{1}{10} \end{bmatrix} \begin{bmatrix} 1 \\ 1 \\ 1 \\ 1 \end{bmatrix} = \frac{3}{10}$$

A 90% confidence interval for $E(Y)$ at this particular setting will be

$$y \pm t_{\alpha/2}\, s\sqrt{a'(X'X)^{-1}a} \quad \text{or} \quad 12.81 \pm 1.943\sqrt{.121833}\,\sqrt{\tfrac{3}{10}} \quad \text{or} \quad 12.81 \pm .37$$

f. Refer to part (e). A 90% prediction interval for growth of bacteria B at time $x_2 = 1$ is

$$y \pm t_{\alpha/2}\, s\sqrt{a'(X'X)^{-1}a} \quad \text{or} \quad 12.81 \pm 1.943\sqrt{.121833}\,\sqrt{1.3} \quad \text{or} \quad 12.81 \pm .78$$

11.81 Let

$$Y' = \begin{bmatrix} Y_1 & Y_2 & Y_3 & \cdots & Y_n \end{bmatrix} \quad \text{and} \quad l' = \begin{bmatrix} 1 & 1 & 1 & \cdots & 1 \end{bmatrix}$$

be two $1 \times n$ vectors. Then we can write

$$\overline{Y} = \begin{bmatrix} \tfrac{1}{n} & \tfrac{1}{n} & \cdots & \tfrac{1}{n} \end{bmatrix} \begin{bmatrix} Y_1 \\ Y_2 \\ Y_3 \\ \vdots \\ Y_n \end{bmatrix} = \tfrac{1}{n} l' Y$$

In matrix form, the equation of interest is $Y = x'\widehat{\beta}$, where

$$x' = \begin{bmatrix} 1 & x_1 & x_2 & \cdots & x_k \end{bmatrix} \quad \text{and} \quad \widehat{\beta}' = \begin{bmatrix} \widehat{\beta}_0 & \widehat{\beta}_1 & \cdots & \widehat{\beta}_k \end{bmatrix}$$

Suppose $Y = \overline{Y}$. Then

$$\overline{Y} = x'\widehat{\beta} = x'(X'X)^{-1}X'Y \quad \text{or} \quad \tfrac{1}{n} l'YY' = x'(X'X)^{-1}X'YY'$$

or

$$\tfrac{1}{n} l' = x'(X'X)^{-1}X' \quad \text{or} \quad \tfrac{1}{n} l'X = x'(X'X)^{-1}X'X \quad \text{or} \quad \tfrac{1}{n} l'X = x'$$

or

$$x' = \begin{bmatrix} 1 & \overline{x}_1 & \overline{x}_2 & \cdots & \overline{x}_k \end{bmatrix}.$$

That is, the point $(\overline{x}_1, \overline{x}_2, \ldots, \overline{x}_k, \overline{Y})$ satisfies the equation $Y = x'\widehat{\beta}$, so that the least squares prediction line must pass through this point.

11.83 a. Let $S_{xy} = \sum (x_i - \overline{x})(y_i - \overline{y})$. Write

$$\widehat{\beta}_1 = \frac{S_{xy}}{S_{xx}} = \frac{S_{xy}}{\sqrt{S_{xx}S_{yy}}}\sqrt{\frac{S_{yy}}{S_{xx}}} = r\sqrt{\frac{S_{yy}}{S_{xx}}}$$

b. The student may verify that if X and Y have a bivariate normal distribution as defined in Section 5.10 of the text, then the conditional distribution of Y, given $X = x$, will be normal with mean

$$E(Y|X = x) = \mu_y + \rho \frac{\sigma_y}{\sigma_x}(x - \mu_x) \quad \text{and variance} \quad V(Y|X = x) = \sigma_y^2(1 - \rho^2)$$

This proof may be found in several texts, one of which is Hogg and Craig, *Introduction to Mathematical Statistics*, in the chapter on "The Bivariate Normal Distribution." For the model given in Exercise 11.21, which gives the expected value of Y_i for the fixed $X_i = x_i$,

$$E(Y_i|X_i = x_i) = \beta_0 + \beta_1 x_i$$

Hence we must have $\beta_1 = \rho\left(\frac{\sigma_y}{\sigma_x}\right)$, $\beta_0 = \mu_y - \beta_1 \mu_x$, and, if $\rho = 0$, $\beta_1 = 0$. In Exercise 11.29 we showed in general that

$$T = \frac{\hat{\beta}_i - \beta_i}{S\sqrt{c_{ii}}}$$

has a t distribution with $n - (k + 1)$ degrees of freedom. In particular, for $k = 1$ and $i = 1$,

$$\frac{\hat{\beta}_1 - \beta_1}{S\sqrt{\frac{1}{S_{xx}}}}$$

has a t distribution with $(n - 2)$ degrees of freedom. Finally, if $\rho = 0$, then $\beta_1 = 0$ and the t statistic becomes

$$T = \frac{\hat{\beta}_1}{S\sqrt{\frac{1}{S_{xx}}}} = \frac{\hat{\beta}_1}{\sqrt{\frac{SSE}{(n-2)}}\sqrt{\frac{1}{S_{xx}}}} = \frac{\hat{\beta}_1\sqrt{(n-2)S_{xx}}}{\sqrt{(1-r^2)S_{yy}}}$$

since

$$SSE = (1 - r^2)S_{yy}.$$

c. Since $\hat{\beta}_1 = r\sqrt{\frac{S_{yy}}{S_{xx}}}$, the t statistic in part (b) becomes

$$T = \frac{r\sqrt{n-2)}}{\sqrt{1-r^2}}$$

and it has a t distribution with $(n - 2)$ degrees of freedom. Since the t distribution depends only on $\nu = n - 2$ and not on the particular value of X_i, the distribution of T will be the same no matter what values of X_i are observed. Hence T has the same distribution unconditionally.

11.85 The necessary computations are

$$\sum x = 741 \qquad\qquad \sum y = 108 \qquad\qquad \sum xy = 10{,}016.3$$

$$\sum x^2 = 68{,}789 \qquad\qquad \sum y^2 = 1459.34 \qquad\qquad S_{xx} = 153.875$$

$$S_{yy} = 1.34 \qquad\qquad S_{xy} = 12.8$$

a. $\quad r = \dfrac{12.8}{\sqrt{153.875(1.34)}} = .89$

b. We need to test

$$H_0: \ \beta_1 = 0 \qquad\qquad \text{vs.} \qquad\qquad H_a: \ \beta_1 \neq 0.$$

Now

$$\widehat{\beta}_1 = \frac{12.8}{153.875} = .0832$$

Also,

$$\text{SSE} = S_{yy} - \widehat{\beta}_1 S_{xy} = 1.34 - .0832(12.8) = .275$$

so

$$s^2 = \frac{.275}{6} = .046.$$

The test statistic is

$$t = \frac{.0832}{\sqrt{\dfrac{.046}{153.875}}} = 4.81$$

For $\alpha = .05$ the rejection region is $|t| > t_{.025,\,6} = 2.447$. Thus, we reject H_0. Siance $t = 4.81 > 3.707$, the p-value is less than $2(.005) = .01$.

CHAPTER 12 CONSIDERATIONS IN DESIGNING EXPERIMENTS

12.1 See Section 12.2 and suppose that $\sigma_1^2 = 9$, $\sigma_2^2 = 25$, $n = 90$. From Example 12.1, $V(\overline{Y}_1 - \overline{Y}_2)$ will be minimized when

$$n_1 = \left(\frac{\sigma_1}{\sigma_1 + \sigma_2}\right)n = \left(\frac{3}{3+5}\right)90 = 33.75 \text{ or } 34 \qquad \text{and} \qquad n_2 = 90 - 34 = 56$$

12.3 From Exercise 12.1 we know that $n_1 = \left(\frac{3}{8}\right)n$ and $n_2 = \left(\frac{5}{8}\right)n$, where n is the combined sample size. The length of a 95% confidence interval for $(\mu_1 - \mu_2)$ is twice the halfwidth, or

$$2(1.96)\sqrt{\frac{\sigma_1^2}{n_1} + \frac{\sigma_2^2}{n_2}}$$

This width must be equal to two units. Hence

$$2 = 2(1.96)\sqrt{\frac{9}{\left(\frac{3}{8}\right)n} + \frac{25}{\left(\frac{5}{8}\right)n}} \qquad \text{or} \qquad 1 = (1.96)^2\left[\frac{\left(\frac{45}{8}\right) + \left(\frac{75}{8}\right)}{\left(\frac{15}{64}\right)n}\right]$$

or

$$1 = (1.96)^2\left[\frac{120(8)}{15n}\right]$$

12.5 Refer to Section 12.2. The maximum quantity of information, or the minimum value for $\sigma_{\widehat{\beta}_1}$, occurs when the six data points are equally divided, with half located at $x = 2$ and half located at $x = 5$. Notice that this allocation of rats provides no information on curvature of the response curve, and it is necessary that the experimenter be almost certain that the response is linear. Again, we are employing the principle of signal amplification, since we are concerned with the selection of treatments (dosages) and the number of experimental units (rats) to be assigned to each dosage.

12.7 Refer to Exercise 12.5. If all the experimental units are assigned to the points $x = 2$ and $x = 5$, the experimenter has no way to determine whether or not the response function is truly linear over the experimental region. By assigning one or two points at $x = 3.5$, the experimenter can check for curvature in the response function.

12.9 **a.** Because the two halves come from the same sample, the percentage of iron ore in each should be similar. If both methods work reasonably well, then each will be measuring similar amounts. Therefore, the observations will be positively correlated.

b. Either analysis would have worked. However, the paired analysis requires fewer assumptions, removes the unwanted variability caused by the inherent differences in the ore samples, and produces a smaller standard error because of the positive correlation discussed in part (a).

12.11 a. $\sigma^2_{(\overline{Y}_1 - \overline{Y}_2)}$ equals $\sigma^2_{\overline{D}}$ if $(2\rho\sigma_1\sigma_2) = 0$. This will happen if $\rho = 0$, i.e., if the two samples are uncorrelated.

b. $\sigma^2_{(\overline{Y}_1 - \overline{Y}_2)}$ is less than $\sigma^2_{\overline{D}}$ if $2\rho\sigma_1 s_2 < 0$. This happens if $\rho < 0$, i.e., if the two samples are negatively correlated.

c. It would be better to implement the paired experiment if the samples are positively correlated and the independent samples experiment when the samples are uncorrelated or negatively correlated. Note that if the samples are uncorrelated and the variances are equal, then independent samples yield about double the degrees of freedom.

12.13 a. A paired-difference test is used since the samples are not independent. The hypothesis of interest is

$$H_0: \ \mu_A - \mu_B = 0 \qquad \text{vs.} \qquad H_a: \ \mu_A - \mu_B \neq 0.$$

The differences are

$$d_i = 2, \ 1, \ 0, \ 3, \ -1, \ 2, \ 4, \ 1.$$

Then

$$\sum d_i = 12 \qquad \text{and} \qquad \sum d_i^2 = 36$$

$$\overline{d} = \frac{\sum d_i}{n} = 1.5 \qquad \text{and} \qquad s_D^2 = \frac{\sum d_i^2 - \dfrac{\left(\sum d_i^2\right)}{n}}{n-1} = \frac{36 - 18}{7} = 2.571$$

The test statistic is

$$t = \frac{\overline{d} - 0}{\dfrac{s_D}{\sqrt{n}}} = \frac{1.5}{\sqrt{\dfrac{2.571}{8}}} = 2.65$$

The rejection region is based on a t distribution with $n - 1 = 7$ degrees of freedom, and is $|t| > 2.365$ with $\alpha = .05$. The null hypothesis is rejected, and we conclude that there is a difference between the two machines.

b. Note that the variation among technicians is very high (some high and some low), while the variation with technicians between machines is not. Hence pairing is important to screen out the variation among technicians.

c. The population of differences is normally distributed with variance σ_D^2. A random sample of n differences has been selected from this population.

12.15 a. $E(Y_{ij}) = \mu_i + E(U_j) + E(\epsilon_{ij}) = \mu_i.$

b. The Y_{ij} are not normally distributed because the sum of a uniform random variable and a normal random variable is not normal.

c. $\begin{aligned} \text{Cov}\,(Y_{2j}, Y_{2j}) &= \text{Cov}\,(\mu_1 + u_j + \epsilon_{ij},\ \mu_2 + u_j + \epsilon_{2j}) \\ &= \text{Cov}\,(u_j, u_j) + \text{Cov}\,(\epsilon_{1j},\ \epsilon_{2j}) + \text{Cov}\,(u_j,\ \epsilon_{2j}) + \text{Cov}\,(u_j,\ \epsilon_{2j}) \\ &= V(u_j) + 0 + 0 + 0 = \tfrac{1}{3} \end{aligned}$

d. $D_j = Y_{1j} - Y_{2j} = \mu_1 + u_j + \epsilon_{1j} - (\mu_2 + u_j + \epsilon_{2j}) = \mu_1 - \mu_2 + \epsilon_{1j} - \epsilon_{2j}.$

Since ϵ_{1j} and ϵ_{2j} are independent normals, D_j is also normal. Finally, for $j \neq j'$,

$\begin{aligned} \text{Cov}\,(D_j, D_{j'}) &= \text{Cov}\,(\mu_1 - \mu_2 + \epsilon_{1j} - \epsilon_{2j},\ \mu_1 - \mu_2 + \epsilon_{1j'} - \epsilon_{2j'}) \\ &= \text{Cov}\,(\epsilon_{1j} - \epsilon_{2j},\ \epsilon_{1j'} - \epsilon_{2j'}) = D \end{aligned}$

since the ϵ_{ij} are independent.

Thus, the D_j are independent, normally distributed random variables.

e. Any choice of nonnormal distribution for the u_j will work as long as $E(u_j) = 0$ and $V(u_j) = \sigma_u^2 > 0$.

12.17 Use Table 12 and see Section 12.4 of the text.

12.19 Randomization avoids the possibility of bias introduced by a nonrandom selection of sample elements. Also, it provides a probabilistic basis for the selection of a sample (see Section 12.4).

12.21 A treatment is a specific combination of factor levels.

12.23 The parts of the experimental design that would increase the accuracy of the experiment are (a) the selection of treatments and (b) choice of the number of experimental units to be assigned to each treatment. The part of the design that would decrease the impact of extraneous sources of variability is the method of assigning treatments to the experimental units.

12.25 The treatments are assigned so that each treatment appears in each row and column exactly once. Hence, for this exercise the assignment of treatments is as follows:

B	A	C
C	B	A
A	C	B

12.27 A random sample is a sample in which every possible sample from the population has an equal probability of being selected.

12.29 Given the model proposed in this exercise, we have the following:

a. $E(Y_{ij}) = E(\mu_i + P_j + \epsilon_{ij}) = \mu_i + E(P_j) + E(\epsilon_{ij}) = \mu_i + 0 + 0 = \mu_i$

b. $\overline{Y}_i$ is the mean of the n observations receiving treatment i. That is,

$$\overline{Y}_i = \frac{\sum_{j=1}^{n} Y_{ij}}{n} \quad \text{and} \quad E(\overline{Y}_i) = \frac{\sum_{j=1}^{n} E(Y_{ij})}{n} = \frac{n\mu_i}{n} = \mu_i$$

Further,

$$V(\overline{Y}_i) = \frac{\sum_{j=1}^{n} V(Y_{ij})}{n^2} = \frac{\sum_{j=1}^{n} V(\mu_i + P_j + \epsilon_{ij})}{n^2} = \frac{\sum_{j=1}^{n} [V(P_j) + V(\epsilon_{ij})]}{n^2}$$

$$= \frac{n\left(\sigma_p^2 + \sigma^2\right)}{n^2} = \frac{\sigma_p^2 + \sigma^2}{n}$$

Notice that P_j and ϵ_{ij} are independent random variables for all i and j.

c. Let $\overline{D} = \overline{Y}_1 - \overline{Y}_2$. Then

$$E(\overline{D}) = E(\overline{Y}_1) - E(\overline{Y}_2) = \mu_1 - \mu_2$$

The variance of $\overline{D}$ can be most easily seen by noting that

$$\overline{D} = \frac{\sum_{j=1}^{n} (Y_{1j} - Y_{2j})}{n} = \frac{\sum_{j=1}^{n} [\mu_1 - \mu_2 + (P_j - P_j) + (\epsilon_{1j} - \epsilon_{2j})}{n}$$

$$= \frac{n}{n}(\mu_1 - \mu_2) + \frac{\sum_{j=1}^{n} \epsilon_{1j}}{n} - \frac{\sum_{j=1}^{n} \epsilon_{2j}}{n} = (\mu_1 - \mu_2) + \overline{\epsilon}_1 - \overline{\epsilon}_2$$

where $\overline{\epsilon}_i$ is the mean of the random errors associated with experimental units receiving treatment i. Then

$$V(\overline{D}) = V(\overline{\epsilon}_1) + V(\overline{\epsilon}_2) = \frac{V(\epsilon_{1j})}{n} + \frac{V(\epsilon_{2j})}{n} = \frac{2\sigma^2}{n}$$

Finally, since $\overline{D}$ is a linear combination of the Y_{ij}, which are normally distributed, $\overline{D}$ is normally distributed.

12.31 In this situation the model must be written as $Y_{ij} = \mu_i + P_{ij} + \epsilon_{ij}$ since the pair effect will differ from one of the $2n$ observations to another. That is, for a fixed pair j, the pair effect P_{ij} may be different for the first member of the pair (P_{1j}) and the second (P_{2j}). Hence

$$\overline{D} = \frac{\sum\limits_{j=1}^{n}(Y_{1j} - Y_{2j})}{n} = \frac{\sum\limits_{j=1}^{n}[\mu_1 - \mu_2 + (P_{1j} - P_{2j}) + (\epsilon_{1j} - \epsilon_{2j})]}{n}$$

and

$$V(\overline{D}) = \frac{1}{n^2}\sum\limits_{j=1}^{n}[V(P_{1j}) + V(P_{2j}) + V(\epsilon_{1j}) + V(\epsilon_{2j})] = \frac{2\sigma_p^2 + 2\sigma^2}{n} > \frac{2\sigma^2}{n}$$

Notice that $V(\overline{D})$ is larger for the completely randomized design, since the unwanted variation due to pairs is not being eliminated.

12.33 The differences along with the calculation of $\overline{d}$ and s_D^2 are shown below.

d_i		
80	$\sum d_i = 450$	Then
90	$\sum d_i^2 = 69{,}900$	$\overline{d} = \frac{450}{4} = 112.5$
230		
50	$n = 4$	$s_D^2 = \dfrac{69{,}900 - \dfrac{(450)^2}{4}}{3} = 6425$

The 90% confidence interval is

$$\overline{d} \pm t_{.05}\frac{s_D}{\sqrt{n}}$$

$$112.5 \pm 2.353\sqrt{\frac{6425}{4}} \quad \text{or} \quad 112.5 \pm 94.104 \quad \text{or} \quad 18.196 < (\mu_1 - \mu_2) < 206.804$$

12.35 There will be nk_1 points at the setting $x = -1$, nk_2 points at $x = 0$, and nk_3 points at $x = 1$. Hence the design matrix will be

$$
X = \begin{bmatrix}
1 & -1 & 1 \\
1 & -1 & 1 \\
\vdots & \vdots & \vdots \\
1 & -1 & 1 \\
1 & 0 & 0 \\
1 & 0 & 0 \\
\vdots & \vdots & \vdots \\
1 & 0 & 0 \\
1 & 1 & 1 \\
1 & 1 & 1 \\
\vdots & \vdots & \vdots \\
1 & 1 & 1
\end{bmatrix}
\begin{array}{l}
\left.\rule{0pt}{40pt}\right\} nk_1 \\[20pt]
\left.\rule{0pt}{40pt}\right\} nk_2 \\[20pt]
\left.\rule{0pt}{40pt}\right\} nk_3
\end{array}
$$

and

$$
X'X = \begin{bmatrix}
n & n(k_3 - k_1) & n(k_1 + k_3) \\
n(k_3 - k_1) & n(k_1 + k_3) & n(k_3 - k_1) \\
n(k_1 + k_3) & n(k_3 - k_1) & n(k_1 + k_3)
\end{bmatrix}
$$

$$
= n \begin{bmatrix}
1 & b & a \\
b & a & b \\
a & b & a
\end{bmatrix} = nA
$$

where $a = k_1 + k_3$ and $b = k_3 - k_1$. The inverse of this matrix can be shown to be

$$
\frac{1}{n\,|A|} \begin{bmatrix}
a^2 - b^2 & 0 & b^2 - a^2 \\
0 & a - a^2 & ab - b \\
b^2 - a^2 & ab - b & a - b^2
\end{bmatrix}
$$

with

$$|A| = a^2 - b^2 - b(ab - ab) + a(b^2 - a^2) = (a-1)(b^2 - a^2) = -k_2(b^2 - a2)$$
$$= -k_2\left(k_3^2 - 2k_1k_3 + k_1^2 - k_1^2 - k_3^2 - 2k_1k_3\right) = 4k_1k_2k_3$$

Hence

$$V(\hat{\beta}_2) = \frac{\sigma^2(a-b^2)}{n4k_1k_2k_3} = \frac{\sigma^2\left[k_1 + k_2 - (k_3 - k_1)^2\right]}{4nk_1k_2k_3}$$

We must minimize

$$Q = \frac{(k_1 + k_3) - (k_3 - k_1)^2}{4k_1k_2k_3} = \frac{(k_1 + k_3) - \left[(k_1+k_3)^2 - 4k_1k_3\right]}{4k_1k_2k_3}$$

$$= \frac{(k_1 + k_3)(1 - k_1 - k_3)}{4k_1k_2k_3} + \frac{4k_1k_3}{4k_1k_2k_3} = \frac{k_1 + k_3}{4k_1k_3} + \frac{1}{k_2} = \frac{k_1 + k_3}{4k_1k_3} + \frac{1}{1 - k_1 - k_3}$$

Differentiating with respect to k_1 and k_2 and setting the resulting equations equal to zero implies

(*) $4k_1^2 = (1 - k_1 - k_3)^2$

$4k_3^2 = (1 - k_1 - k_3)^2$

Hence, since k_1, k_2, k_3 are positive, $k_1 = k_3$. From (*), noting that $1 - 2k_1 = 1 - k_1 - k_3 = k_2 > 0$,

$$4k_1^2 = (1 - 2k_1)^2$$

$$2k_1 = 1 - 2k_1$$

$$k_1 = \tfrac{1}{4}$$

It follows that $k_1 = k_3 = \tfrac{1}{4}$ and $k_2 = 1 - k_1 - k_3 = \tfrac{1}{2}$.

CHAPTER 13 THE ANALYSIS OF VARIANCE

13.1 **a.** & **c.** The hypothesis to be tested is

$$H_0: \ \mu_1 = \mu_2 \qquad \text{vs.} \qquad H_a: \ \mu_1 \neq \mu_2$$

The analysis of variance F test will be used. A completely randomized design has been employed. We assume that independent random samples of size $n_1 = n_2 = 8$ have been drawn from two normal populations with means μ_1 and μ_2, respectively, and with common variance σ^2. The object of the analysis of variance is to partition the total sum of squares into two parts, the sum of squares for treatment and error. Then

$$F = \frac{\text{MST}}{\text{MSE}}$$

will provide a test statistic to test the null hypothesis $H_0: \ \mu_1 = \mu_2$ against the alternative that this equality does not hold. In order to perform the analysis of variance, we must calculate the following quantities:

$$\bar{y}_1 = 1.875 \qquad \bar{y}_2 = 2.625 \qquad \sum_{ij} y_{ij}^2 = 94 \qquad \sum_{ij} y_{ij} = 36$$

Using the formulas given in this section, calculate

$$\text{SST} = \frac{n_1}{2}(\bar{y}_1 - \bar{y}_2)^2 = \frac{8}{2}(1.875 - 2.625)^2 = 4(.5625) = 2.25$$

$$\text{TSS} = \sum_{i} \sum_{j} y_{ij}^2 - \frac{\left(\sum_{i} \sum_{j} y_{ij}\right)^2}{16} = 94 - \frac{(36)^2}{16} = 94 - 81 = 13$$

Then

$$\text{SSE} = \text{TSS} - \text{SST} = 13 - 2.25 = 10.75$$

For the test of the hypothesis of equality of means, the appropriate mean squares are

$$\text{MST} = \frac{\text{SST}}{1} = 2.25 \qquad \text{and} \qquad \text{MSE} = \frac{\text{SSE}}{2n_1 - 2} = .7679$$

and

$$F = \frac{\text{MST}}{\text{MSE}} = \frac{2.25}{.7679} = 2.93$$

Notice that the critical value for rejection with $\alpha = .05$, based on 1 and 14 degrees of freedom, is $F = 4.60$. Hence the null hypothesis is not rejected.

There is not sufficient evidence to conclude that there is a difference in mean reaction times for the two stimuli, using a 5% significance level (p-value $> .10$).

b.   The value MSE $= .7679$ is the same as s^2. The t statistic is

$$t = \frac{\overline{y}_1 - \overline{y}_2}{\sqrt{s^2\left[\left(\frac{1}{n_1}\right)+\left(\frac{1}{n_2}\right)\right]}} = \frac{1.875 - 2.625}{\sqrt{.7679\left[\left(\frac{1}{8}\right)+\left(\frac{1}{8}\right)\right]}} = -1.712$$

Since $|-1.712| < 2.145 = t_{.025,14}$, we do not reject H_0. Note that $t^2 = F$, i.e., $(-1.712)^2 = 2.93$.

13.3   See Section 13.3 of the text.

13.5   Since W has a χ^2 distribution with r degrees of freedom, the moment-generating function is

$$M_W(t) = (1 - 2t)^{-r/2} = E\left(e^{tw}\right) = E\left(e^{t(u+v)}\right) = E\left(e^{tu}e^{tv}\right)$$
$$= E\left(e^{tu}\right)E\left(e^{tv}\right) \qquad \text{since } u \text{ and } v \text{ are independent}$$
$$= E\left(e^{tu}\right)(1 - 2t)^{-s/2} \qquad \text{since } v \text{ has a } \chi^2 \text{ distribution with } s \text{ degrees of freedom.}$$

Therefore,

$$E\left(e^{tu}\right) = \frac{(1 - 2t)^{-r/2}}{(1 - 2t)^{-s/2}} = (1 - 2t)^{-(r-s)/2}$$

By Theorem 6.1, u has a χ^2 distribution with $r - s$ degrees of freedom.

13.7   This is similar to Example 13.2 in the text. The null hypothesis is H_0: $\mu_1 = \mu_2 = \mu_3 = \mu_4$ against the alternative that at least one of the above equalities does not hold. The following calculations are necessary.

(1) $\text{CM} = \dfrac{\left(\sum\limits_i \sum\limits_j y_{ij}\right)^2}{12} = 58.08$

(2) $\text{TSS} = \sum\limits_i \sum\limits_j y_{ij}^2 - \text{CM} = 58.115 - \text{CM} = .035$

(3) $\text{SST} = \sum\limits_i \dfrac{y_{i*}^2}{n_i} - \text{CM} = \dfrac{(6.75)^2 + (6.50)^2 + (6.50)^2 + (6.65)^2}{3} - \text{CM}$

$\qquad = 58.095 - \text{CM} = .015$

There are four treatments and hence 3 degrees of freedom for treatments, $(n - 1) = 11$ total degrees of freedom, and $(11 - 3) = 8$ degrees of freedom for error. The ANOVA table is shown below. In order to determine whether or not the four types of concrete differ in average strength, we run an F test. The critical value of F with $\alpha = .05$ and 3 and 8 degrees of freedom is $F = 4.07$. The computed value of F is

$$F = \frac{MST}{MSE} = \frac{.0050}{.0025} = 2$$

Source	d.f.	SS	MS
Treatments	3	.015	.0050
Error	8	.020	.0025
Total	11	.035	

which does not exceed the critical value. Therefore, the null hypothesis of equality cannot be rejected and we cannot say that the strengths of the four types of concrete are significantly different.

13.9 H_0: $\mu_1 = \mu_2 = \mu_3 = 0$ vs. H_a: One or more of μ_i's differ

a. $y_{1.} = 14(.93) = 13.02$; $y_{2.} = 14(1.21) = 16.94$; $y_{3.} = 14(.92) = 12.88$

b. Total $= y_{1.} + y_{2.} + y_{3.} = 42.84$

c. $CM = \dfrac{(42.84)^2}{42} = 43.6968$

d. $SST = \dfrac{(13.02)^2 + (16.94)^2 + (12.88)^2}{14} - CM = .7588$

e. $s_1^2 = 14(.04)^2 = .0224$; $s_2^2 = 14(.03)^2 = .0126$; $s_3^2 = 14(.04)^2 = .0224$

f. & g. $SSE = \displaystyle\sum_{i=1}^{p} (n_i - 1)s_1^2 = 13(.0224) + 13(.0126) + 13(.0224) = .7462$

h.
Source	df	SS	MS	F
Treatments	2	.7588	.3794	19.83
Error	39	.7462	.019133	
Total	41			

i. The test statistic is

$$F = \frac{MST}{MSE} = \frac{.3794}{.019133} = 19.83$$

and the rejection region with 2 and 39 df is approximately $F > F_{.05} = 3.23$. The null hypothesis is rejected and there is a difference between the means.

p-value < 0.005.

13.11 Using the pooled sum of squares formula for SSE, we have

$$SSE = \sum (n_i - 1)s_i^2 = 44(.7)^2 + 101(.64)^2 + 17(.9)^2 = 76.6996$$

The treatment totals are

$$y_{1.} = 45(4.59) = 206.55 \quad y_{2.} = 102(4.88) = 497.76 \quad y_{3.} = 18(6.24) = 112.32$$

so that

$$CM = \frac{\left[\sum\sum y_{ij}\right]^2}{n} = \frac{(816.63)^2}{165} = 4041.725$$

$$SST = \sum \frac{y_{i.}^2}{n_i} - CM = \sum n_i y_{i.}^2 - CM$$
$$= 45(4.59)^2 + 102(4.88)^2 + 18(6.24)^2 - CM = 4078.010 - 4041.725 = 36.285$$

The ANOVA table is shown below. Then

$$F = \frac{MST}{MSE} = \frac{18.143}{.4735} = 38.316$$

Source	d.f.	SS	MS
Treatments	2	36.285	18.143
Error	162	76.6996	0.4735
Total	164		

From Table 7, $F = 38.316 > 7.88 = F_{.005}$ for 2 and ∞ degrees of freedom. Thus, the p-value is less than .005. We reject H_0 at the $\alpha = .05$ level (p-value $< .005$).

13.13 We have a completely randomized design with four treatments. The analysis is as follows:

(1) $$CM = \frac{\left[\sum_i \sum_j y_{ij}\right]^2}{n} = \frac{(110.6)^2}{19} = 643{,}8084$$

(2) $$TSS = \sum_i \sum_j y_{ij}^2 - CM = 652.26 - 643.8084 = 8.4516$$

(3) $$SST = \sum_i \frac{y_{1.}^2}{n_i} - CM$$
$$= \frac{(30.4)^2}{5} + \frac{(32.2)^2}{5} + \frac{(23.9)^2}{5} + \frac{(24.1)^2}{4} - 643.808 = 7.8361$$

(4) $$SSE = TSS - SST = 8.452 - 7.836 = .616$$

The ANOVA table is shown below. The F statistic is

$$F = \frac{MST}{MSE} = \frac{2.61203}{.04103} = 63.66$$

Source	d.f.	SS	MS
Treatments	2	7.8361	2.61203
Error	15	.6155	.04103
Total	18	8.4516	

The critical value for F with $\alpha = .005$ and with 3 and 15 degrees of freedom is $F = 6.48 < 63.66$. Thus, the p-value $< .005$. We conclude that there is a difference in mean dissolved oxygen content for the four locations.

13.15 $E(\overline{Y}_{i.}) = E\left(\frac{1}{n_i} \sum_{j=1}^{n_i} Y_{ij}\right) = E\left(\frac{1}{n_i} \sum_{j=1}^{n_i} (\mu + \tau_i + \epsilon_{ij})\right)$

$\qquad = E\left(\frac{1}{n_i} \sum_{j=1}^{n_i} (\mu_i + \epsilon_{ij})\right)$ \qquad\qquad since $\mu_i = \mu + \tau_i$,

$\qquad = \frac{1}{n_i} \sum_{j=1}^{n_i} E(\mu_i) + \frac{1}{n_i} \sum_{j=1}^{n_i} E(\epsilon_{ij})$

$\qquad = \mu_i + 0 = \mu_i,$ \qquad\qquad the mean of ith population.

Similarly,

$$V(\overline{Y}_{i.}) = V\left(\frac{1}{n_i} \sum_{j=1}^{n_i} Y_{ij}\right) = V\left(\mu + \tau_i + \frac{1}{n_i} \sum_{j=1}^{n_i} \epsilon_{ij}\right) = V\left(\frac{1}{n_i} \sum_{j=1}^{n_i} \epsilon_{ij}\right)$$

$$= \frac{1}{n_i^2}\left(\sum_{j=1}^{n_i} V(\epsilon_{ij}) + 2 \sum_{j < j'} \sum \text{Cov}(\epsilon_{ij}, \epsilon_{ij'})\right)$$

$$= \frac{1}{n_i^2}\left(\sum_{j=1}^{n_i} \sigma^2 + 0\right) \qquad\qquad \text{since the } \epsilon_{ij} \text{ are independent,}$$

$$= \frac{\sigma^2}{n_i 2}$$

13.17 a τ_i is defined as $\mu_i - \mu$.

Thus, $\tau_1 = \tau_2 = \ldots = \tau_k$ can be written as

$\mu_1 - \mu = \mu_2 - \mu = \ldots = \mu_k - \mu = 0$

Now, $\mu_i - \mu = 0 \Rightarrow \mu_i = \mu$ for all $i = 1, \ldots, k$.

This implies $\mu_1 = \ldots = \mu_k$.

Conversely, if $\mu_1 = \ldots = \mu_k$,

$\mu + \tau_1 = \ldots = \mu + \tau_k$

This simplifies to $\tau_1 = \ldots = \tau_k$

Since $\sum_{i=1}^{k} \tau_i = 0$, $\tau_1 = \ldots = \tau_k = 0$

b. Consider $\mu_i = \mu + \tau_i$ and $\mu_{i'} = \mu + \tau_{i'}$

If $\mu_i \neq \mu_{i'}$, then $\mu + \tau_i \neq \mu + \tau_{i'}$, which simplifies to $\tau_i \neq \tau_{i'}$.

Since $\sum_{i=1}^{k} \tau_i = 0$, this implies at least one $\tau_i \neq 0$.

Conversely, we can consider

$$\mu = \mu_i - \tau_i$$
$$\mu = \mu_{i'} - \tau_{i'}$$

Thus, $\mu_i - \tau_i = \mu_{i'} - \tau_{i'}$.

Since $\sum_{i=1}^{k} \tau_i = 0$ and if $\tau_i \neq 0$ for at least one i, then for the preceding to hold,

$$\mu_i \neq \mu_{i'}.$$

13.19 a. The 95% confidence interval is

$$(\overline{Y}_{1.} - \overline{Y}_{4.}) = t_{.025}\, S_{14}\sqrt{\tfrac{1}{6}+\tfrac{1}{4}},$$

where

$$S_{14}^2 = \frac{(n_1-1)S_1^2 + (n_4-1)S_4^2}{n_1+n_4-2} = \frac{5(66.67)+3(33.58)}{8} = 54.26,$$

and $t_{.025}$ is based upon 8 degrees of freedom. This gives

$$-12.08 \pm (2.306)\sqrt{54.26}\left(\sqrt{\tfrac{1}{6}+\tfrac{1}{4}}\right) \quad \text{or} \quad -12.08 \pm 10.96 = (-23.04,\ -1.12),$$

suggesting $\mu_4 > \mu_1$.

b. The interval in part (a) is longer than the interval given in Example 13.4, with lengths 21.92 and 21.46, respectively.

c. The major reason the interval in part (a) is longer is the loss of 11 degrees of freedom resulting from excluding data on Techniques 2 and 3 when estimating σ^2. The $t_{.025}$ value is 2.306 based on 8 degrees of freedom compared to 2.093 based on 19 degrees of freedom.

13.21 Need to consider

$$(\overline{Y}_{1.} - \overline{Y}_{2.}) \pm (t_{.025,39})s\sqrt{\tfrac{1}{n_1}+\tfrac{1}{n_2}} \quad \text{or} \quad s = \sqrt{\text{MSE}} = .1383.$$

We give as our 95% estimate interval

$$(.93 - 1.21) \pm 1.96(.1383)\sqrt{\tfrac{2}{14}} \quad \text{or} \quad -.28 \pm .102 \quad \text{or} \quad (-.382,\ -.178)$$

At the 95% confidence level, we would conclude that there is a significant difference between the mean bone densities for the two groups of women since the confidence interval formed contains all negative values. This suggests that $\mu_2 > \mu_1$.

13.23 Confidence intervals are calculated as in Exercise 13.22.

a. $\bar{y}_A \pm t_{.025,8}\dfrac{s}{\sqrt{n_A}}$ or $76 \pm 2.306\dfrac{\sqrt{62.233}}{\sqrt{5}}$ or 76 ± 8.142

or

$67.86 < \mu_A < 84.14.$

b. $\bar{y}_B \pm t_{.025,8}\dfrac{s}{\sqrt{n_B}}$ or $66.33 \pm 2.306\dfrac{\sqrt{62.233}}{\sqrt{3}}$ or 64.33 ± 10.51

or

$55.82 < \mu_B < 76.84.$

c. $(\bar{y}_A - \bar{y}_B) \pm t_{.025,8}\, s\sqrt{\dfrac{1}{n_A} + \dfrac{1}{n_B}}$ or $9.667 \pm 2.306(7.895)\sqrt{\dfrac{1}{3} + \dfrac{1}{5}}$

or

9.667 ± 13.295 or $-3.628 < \mu_A - \mu_B < 22.962$

13.25 a. A 95% confidence interval for μ_B is

$\bar{y}_B \pm t_{.025,162}\dfrac{s}{\sqrt{n_B}}$ or $6.24 \pm 1.96\sqrt{\dfrac{.4735}{18}}$ or $6.24 \pm .318$

b. A 95% confidence interval for $\mu_s - \mu_L$ is

$(\bar{y}_s - \bar{y}_L) \pm t_{.05,162}\dfrac{s}{\sqrt{\dfrac{1}{n_s} + \dfrac{1}{n_L}}}$

or

$(4.59 - 4.58) \pm 1.96\sqrt{.4735}\sqrt{\dfrac{1}{45} + \dfrac{1}{102}}$ or $-.29 \pm .241$

c. Probably not. The driving habits of people vary from town to town. Thus, the vehicles sampled do not represent a random sample of vehicles from all towns.

13.27 a. We wish to test H_0: $\mu_1 = \mu_2 = \mu_3 = \mu_4$ against the general alternative. Calculate

$$CM = \frac{(49)^2}{16} = 150.0625$$

$$TSS = 183 - CM = 32.9375$$

$$SST = \frac{(8)^2 + (12)^2 + (13)^2 + (16)^2}{4} - CM = 8.1875$$

The ANOVA table is shown below. Then

$$F = \frac{MST}{MSE} = 1.32$$

Source	d.f.	SS	MS
Treatments	3	8.1875	2.7292
Error	12	24.7500	2.0625
Total	15	32.9375	

which is compared to $F_{.05} = 3.49$ with 3 and 12 degrees of freedom. We do not reject H_0.

b. $(\bar{y}_{4.} - \bar{y}_{1.}) \pm t_{.025,12} \sqrt{s^2\left(\frac{1}{n_4} + \frac{1}{n_1}\right)}$ or $2 \pm 2.179\sqrt{2.0625\left(\frac{2}{4}\right)}$

or

2 ± 2.21

or

$-.21 < \mu_4 - \mu_1 \leq 4.21.$

Intervals constructed in this manner will enclose $(\mu_4 - \mu_1)$ 95% of the time in repeated sampling. Hence we are fairly confident that this particular interval encloses $(\mu_4 - \mu_1)$.

13.29 $(\bar{y}_{2.} - \bar{y}_{3.}) \pm t_{\alpha/2,15} \, s\sqrt{\frac{1}{n_2} + \frac{1}{n_3}}$ or $(6.44 - 4.78) \pm 2.131 \, s\sqrt{\frac{1}{5} + \frac{1}{5}}$

or

$1.66 \pm 2.131(.2025)\sqrt{\frac{2}{5}}$ or $1.66 \pm .273$ or $1.39 < \mu_2 - \mu_3 < 1.93$

13.31 a. The 90% confidence interval for $(\mu_1 - \mu_4)$ is

$$(\bar{y}_{1.} - \bar{y}_{4.}) \pm t_{.05,36}\sqrt{MSE\left(\frac{1}{n_1} + \frac{1}{n_4}\right)}$$

or

$$(30.9 - 28.2) \pm 1.645\sqrt{25.986\left(\frac{2}{10}\right)}$$

or

2.7 ± 3.75 or $-1.05 < \mu_1 - \mu_4 < 6.45$

b. A 90% confidence interval for μ_2 is

$$\bar{y}_{2.} \pm t_{.05,36}\sqrt{\frac{MSE}{n_2}}$$ or $27.5 \pm 1.645\sqrt{\frac{25.9861}{10}}$ or 27.5 ± 2.65

or

$24.85 < \mu_2 < 30.15.$

13.33 **a.** $\frac{1}{bk} \sum_{i=1}^{k} \sum_{j=1}^{b} E(Y_{ij}) = \frac{1}{bk} \sum_{i=1}^{k} \sum_{j=1}^{b} (\mu + \tau_i + \beta_j)$

$$= \frac{1}{bk}\left(bk\mu + b\sum_{i=1}^{k}\tau_i + k\sum_{j=1}^{k}\beta_j\right)$$

$$= \frac{1}{bk}(bk\mu + 0 + 0) = \mu$$

b. μ is the overall mean for the $n = bk$ Y_{ij}.

13.35 **a.** $E(\overline{Y}_{i\,.} - \overline{Y}_{i'\,.}) = E(\overline{Y}_{i\,.}) - E(\overline{Y}_{i'\,.}) = \mu_i - \mu_{i'}$.

Recall that $\mu_i = \mu + \tau_i$; $\mu_{i'} = \mu + \tau_{i'}$

Thus,

$$\mu_i - \mu_{i'} = (\mu + \tau_i) - (\mu + \tau_{i'}) = \tau_i - \tau_{i'}.$$

b. $V(\overline{Y}_{i\,.} - \overline{Y}_{i'\,.}) = V(\overline{Y}_{i\,.}) + V(\overline{Y}_{i\,.}) - 2\,\text{Cov}\,(\overline{Y}_{i\,.}, \overline{Y}_{i'\,.})$

$$= \frac{\sigma^2}{b} + \frac{\sigma^2}{b} - 0 \qquad \text{from Exercise 13.34 and since } \overline{Y}_{i\,.}, \overline{Y}_{i'\,.}$$
$$\text{are independent.}$$

$$= \frac{2\sigma^2}{b}$$

13.37 **a.** A summary of the data is

$Y_1\,. = 9.32$	$Y_2\,. = 9.45$	$Y_{.1} = 2.27$
$Y_{.2} = 3.45$	$Y_{.3} = 2.14$	$Y_{.4} = 3.73$
$Y_{.5} = 2.93$	$Y_{.6} = 4.35$	

$$\sum_{i=1}^{2} \sum_{j=1}^{6} y_{ij} = 18.77$$

$$\sum_{i=1}^{2} \sum_{j=1}^{6} y_{ij}^2 = 31.1013$$

$$\text{CM} = \frac{(18.77)^2}{12} = 29.3594$$

Total SS $= 31.1013 - 29.3594 = 1.7419$

$$\text{SST} = \frac{(9.32)^2 + (9.45)^2}{6} - 29.3594 = .0014$$

$$\text{SSB} = \frac{(2.27)^2 + (2.45)^2 + (2.14)^2 + (3.73)^2 + (2.93)^2 + (4.25)^2}{2} - 29.3594$$

$$= 1.7382$$

$$\text{SSE} = 1.7419 - .0014 - 1.7382 = .0023$$

The ANOVA table is

Source	df	SS	MS
Computer	1	.0014	.0014
Program	5	1.7382	.3476
Error	5	.0023	.00045
Total	11	1.7419	

To test H_0: $\mu_1 = \mu_2$, we use

$$F = \frac{\text{MST}}{\text{MSE}} = 3.05$$

Since $3.05 < 6.61 = F_{.05}$ with 1 and 5 degrees of freedom, we fail to reject H_0. Thus, we see no evidence of a difference in mean CPU time between computer 1 and computer 2.

This decision is the same as the one reached in Exercise 12.10(a).

b. The *p*-value is greater than .10. This is consistent with Exercise 12.10(b).

c. Except for roundoff, $S_D^2 = 2\text{MSE}$.

13.39 The factor of interest is "soil preparation," and the blocking factor is "locations." A randomized block design is employed and the analysis of variance is as shown below.

(1) $\text{CM} = \dfrac{(162)^2}{12} = 2187$

(3) $\text{SST (preparations)} = \dfrac{8900}{4} - \text{CM} = 38$

(2) $\text{TSS} = 2298 - \text{CM} = 111$

(4) $\text{SSB (locations)} = \dfrac{6746}{3} - \text{CM} = 61.67$

The ANOVA table is shown at the right.

Source	d.f.	SS	MS
Blocks	3	61.67	20.56
Treatments	2	38.00	19.00
Error	6	11.33	1.89
Total	11	111.00	

a. The F statistic to detect a difference owing to soil preparations is

$$F = \frac{\text{MST}}{\text{MSE}} = 10.05$$

Then the critical value of F, based on 2 and 6 degrees of freedom, is $F_{.05} = 5.14$, and we conclude that there is a significant effect owing to soil preparation.

b. The F statistic to detect a difference owing to locations is

$$F = \frac{\text{MSB}}{\text{MSE}} = 10.88$$

Then the critical value of F, based on 3 and 6 degrees of freedom, is $F_{.05} = 4.76$, and we conclude that there is a significant effect owing to locations.

13.41 Using a randomized block design with locations as blocks, recall that $CM = 150.0625$, $TSS = 32.9375$, and $SST = 8.1875$. Then

$$SSB = \frac{(15)^2 + (8)^2 + (14)^2 + (12)^2}{4} - CM = 7.1875$$

SSE will be obtained by subtraction; see the ANOVA table. To test for a difference in treatment means, the test statistic is

$$F = \frac{MST}{MSE} = 1.40$$

Source	d.f.	SS	MS
Treatments	3	8.1875	2.729
Blocks	3	7.1875	2.396
Error	9	17.5625	1.95139
Total	15	32.9375	

which is compared to $F_{.05} = 3.86$ with 3 and 9 degrees of freedom. There is no evidence of significant treatment differences. Notice that the test for block differences yields $F = \frac{MST}{MSE} = 1.23$, which is not significant. Blocking was not worthwhile.

13.43 A summary of the data is

$$y_{1 \cdot} = 497.7 \qquad y_{2 \cdot} = 531.3 \qquad y_{3 \cdot} = 491.3$$

$$k = 3 \qquad y_{\cdot 1} = 211.1 \qquad y_{\cdot 2} = 202.7$$

$$y_{\cdot 3} = 233.1 \qquad y_{\cdot 4} = 218.1 \qquad y_{\cdot 5} = 220.5$$

$$y_{\cdot 6} = 205.3 \qquad y_{\cdot 7} = 229.5 \qquad b = 7$$

$$\sum y = 1520.3 \qquad \sum y^2 = 110{,}587.13$$

$$CM = \frac{(1520.3)^2}{21} = 110{,}062.48$$

$$TSS = 110{,}587.13 - \frac{(1520.3)^2}{21} = 524.650$$

$$SST = \frac{(497.7)^2}{7} + \frac{(531.3)^2}{7} + \frac{(491.3)^2}{7} - \frac{(1520.3)^2}{21} = 131.901$$

$$SSB = \frac{(211.1)^2}{3} + \ldots + \frac{(229.5)^2}{3} - \frac{(1520.3)^2}{21} = 268.290$$

$$SSE = 524.650 - 131.901 - 268.290 = 124.459.$$

The ANOVA table is given on the right. To test H_0: $\mu_1 = \mu_2 = \mu_3$, we use

$$F = \frac{\text{MST}}{\text{MSE}} = 6.36$$

Source	d.f.	SS	MS
Treatments	2	131.901	65.9505
Blocks	6	268.90	44.8167
Error	12	124.459	10.3716
Total	20	524.65	

Since $6.36 > 3.89 = F_{.05}$ with 2 and 12 degrees of freedom, we reject H_0.

13.45 Some preliminary results will be necessary in order to obtain the solution.

(1) $E\left(Y_{ij}^2\right) = V(Y_{ij}) + [(E(Y_{ij})]^2 = \sigma^2 + (\mu + \tau_i + \beta_j)^2$

(2) With $\overline{Y}_{..} = \dfrac{\sum\limits_i \sum\limits_j Y_{ij}}{bk} = \mu + \dfrac{\sum\limits_i \sum\limits_j \epsilon_{ij}}{bk}$, then

$$E\left(\overline{Y}_{..}\right) = \mu \qquad V\left(\overline{Y}_{..}\right) = \frac{1}{(bk)^2} \sum_i \sum_j V(\epsilon_{ij}) = \frac{\sigma^2}{bk}$$

$$E\left(\overline{Y}_{..}^2\right) = \frac{\sigma^2}{bk} + \mu^2$$

(3) With $\overline{Y}_{.j} = \dfrac{\sum\limits_i Y_{ij}}{k} = \mu + \beta_j = \dfrac{\sum\limits_i \epsilon_{ij}}{k}$, then

$$E\left(\overline{Y}_{.j}\right) = \mu + \beta_j \qquad V\left(\overline{Y}_{.j}\right) = \frac{\sigma^2}{k} \qquad E\left(\overline{Y}_{.j}^2\right) = \frac{\sigma^2}{k} + (\mu + \beta_j)^2$$

(4) $E\left(\overline{Y}_{i.}^2\right) = V(\overline{Y}_{i.}) + [E(\overline{Y}_{i.})]^2 = \dfrac{\sigma^2}{b} + (\mu + \tau_i)^2$

a. $E(\text{MST}) = \dfrac{b}{k-1} E\left[\sum_i (\overline{Y}_{i.} - \overline{Y}_{..})^2\right] = \dfrac{b}{k-1} E\left[\sum_i \overline{Y}_{i.}^2 - k\overline{Y}_{..}^2\right]$

which can be seen by expanding $\sum\limits_i (\overline{Y}_{i.} - \overline{Y}_{..})^2$. Then

$$E(\text{MST}) = \frac{b}{k-1}\left[\sum_i E\left(\overline{Y}_{i.}^2\right) - kE(\overline{Y}_{..}^2)\right]$$

$$= \frac{b}{k-1}\left[\sum_i \left(\frac{\sigma^2}{b} + \mu^2 + 2\mu\tau_i + \tau_i^2\right) - k\left(\frac{\sigma^2}{bk} + \mu^2\right)\right]$$

$$= \frac{b}{k-1}\left[\frac{(k-1)\sigma^2}{b} + \sum_i \tau_i^2\right] = \sigma^2 + \frac{b}{k-1}\sum_i \tau_i^2$$

b. $E(\text{MSB}) = \frac{k}{b-1}\left[\sum_j E\left(\overline{Y}^2_{.j}\right) - bE\left(\overline{Y}^2_{..}\right)\right]$

$$= \frac{k}{b-1}\left[\sum_j \left(\frac{\sigma^2}{k} + \mu^2 + 2\mu\beta_j + \beta_j^2\right) + \frac{b\sigma^2}{bk} - b\mu^2\right]$$

$$= \frac{k}{b-1}\left[\frac{(b-1)\sigma^2}{k} + \sum_j \beta_j^2\right] = \sigma^2 + \frac{k}{b-1}\sum_j \beta_j^2$$

c. Recall that $\text{TSS} = \sum_i \sum_j Y_{ij}^2 - bk\overline{Y}^2_{..}$, so that

$$E(\text{TSS}) = \sum_i \sum_j \left(\sigma^2 + \mu2 + \tau_i^2 + \beta_j^2\right) - bk\left(\frac{\sigma^2}{bk} + \mu^2\right)$$

$$= (bk-1)\sigma^2 + b\sum_i \tau_i^2 + k\sum_j \beta_j^2$$

By the additivity property,

$$E(\text{SSE}) = E(\text{TSS}) - E(\text{SST}) - E(\text{SSB})$$

$$= (bk-1)\sigma^2 - (k-1)\sigma^2 - (b-1)\sigma^2 + b\sum_i \tau_i^2 - b\sum_i \tau_i^2$$

$$\quad + k\sum_j \beta_j^2 - k\sum_j \beta_j^2$$

$$= (bk - k - b + 1)\sigma^2$$

Finally, since

$$\text{MSE} = \frac{\text{SSE}}{bk - k - b + 1} \qquad \text{then} \qquad E(\text{MSE}) = \sigma^2$$

13.47 $(\overline{Y}_A - \overline{Y}_B) \pm t_{.025,6}\sqrt{\text{MSE}\left(\frac{2}{b}\right)}$ or $(7-5) \pm 2.447\sqrt{2\left(\frac{2}{3}\right)}$ or 2 ± 2.83

or

$$-.83 < \mu_A - \mu_B < 4.83.$$

13.49 $(\overline{Y}_B - \overline{Y}_D) \pm t_{.025,12}\sqrt{\text{MSE}}\sqrt{\frac{2}{b}}$ or $\left(\frac{10.756}{4} - \frac{10.175}{4}\right) \pm 2.179\sqrt{.0135}\sqrt{\frac{2}{4}}$

or

$$.145 \pm .179.$$

13.51 $(\overline{Y}_{1.} - \overline{Y}_{2.}) \pm t_{.055,12}\sqrt{\text{MSE}}\sqrt{\frac{2}{b}}$ or $(71.1 - 75.9) \pm 3.055\sqrt{10.3716}\sqrt{\frac{2}{7}}$

or

$$-4.8 \pm 5.259.$$

13.53 It is necessary to have $\frac{2\sigma}{\sqrt{n_A}} \leq 10$. Estimating σ with $\sqrt{MSE}$, we solve

$$\frac{2\sqrt{62.333}}{\sqrt{n_A}} \leq 10 \qquad \text{or} \qquad n_A \geq 2.49$$

Hence, $n = 3$ observations are necessary.

13.55 It is necessary to have

$$2\sigma_{\overline{Y}_{i\cdot} - \overline{Y}_{i'\cdot}} \leq 1 \qquad \text{or} \qquad 2\sigma\sqrt{\frac{2}{b}} \leq 1$$

Using $\sigma = \sqrt{MSE}$, solve

$$2\sqrt{1.89\left(\frac{2}{b}\right)} \leq 1 \qquad \text{or} \qquad b \geq 15.12$$

Hence, $b = 16$ locations must be used.

Thus, $3b = 48$ is the total number of observations required in the entire experiment.

13.57 There are three intervals to construct, so each interval should have confidence coefficient $1 - \left(\frac{.05}{3}\right) = .9833$. Since MSE = .4735 with 162 degrees of freedom, we may use a Z-multiplier. Now, $1 - .9833 = .01667$ and $\frac{.01667}{2} = .00833$. Thus, we will use $Z_{.0083} = 2.39$. The confidence intervals are of the form

$$(\overline{y}_{i\cdot} - \overline{y}_{j\cdot}) \pm 2.39\sqrt{MSE\left(\frac{1}{n_i} + \frac{1}{n_j}\right)}$$

For the pairs (i, j) of $(1, 2)$, $(1, 3)$, and $(2, 3)$, they are

$$(1, 2): \ -.29 \pm 2.39\sqrt{.4735\left(\frac{1}{45} + \frac{1}{102}\right)} \qquad \text{or} \qquad -.29 \pm .294$$

$$(1, 3): \ -1.65 \pm 2.39\sqrt{.4735\left(\frac{1}{45} + \frac{1}{18}\right)} \qquad \text{or} \qquad -1.65 \pm .459$$

$$(2, 3): \ -1.36 \pm 2.39\sqrt{.4735\left(\frac{1}{102} + \frac{1}{18}\right)} \qquad \text{or} \qquad -1.36 \pm .420$$

The simultaneous coverage rate of intervals constructed in this manner is .95.

13.59 There are three intervals to construct, so that each interval should have confidence coefficient $1 - \left(\frac{\alpha}{3}\right) = 1 - \left(\frac{.10}{3}\right) = .97$. Since no value of t is given with area .01667 to its right, we choose to use $t_{.01}$ so that the overall confidence coefficient will be .94 rather than .85, which would occur if $t_{.025}$ were used. The three intervals all have half width $3.143\sqrt{1.89\left(\frac{2}{4}\right)} = 3.06$. The three intervals are shown below.

(1, 2):	-3.5 ± 3.06	or	-6.56 to $-.44$
(1, 3):	$.5 \pm 3.06$	or	-2.56 to 3.56
(2, 3):	4.0 ± 3.06	or	$.94$ to 7.06

13.61   **a.** $\beta_0 + \beta_3$ is the mean response to treatment A in block III.
       b. $\beta_3 = $ difference in mean responses to chemicals A and D in block III.

13.63 The complete model is

$$Y_{ij} = \beta_0 + \beta_1 x_1 + \beta_3 x_2 + \beta_3 x_3 + \beta_4 x_4 + \beta_5 x_5 + \epsilon$$

where

$$x_1 = \begin{cases} 1, & \text{if block 1} \\ 0, & \text{otherwise} \end{cases} \qquad x_2 = \begin{cases} 1, & \text{if block 2} \\ 0, & \text{otherwise} \end{cases} \qquad x_3 = \begin{cases} 1, & \text{if treatment 1} \\ 0, & \text{otherwise} \end{cases}$$

$$x_4 = \begin{cases} 1, & \text{if treatment 2} \\ 0, & \text{otherwise} \end{cases} \qquad x_5 = \begin{cases} 1, & \text{if treatment 3} \\ 0, & \text{otherwise} \end{cases}$$

Then for the complete model,

$$Y = \begin{bmatrix} 5 \\ 3 \\ 8 \\ 4 \\ 9 \\ 8 \\ 13 \\ 6 \\ 7 \\ 4 \\ 9 \\ 8 \end{bmatrix}
\qquad
X = \begin{bmatrix} 1 & 1 & 0 & 1 & 0 & 0 \\ 1 & 1 & 0 & 0 & 1 & 0 \\ 1 & 1 & 0 & 0 & 0 & 1 \\ 1 & 1 & 0 & 0 & 0 & 0 \\ 1 & 0 & 1 & 1 & 0 & 0 \\ 1 & 0 & 1 & 0 & 1 & 0 \\ 1 & 0 & 1 & 0 & 0 & 1 \\ 1 & 0 & 1 & 0 & 0 & 0 \\ 1 & 0 & 0 & 1 & 0 & 0 \\ 1 & 0 & 0 & 0 & 1 & 0 \\ 1 & 0 & 0 & 0 & 0 & 1 \\ 1 & 0 & 0 & 0 & 0 & 0 \end{bmatrix}$$

$$X'X = \begin{bmatrix} 12 & 4 & 4 & 3 & 3 & 3 \\ 4 & 4 & 0 & 1 & 1 & 1 \\ 4 & 0 & 4 & 1 & 1 & 1 \\ 3 & 1 & 1 & 3 & 0 & 0 \\ 3 & 1 & 1 & 0 & 3 & 0 \\ 3 & 1 & 1 & 0 & 0 & 3 \end{bmatrix}
\qquad
(X'X)^{-1} = \begin{bmatrix} \frac{1}{2} & -\frac{1}{4} & -\frac{1}{4} & -\frac{1}{3} & -\frac{1}{3} & -\frac{1}{3} \\ -\frac{1}{4} & \frac{1}{2} & \frac{1}{4} & 0 & 0 & 0 \\ -\frac{1}{4} & \frac{1}{4} & \frac{1}{2} & 0 & 0 & 0 \\ -\frac{1}{3} & 0 & 0 & \frac{2}{3} & \frac{1}{3} & \frac{1}{3} \\ -\frac{1}{3} & 0 & 0 & \frac{1}{3} & \frac{2}{3} & \frac{1}{3} \\ -\frac{1}{3} & 0 & 0 & \frac{1}{3} & \frac{1}{3} & \frac{2}{3} \end{bmatrix}$$

$$X'Y = \begin{bmatrix} 84 \\ 20 \\ 36 \\ 21 \\ 15 \\ 30 \end{bmatrix}
\qquad
\widehat{\beta} = \begin{bmatrix} 6 \\ -2 \\ 2 \\ 1 \\ -1 \\ 4 \end{bmatrix}$$

and $\text{SSE}_2 = 674 - 662 = 12$, with $12 - 6 = 6$ degrees of freedom. The reduced model is $Y_{ij} = \beta_0 + \beta_1 x_1 + \beta_2 x_2 + \epsilon$, with x_1 and x_2 as defined in the complete model. The X matrix has columns 4, 5, and 6 deleted, and

$$X'X = \begin{bmatrix} 12 & 4 & 4 \\ 4 & 4 & 0 \\ 4 & 0 & 4 \end{bmatrix} \quad (X'X)^{-1} = \begin{bmatrix} \frac{1}{4} & -\frac{1}{4} & -\frac{1}{4} \\ -\frac{1}{4} & \frac{1}{2} & \frac{1}{4} \\ -\frac{1}{4} & \frac{1}{4} & \frac{1}{2} \end{bmatrix} \quad X'Y = \begin{bmatrix} 84 \\ 20 \\ 36 \end{bmatrix}$$

$$\hat{\beta} = \begin{bmatrix} 7 \\ -2 \\ 2 \end{bmatrix}$$

and $\text{SSE}_1 = 674 - 620 = 54$ with $12 - 3 = 9$ degrees of freedom. Then

$$F = \frac{S_3^2}{S_2^2} = \frac{\frac{54 - 12}{3}}{\frac{12}{6}} = 7.00$$

which is compared to $F_{.05} = 4.76$ with 3 and 6 degrees of freedom. H_0 is rejected; there is a difference in the treatment means.

13.65 **a.** Experimental units are patches of skin, while the three people act as blocks.

b. Incorporate the given sums of squares into an ANOVA table, as shown at the right. Then

$$F = \frac{\text{MST}}{\text{MSE}} = \frac{.59}{.56} = 1.05$$

Source	d.f.	SS	MS
Treatments	2	1.18	.59
Blocks	2	.78	.39
Error	4	2.24	.56
Total	8	4.2	

which is compared to $F_{.05} = 6.94$ with 2 and 4 degrees of freedom. H_0 is not rejected; there is no significant difference in treatment means.

13.67 A 95% confidence interval for $(\mu_A - \mu_B)$ is given by

$$(\bar{y}_A - \bar{y}_B) \pm t_{.025,6}\ s\sqrt{\frac{1}{n_A} + \frac{1}{n_B}} \quad \text{or} \quad (2.25 - 2.166) \pm 2.447\sqrt{.000833}\ \sqrt{\frac{1}{3} + \frac{1}{3}}$$

or

$$.08 \pm .06 \quad \text{or} \quad .02 < \mu_A - \mu_B < .14$$

Thus, in repeated sampling, confidence intervals constructed in this manner will enclose the mean difference $(\mu_A - \mu_B)$ 95% of the time. Notice that this confidence interval is not as wide as the interval found in Exercise 13.22(b). That is, by reducing SSE, the width of the confidence interval has been reduced and a more precise bound on $(\mu_A - \mu_B)$ has been obtained.

13.69 **a.** This is similar to Exercise 13.45. It is necessary to have

$$2\sigma_{\overline{Y}} \leq 10 \qquad \text{or} \qquad 2\frac{\sigma}{\sqrt{n}} \leq 10 \qquad \text{or} \qquad \frac{2(20)}{\sqrt{n}} \leq 10$$

so that $\sqrt{n} \geq 4$ and $n \geq 16$.

b. When 16 patients are assigned to each of the 9 treatments, there are $(n_1 + n_2 + \ldots + n_9 - 9)$ degrees of freedom for estimating σ^2. In this case there are $16(9) - 9 = 135$ degrees of freedom.

c. With 16 replications for each treatment, an approximate half width of the confidence interval for the difference in mean response for two treatments is

$$2\sigma\sqrt{\frac{1}{n_1} + \frac{1}{n_2}} = 2(20)\sqrt{\frac{1}{16} + \frac{1}{16}} = 14.14$$

13.71 Refer to Exercise 13.70. If the experiment is run as a completely randomized design, the block factor (brand of gasoline) has a negligible effect on the response. Hence the sum of squares for blocks (which was calculated in Exercise 13.61) will not be removed from the sum of squares for error. The calculations for SST, TSS, and CM remain the same, and the ANOVA table is as shown below.

a. For a test of significance of treatments, the test statistic is

$$F = \frac{\text{MST}}{\text{MSE}} = 7.3287$$

Source	d.f.	SS	MS
Treatments	2	15.4696	7.7348
Error	6	6.3324	1.0554
Total	8	21.8020	

The critical value of F, based on 2 and 6 degrees of freedom, is 5.14. Notice that the observed value of the test statistic exceeds the critical value of F. Hence we reject the null hypothesis of no difference between treatments and conclude that cars have a significant effect on gasoline mileage.

b. Notice that we are able to reject the null hypothesis in this case but were unable to do so in Exercise 13.70. The reason for this is that the drop in SSE (due to the isolation of block sum of squares in the first case) was not sufficient to compensate for the loss of degrees of freedom for estimating σ^2. The loss of 2 degrees of freedom caused the critical value of F to be substantially increased, and the information gained by blocking was not sufficient to overcome this loss.

c. The randomized block design randomly assigns treatments to experimental units within each block, while the completely randomized design randomly assigns the treatments to all experimental units. Thus, when we have data from an RBD, the randomization is wrong for the completely randomized design and the results do not fit the model.

13.73 a. A rearrangement of the results gives

Levels of Digitalis	Dogs			
	1	2	3	4
A	1342	1140	1029	1150
B	1608	1387	1296	1319
C	1881	1698	1549	1579

The analysis is as follows:

(1) $CM = \dfrac{\left[\sum_i \sum_j y_{ij}\right]^2}{n} = \dfrac{(16,978)^2}{12} = 24,021,040.333$

(2) $TSS = \sum_i \sum_j y_{ij}^2 - CM = 24,724,722 - 24,021,040.333 = 703,681.667$

(3) $SST \text{ (digitalis)} = \dfrac{\sum_i Y_{i.}^2}{n_T} - CM = \dfrac{(4661)^2}{4} + \dfrac{(5610)^2}{4} + \dfrac{(6707)^2}{4} - 24,021,040$

$\qquad = 524,177,167$

(4) $SSB \text{ (dogs)} = \dfrac{\sum_j Y_{.j}^2}{n_B} - CM$

$\qquad = \dfrac{(4831)^2}{3} + \dfrac{(4225)^2}{3} + \dfrac{(3874)^2}{3} + \dfrac{(4048)^2}{3} - 24,021,040.33$

$\qquad = 173,415$

(5) $SSE = TSS - SST - SSB = 6089.5$

The ANOVA table is shown below.

Source	d.f.	SS	MS	F
Treatments	2	524,177.167	262,088.58	258.237
Blocks	3	173,415.00	57,805.00	56.95
Error	6	6,089.5	1,014.9167	
Total	11	703,681.667		

b. There are $n - k - b + 1 = 12 - 3 - 4 + 1 = 6$ degrees of freedom associated with SSE.

c. To test the null hypothesis that there is no difference in mean uptake of calcium for the three levels of digitalis, we use an F test, where

$$F = \frac{MST}{MSE} = \frac{262,088.58}{1014.9167} = 258.237$$

The critical value of F, based on 2 and 6 degrees of freedom, is $F_{.05} = 5.14$. Since the test statistic is in the rejection region, we reject the null hypothesis and conclude that at least one of the levels of digitalis causes a different level of calcium in the heart muscle of dogs.

d. To test the null hypothesis that there is no difference in the mean uptake in calcium for the four heart muscles, we use an F test where

$$F = \frac{\text{MSB}}{\text{MSE}} = \frac{57,805.00}{1014.9167} = 56.95.$$

The critical value of F, based on 3 and 6 degrees of freedom, is $F_{.05} = 4.76$. Since the test statistic is in the rejection region, we reject the null hypothesis and conclude that at least one of the heart muscles differs in mean calcium uptake.

e. The standard deviation of the difference between the mean calcium uptake for two levels of digitalis is

$$s\sqrt{\frac{1}{n_i} + \frac{1}{n_j}} = \sqrt{\text{MSE}}\,\sqrt{\frac{1}{4} + \frac{1}{4}} = (1014.9167)\sqrt{.5} = 22.53.$$

f. A 95% confidence interval for the difference in mean response between treatments A and B is

$$(\bar{y}_A - \bar{y}_B) \pm t_{\alpha/2,6}\, s\sqrt{\frac{1}{n_A} + \frac{1}{n_B}} \qquad \text{or} \qquad (1165.25 - 1402.5) \pm (2.447)(22.53)$$

or

$$-237.25 \pm 55.13 \qquad \text{or} \qquad -292.38 < \mu_A - \mu_B < -182.12.$$

13.75 a. The design is completely randomized with five treatments, containing 4, 7, 6, 5, and 5 measurements, respectively. The analysis is as follows:

(1) $CM = \dfrac{(20.6)^2}{27} = 15.717$

(2) $TSS = 17,500 - CM = 1.783$

(3) $SST = \dfrac{(2.5)^2}{4} + \dfrac{(4.7)^2}{7} + \dfrac{(6.4)^2}{6} + \dfrac{(4.6)^2}{5} + \dfrac{(2.4)^2}{5} - CM = 1.212$

The ANOVA table is shown at the right. The F test to detect differences in mean reaction time to five stimuli is

Source	d.f.	SS	MS
Treatments	4	1.212	.303
Error	22	.571	.02596
Total	26	1.783	

$$F = \frac{\text{MST}}{\text{MSE}} = 11.68$$

The critical value of F for $\alpha = .05$, with 4 and 22 degrees of freedom, is $F = 2.82$, and we conclude that there is a significant difference owing to treatments. Since $F = 11.68 > 5.02$, the p-value is less than .005.

b. The hypothesis of interest is

$$H_0: \ \mu_A - \mu_D = 0 \qquad \text{vs.} \qquad H_a: \ \mu_A - \mu_D \neq 0,$$

and the test statistic is

$$t = \frac{\bar{y}_A - \bar{y}_D}{s\sqrt{\frac{1}{n_A} + \frac{1}{n_D}}} = \frac{.625 - .920}{\sqrt{.02596}\,\sqrt{\frac{1}{4} + \frac{1}{5}}} = -2.73$$

The critical value of t, with $\alpha = .05$ and 22 degrees of freedom, is $t_{.025,\,22} = 2.074$ and the rejection region is $|t| > 2.074$. Hence the null hypothesis is rejected, and we conclude that there is a difference between stimuli A and D. Since $2.508 < |t| < 2.819$, $2(.005) < p\text{-value} < 2(.01)$.

13.77 We will construct four confidence intervals, so each interval should have confidence coefficient $1 - \left(\frac{.05}{4}\right) = .9875 = .99$. We will use the t value $t_{.005,\,12} = 3.055$. The intervals all have half width $3.055\sqrt{.0135}\,\sqrt{\frac{2}{4}} = .251$. The intervals for the various varieties are

$$\mu_A - \mu_D: \ .320 \pm .251 \qquad\qquad \mu_B - \mu_D: \quad .145 \pm .251$$

$$\mu_C - \mu_D: \ .023 \pm .251 \qquad\qquad \mu_E - \mu_D: \ -.124 \pm .251$$

13.79 a. Y_{ij} and $Y_{ij'}$ are independent if $\text{Cov}(Y_{ij}, Y_{ij'}) = 0$. Refer to Section 5.10 on the bivariate normal distribution. Using Theorem 5.12,

$$
\begin{aligned}
\text{Cov}(Y_{ij}, Y_{ij'}) &= \text{Cov}(\mu + \tau_i + \beta_j + \epsilon_{ij}, \ \mu + \tau_i + \beta_{j'} + \epsilon_{ij'}) \\
&= \text{Cov}(\beta_j + \epsilon_{ij}, \ \beta_{j'} + \epsilon_{ij'}) \\
&= \text{Cov}(\beta_j, \beta_{j'}) + \text{Cov}(\epsilon_{ij}, \beta_{j'}) + \text{Cov}(\beta_j, \epsilon_{ij'}) \\
&\quad + \text{Cov}(\epsilon_{ij}, \epsilon_{ij'}) \\
&= 0 + 0 + 0 + 0 = 0
\end{aligned}
$$

Since the β_j's are independent, the ϵ_{ij}'s are independent, and the β_i and ϵ_{ij} are independent of each other.

$$
\begin{aligned}
\text{Cov}(Y_{ij}, Y_{i'j'}) &= \text{Cov}(\mu + \tau_i + \beta_j + \epsilon_{ij}, \ \mu + \tau_{i'} + \beta_{j'} + \epsilon_{i'j'}) \\
&= \text{Cov}(\beta_j + \epsilon_{ij}, \ \beta_{j'} + \epsilon_{i'j'}) \\
&= \text{Cov}(\beta_j, \beta_{j'}) + \text{Cov}(\epsilon_{ij}, \beta_{j'}) + \text{Cov}(\beta_j, \epsilon_{i'j'}) \\
&\quad + \text{Cov}(\epsilon_{ij}, \epsilon_{i'j'}) \\
&= 0 + 0 + 0 + 0 = 0
\end{aligned}
$$

b. $\text{Cov}\,(Y_{ij},\,Y_{i'j}) = \text{Cov}\,(\mu + \tau_i + \beta_j + \epsilon_{ij},\ \mu + \tau_{i'} + \beta_j + \epsilon_{i'j})$

$\qquad\qquad\qquad = \text{Cov}\,(\beta_j + \epsilon_{ij},\ \beta_j + \epsilon_{i'j})$

$\qquad\qquad\qquad = \text{Cov}\,(\beta_j,\ \beta_j) + \text{Cov}\,(\epsilon_{ij},\ \beta_{i'j}) + \text{Cov}\,(\beta_j,\ \epsilon_{i'j}) + \text{Cov}\,(\epsilon_{ij},\ \beta_j)$

$\qquad\qquad\qquad = V(\beta_j) + 0 + 0 + 0 = \sigma_B^2.$

c. When $\sigma_B^2 = 0$, $\text{Cov}\,(Y_{ij},\,Y_{i'j}) = 0$.

13.81 **a.** $\overline{Y}_{\cdot j} = \frac{1}{k}\sum\limits_{i=1}^{k} Y_{ij} = \frac{1}{k}\sum\limits_{i=1}^{k}\,(\mu + \tau_i + \beta_j + \epsilon_{ij}) = \mu + \frac{1}{k}\sum\limits_{i=1}^{k}\tau_i + \beta_j + \frac{1}{k}\sum\limits_{i=1}^{k}\epsilon_{ij}$

$\qquad\qquad\quad = \mu + \beta_j + \overline{\epsilon}_{\cdot j}$

$\qquad\qquad E(\overline{Y}_{\cdot j}) = E(\mu + \beta_j + \epsilon_{\cdot j}) = \mu + E(\beta_j) + E(\overline{\epsilon}_{\cdot j}) = \mu + 0 + 0 = \mu$

$\qquad\qquad V(\overline{Y}_{\cdot j}) = V(\mu + \beta_j + \epsilon_{\cdot j}) = \mu + V(\beta_j) + \frac{1}{k^2}\sum\limits_{i=1}^{k} V(\epsilon_{ij}) = \sigma_B^2 + \frac{\sigma_\epsilon^2}{k}$

b. $E(\text{MST}) = \sigma^2 + \left(\dfrac{b}{k-1}\right)\sum\limits_{i=1}^{k}\tau_i^2$ as calculated in Exercise 13.45, since the block effects cancel when we are making treatment comparisons.

c. $E(\text{MSB}) = k\left[\dfrac{\sum\limits_{j=1}^{k}\,(\overline{Y}_{\cdot j} - \overline{Y})^2}{b-1}\right]$

$\qquad\qquad\qquad = k\left[\text{an unbiased estimator of } V(\overline{Y}_{\cdot j}\right]$

$\qquad\qquad\qquad = kV(\overline{Y}_{\cdot j})$

$\qquad\qquad\qquad = k\left[\sigma_B^2 + \dfrac{\sigma_\epsilon^2}{k}\right]$ $\qquad\qquad$ from part (a)

$\qquad\qquad\qquad = \sigma_\epsilon^2 + k\sigma_B^2$

d. $E[\text{MSE}] = \sigma_\epsilon^2$ $\qquad\qquad$ (similar to derivation in Exercise 13.45(c))

13.83 a. The vector AY can be displayed as

$$AY = \begin{bmatrix} \dfrac{\sum\limits_i Y_i}{\sqrt{n}} \\[2.5ex] \dfrac{Y_1 - Y_2}{\sqrt{2}} \\[2.5ex] \dfrac{Y_1 + Y_2 - 2Y_3}{\sqrt{2 \cdot 3}} \\[2.5ex] \vdots \\[2.5ex] \dfrac{[(Y_1 + Y_2 + \ldots + Y_{n-1}) - (n-1)Y_n]}{\sqrt{n(n-1)}} \end{bmatrix} = \begin{bmatrix} \sqrt{n}\,\overline{Y} \\[1ex] U_1 \\[1ex] U_2 \\[1ex] \vdots \\[1ex] U_{n-1} \end{bmatrix}$$

where $U_1, U_2, \ldots, U_{n-1}$ are linear functions of $Y_1, Y_2, \ldots, Y_n$. Then

$$\sum_{i=1}^{n} Y_i^2 = Y'Y = Y'A'AY = n\overline{Y}^2 + \sum_{i=1}^{n-1} U_i^2.$$

b. Write $L_i = \sum\limits_{j=1}^{n} a_{ij} Y_j$ to be a linear function of $Y_1, Y_2, \ldots, Y_n$. Then two

such linear functions, say L_i and L_k, will be pairwise orthogonal if and only if

$\sum\limits_{j=1}^{n} a_{ij} a_{kj} = 0$, and hence L_i and L_k will be independent if the Y_j are normal

(see Exercise 5.81). Let $L_1, L_2, \ldots, L_n$ be the n linear functions defined by

$\sqrt{n}\,\overline{Y}, U_1, U_2, \ldots, U_{n-1}$. The constants a_{ij}, $j = 1, 2, \ldots, n$, are the elements of the ith row of the matrix A. Moreover, if any two rows of the matrix A are multiplied together, the result is zero. Consider the row vectors

$$a_1' = \begin{bmatrix} \dfrac{1}{\sqrt{n}} & \dfrac{1}{\sqrt{n}} & \cdots & \dfrac{1}{\sqrt{n}} \end{bmatrix}$$

and

$$a_i' = \begin{bmatrix} \dfrac{1}{\sqrt{i(i-1)}} & \dfrac{1}{\sqrt{i(i-1)}} & \cdots & \dfrac{-(i-1)}{\sqrt{i(i-1)}} & 0 & \cdots & 0 \end{bmatrix}$$

for $i = 2, \ldots, n$,

$$a_1' a_i = \frac{i-1}{\sqrt{n}\,\sqrt{i(i-1)}} - \frac{i-1}{\sqrt{n}\,\sqrt{i(i-1)}} = 0$$

$$a_i' a_j = \frac{j-1}{\sqrt{i(i-1)(j)(j-1)}} - \frac{j-1}{\sqrt{i(i-1)(j)(j-1)}} = 0$$

Thus, $L_1, L_2, \ldots, L_n$ are independent linear functions of $Y_1, Y_2, \ldots, Y_n$.

c. $\displaystyle\sum_{i=1}^{n} (Y_i - \overline{Y})^2 = \sum_{i=1}^{n} Y_i^2 - n\overline{Y}^2 = n\overline{Y}^2 + \sum_{i=1}^{n-1} U_i^2 - n\overline{Y}^2 = \sum_{i=1}^{n-1} U_i^2$

Since U_i is independent of $\sqrt{n}\,\overline{Y}$ for $i = 1, 2, \ldots, n-1$, so is U_i^2 and also $\displaystyle\sum_{i=1}^{n-1} U_i^2$. Thus,

$$\sum_{i=1}^{n-1} U_i^2 = \sum_{i=1}^{n} (Y_i - \overline{Y})^2 \text{ is independent of } \overline{Y}.$$

d.   Write

$$\frac{\displaystyle\sum_{i=1}^{n} (Y_i - \mu)^2}{\sigma^2} = \frac{\displaystyle\sum_{i=1}^{n} (Y_i - \overline{Y} + \overline{Y} - \mu)^2}{\sigma^2} = \frac{\displaystyle\sum_{i=1}^{n} (Y_i - \overline{Y})^2}{\sigma^2} + \frac{n(\overline{Y} - \mu)^2}{\sigma^2}$$

$$= X_1 + X_2$$

X_1 and X_2 are independent from part (c).

(1) Since Y_i are normal, $i = 1, 2, \ldots, n$, $\dfrac{(Y_i - \mu)}{\sigma}$ is standard normal, and $\left(\dfrac{Y_i - \mu}{\sigma}\right)^2$ has a χ^2 distribution with 1 degree of freedom.

(2) Since $Y_1, Y_2, \ldots, Y_n$ are independent,

$$\sum_{i=1}^{n} \left(\frac{Y_i - \mu}{\sigma}\right)^2 = \frac{\displaystyle\sum_{i=1}^{n} (Y_i - \mu)^2}{\sigma^2} = X_1 + X_2$$

has a χ^2 distribution with n degrees of freedom.

(3) $\overline{Y}$ has a normal distribution with mean μ and variance $\dfrac{\sigma^2}{n}$, so that $\sqrt{n}\,\dfrac{\overline{Y} - \mu}{\sigma}$ has a standard normal distribution and $X_2 = \dfrac{n(\overline{Y} - \mu)^2}{\sigma^2}$ has a χ^2 distribution with 1 degree of freedom. Now consider the distribution of X_1 using moment-generating functions. Since X_1 and X_2 are independent,

$$m_{X_1 + X_2}(t) = m_{X_1}(t) m_{X_2}(t) \quad \text{or} \quad \frac{1}{(1 - 2t)^{n/2}} = m_{X_1}(t) \left[\frac{1}{(1 - 2t)^{1/2}}\right]$$

or

$$m_{X_1}(t) = \frac{1}{(1 - 2t)^{(n-1)/2}}$$

and X_1 is evidently distributed as a χ^2 random variable with $(n - 1)$ degrees of freedom.

CHAPTER 14 ANALYSIS OF CATEGORICAL DATA

14.1 One thousand cars were each classified according to the lane that they occupied (1 through 4). The objective is to determine whether or not some lanes were preferred over others. This is a multinomial experiment with $k = 4$ cells. If no lane is preferred over another, the probability that a car will be driven in lane i,

$i = 1, 2, 3, 4$, is $\frac{1}{4}$. The null hypothesis is then

$$H_0: \ p_1 - p_2 = p_3 = p_4 = \frac{1}{4}$$

Notice that the hypothesis to be tested is a test of specified numerical values for the probabilities rather than a test of their relationship to one another. Hence no degrees of freedom are lost for estimating cell probabilities. The test statistic is

$$X^2 = \sum_{i=1}^{k} \frac{[n_i - E(n_i)]^2}{E(n_i)}$$

which, when n is large, will possess an approximate chi-square distribution in repeated sampling. The values of n_i are the actual counts observed in the experiment, and

$$E(n_i) = np_i = 1000\left(\frac{1}{4}\right) = 250$$

A table of observed and expected cellcounts is shown at the right. Then

	Lane 1	Lane 2	Lane 3	Lane 4
n_i	294	276	238	192
$E(n_i)$	250	250	250	250

$$X^2 = \frac{(294 - 250)^2}{250} + \frac{(276 - 250)^2}{250} + \frac{(238 - 250)^2}{250} + \frac{(192 - 250)^2}{250}$$

$$= \frac{6120}{250} = 24.48$$

To obtain the rejection region for this test, the degrees of freedom associated with X^2 must be determined. The number of degrees of freedom is equal to the number of cells, k, less 1 degree of freedom for each linearly independent restriction placed on $n_1, n_2, \ldots, n_k$. For this example, $k = 4$ and one degree of freedom is lost because of the restriction that $\sum_i n_i = n$. Hence X^2 has $(k - 1) = (4 - 1) = 3$ degrees of freedom and the appropriate upper-tailed rejection region is

$$X^2 \geq X^2_{3(.005)} = 7.81$$

Thus the conclusion is to reject the null hypothesis, with a probability of error equal to $\alpha = .05$. Remember that a one-tailed test is employed, using the upper-tail values of X^2, because large deviations of the observed cell counts will tend to contradict H_0. Hence we will reject the null hypothesis when X^2 is large. Since $24.48 > 12.8381 = X^2_{3(.005)}$, the p-value is less than .005.

14.3   **a.**   Let p denote the true proportion of heart attacks occurring on Monday. The hypothesis to be tested is

$$H_0: \ p = \tfrac{1}{7} \qquad \text{vs.} \qquad H_a: \ p > \tfrac{1}{7}$$

From Section 8.3 and 10.3, the observed value of the test statistic is

$$z = \frac{\widehat{p} - p_0}{\sqrt{\dfrac{p_0(1 - p_0)}{n}}} = \frac{.18 - \tfrac{1}{7}}{\sqrt{\dfrac{\left(\tfrac{1}{7}\right)\left(\tfrac{6}{7}\right)}{200}}} = 1.50.$$

From Table IV, Appendix 3, we see that $P(z > 1.645) = .05$. Hence, we take $z > 1.645$ as the rejection region. Since the observed value of t is not in the rejection region, there is not sufficient evidence of heart attacks being more likely to occur on Monday than any other day of the week.

b.   The test is suggested by the data. Our hypothesis should test a question of interest formulated before the experiment is conducted. Data is then gathered to support or refute the given hypothesis.

c.   Monday is popularly known as the most stressful workday of the week; one has a long five days until the weekend. One might wish to investigate if this extra stress results in a disproportionate amount of heart attacks.

14.5   Similar to previous exercises. The null hypothesis to be tested is

$$H_0: \ p_1 = .69; \ p_2 = .21; \ p_3 = .07; \ p_4 = .03$$

against the alternative that at least one of these probabilities is incorrect. The observed and expected cell counts are shown below.

User	1	2	3	4
n_i	102	32	12	4
$E(n_i)$	103.5	31.5	10.5	4.5

The test statistic is

$$X^2 = \frac{(02 - 103.5)^2}{103.5} + \cdots + \frac{(4 - 4.5)^2}{4.5} = .2995$$

and the p-value with $k - 1 = 3$ d.f. is p-value $> .95$. The null hypothesis is not rejected and we cannot conclude that the figures given are inaccurate.

14.7   This is similar to Example 14.2. Using $\overline{Y}$ to estimate the Poisson parameter λ, calculate

$$\overline{y} = \frac{\sum y_i f_i}{n} = \frac{0(56) + 1(104) + \ldots + 10(2) + 11(0) + 19(1)}{400} = \frac{976}{400} = 2.44$$

The expected cell counts are the estimated as

$$\widehat{E}(n_i) = n\widehat{p}_i = 400 \; \frac{e^{-2.44}(2.44)^{y_i}}{y_i!}$$

Notice that when $Y = 7$, the expected cell count drops below 5. Hence the final group is $Y \geq 7$. The observed and estimated expected cell counts are shown in the table at the right.

No. of Colonies	n_i	$\widehat{p}_i$	$\widehat{E}(n_i)$
0	56	.087	34.86
1	104	.2127	85.07
2	80	.2595	103.73
3	62	.2110	84.41
4	42	.1287	51.49
5	27	.0628	25.13
6	9	.0255	10.22
7 or more	20		$400 - 394.96$ $= 5.04$

Then

$$X^2 = \frac{(56 - 34.86)^2}{34.86} + \ldots + \frac{(20 - 5.04)^2}{5.04} = 69.42.$$

The rejection region, based on $k - 2 = 6$ degrees of freedom (see Example 14.2), is $X^2 \geq 12.59$, and the null hypothesis is rejected. The data do not fit the Poisson distribution.

14.9   a.   The table of estimated expected cell counts is

	JAS Score		
3-year follow-up	Less than -5	-5 to 5	Greater than 5
Died	17.09	16.24	15.67
Alive	162.91	154.76	149.33

The test statistic is

$$X^2 = \frac{(21 - 17.09)^2}{17.09} + \frac{(17 - 16.24)^2}{16.24} + \ldots + \frac{(154 - 149.33)^2}{149.33} = 2.56$$

with $(r - 1)(c - 1) = 2$ d.f., since

$$X^2 = 2.56 < 5.99 = \chi^2_{.05},$$

we do not reject H_0 at the $\alpha = .05$ level. There is insufficient evidence to indicate a dependence between mortality rate and level of Type A behavior.

b. Since $X^2 = 2.56 < 4.61 = \chi^2_{.10}$, $p > .10$. Thus the results are not significant.

14.11 **a.** $X^2 = \sum\limits_{j=1}^{c} \sum\limits_{i=1}^{r} \dfrac{[n_{ij} - E(\hat{n}_{ij})]^2}{E(\hat{n}_{ij})} = \sum\limits_{j=1}^{c} \sum\limits_{i=1}^{r} \left[\dfrac{\left(n_{ij} - \dfrac{r_i c_j}{n}\right)^2}{\dfrac{r_i c_j}{n}} \right]$

$= n \sum\limits_{j=1}^{c} \sum\limits_{i=1}^{r} \left[\dfrac{n_{ij}^2 - \dfrac{2 n_{ij} r_i c_j}{n} + \dfrac{r_i^2 c_j^2}{n^2}}{r_i c_j} \right]$

$= n \left[\sum\limits_{j=1}^{c} \sum\limits_{i=1}^{r} \dfrac{n_{ij}^2}{r_i c_j} - 2 \sum\limits_{j=1}^{c} \sum\limits_{i=1}^{r} \dfrac{n_{ij}}{n} + \sum\limits_{j=1}^{c} \sum\limits_{i=1}^{r} \dfrac{r_i c_j}{n^2} \right]$

$= n \left[\sum\limits_{j=1}^{c} \sum\limits_{i=1}^{r} \dfrac{n_{ij}^2}{r_i c_j} - \dfrac{2n}{n} + \dfrac{\left(\sum\limits_{j=1}^{c} c_j\right)\left(\sum\limits_{i=1}^{r} r_i\right)}{n^2} \right]$

$= n \left[\sum\limits_{j=1}^{c} \sum\limits_{i=1}^{r} \dfrac{n_{ij}^2}{r_i c_j} - \dfrac{2n}{n} + \dfrac{n^2}{n^2} \right] = n \left[\sum\limits_{j=1}^{c} \sum\limits_{i=1}^{r} \dfrac{n_{ij}^2}{r_i c_j} - 1 \right]$

b. When every entry in the contingency is multiplied by the same $k > 0$,

$$X^2 = kn \left[\sum\limits_{j=1}^{c} \sum\limits_{i=1}^{r} \dfrac{(kn_{ij})^2}{(kr_i)(kc_j)} - 1 \right] = kn \left[\sum\limits_{j=1}^{c} \sum\limits_{i=1}^{r} \dfrac{n_{ij}^2}{r_i c_j} - 1 \right]$$

Thus, if the pattern of responses is the same, then the X^2 will be increased k times.

14.13 **a.-b.** The MINITAB printouts below are used to analyze the data for the two contingency tables. The observed values of the test statistics are $X^2 = 19{,}043$ and $X^2 = 60.139$, for faculty and student responses, respectively. The rejection region, with $(3)(2) = 6$ d.f., is $X^2 > 16.81$ with $\alpha = .01$, and H_0 is rejected for both cases.

MTB > CHISQ C1-C3
Expected counts are printed below observed counts

	C1	C3	C3	Total
1	4	0	0	4
	1.53	1.47	1.00	
2	15	12	3	30
	11.50	11.00	7.50	
3	2	7	7	16
	6.13	5.87	4.00	
4	2	3	5	10
	3.83	3.67	2.50	
Total	23	22	15	60

ChiSq = 3.968 + 1.467 + 1.000 +
        1.065 + 0.091 + 2.700 +
        2.786 + 0.219 + 2.250 +
        0.877 + 0.121 + 2.500 = 19.043

df = 6

7 cells with expected counts less than 5.0

MTB > CHISQ C4-C6

Expected counts are printed below observed counts

	C4	C5	C6	Total
1	19	6	2	27
	6.88	9.56	10.57	
2	19	41	27	87
	22.16	30.80	34.04	
3	3	7	31	41
	10.44	14.52	16.04	
4	0	3	3	6
	1.53	2.12	2.35	
Total	41	57	63	161

ChiSq = 21.379 + 1.325 + 6.944 +
         0.449 + 3.377 + 1.457 +
         5.303 + 3.891 + 13.943 +
         1.528 + 0.361 + 0.181 = 60.139

df = 6

3 cells with expected counts less than 5.0

c. Expected values of some cell counts are less than 5. Thus, the χ^2 approximation to the test statistic X^2 may not be valid.

14.15 a. The estimated expected cell counts are calculated as

$$\widehat{E}(n_{ij}) = \frac{r_i c_j}{n},$$

and are shown in parentheses in the table below.

Results	Life Threatening Complications		Total
	No	Yes	
Negative	166 (151.69)	1 (15.31)	135
Positive	260 (274.31)	42 (27.69)	302
Total	426	43	469

Then

$$X^2 = \sum_{ij} \frac{\left[n_{ij} - \widehat{E}(n_{ij}) \right]^2}{\widehat{E}(n_{ij})} = \frac{(166 - 151.69)^2}{151.69} + \ldots + \frac{(42 - 27.69)^2}{27.69} = 22.94$$

With $\alpha = .05$ and $1(1) = 1$ d.f., a one-tailed rejection region is found using Table 5 to be $X^2 > \chi^2_{.05} = 3.84146$.

Since $X^2 = 22.87$, H_0 is rejected. The probability of having life-threatening myocardial infarctions is dependent upon the electrocardiogram results.

b. The value $X^2 = 22.87$ is greater than $7.88 = \chi^2_{.005}$, so that the p-value $< .005$.

14.17 Three different contingency tables are given and the tests proceed as in previous exercises. Tables are provided for each situation.

a. 20 (13.44)   4 (10.56) $X^2 = 13.99$; rejection region is $X^2 \geq 3.84$; reject
        8 (14.56)   18 (11.44) H_0: species segregate.

b. 4 (10.56)   20 (13.44) $X^2 = 13.99$; rejection region is $X^2 \geq 3.84$; reject
        18 (11.44)   18 (14.56) H_0: species are overly mixed.

c. 20 (18.24   4 (5.76) $X^2 = 1.36$; rejection region is $X^2 \geq 3.84$; do not
        18 (19.76)   8 (6.24)   reject H_0.

14.19 Refer to Section 10.3. The two-tailed z test was used to test a hypothesis

$$H_0: \; p_1 - p_2 \qquad \text{vs.} \qquad H_a: \; p_1 \neq p_2.$$

The test statistic was

$$Z = = \frac{\hat{p}_1 - \hat{p}_2}{\sqrt{\hat{p}\,\hat{q}\left[\left(\frac{1}{n_1}\right) + \left(\frac{1}{n_2}\right)\right]}} \quad \text{and} \quad Z^2 = \frac{(\hat{p}_1 - \hat{p}_2)^2}{\hat{p}\,\hat{q}\left(\frac{n_1 + n_2}{n_1 n_2}\right)} = \frac{n_1 n_2 (\hat{p}_1 - \hat{p}_2)^2}{(n_1 + n_2)\hat{p}\,\hat{q}}$$

Notice that

$$\hat{p} = \frac{y_1 + y_2}{n_1 + n_2} = \frac{n_1 \hat{p}_1 + n_2 \hat{p}_2}{n_1 + n_2}$$

Now consider the chi-square test statistic used in Exercise 14.17. The hypothesis to be tested is

$$H_0: \text{ Independence of classification} \quad \text{vs.} \quad H_a: \text{ dependence of classification}$$

That is, the null hypothesis asserts that the percentage of patients who show improvement is independent of whether or not they have been treated with the serum. If the null hypothesis is true, then $p_1 = p_2$. Hence the two tests are designed to test the same hypothesis.

In order to show that Z^2 is equivalent to X^2, it is necessary to rewrite the chi-square test statistic in terms of quantities $\hat{p}_1$, $\hat{p}_2$, n_1, and n_2.

(1) Consider n_{11}, the observed number of treated patients who have improved. Since $\hat{p}_1 = \frac{n_{11}}{n_1}$, we have $n_{11} = n_1 \hat{p}_1$. Similarly, $n_{21} = n_1 \hat{q}_1$, $n_{12} = n_2 \hat{p}_2$, $n_{22} = n_2 \hat{q}_2$.

(2) The estimated expected cell counts are calculated under the assumption that the null hypothesis is true. Consider

$$\hat{E}(n_{11}) = \frac{r_1 c_1}{n} = \frac{(n_{11} + n_{12})(n_{11} + n_{21})}{n_1 + n_2} = \frac{(y_1 + y_2)(n_{11} + n_{21})}{n_1 + n_2} = n_1 \hat{p}$$

Similarly, $\hat{E}(n_{21}) = n_1 \hat{q}$, $\hat{E}(n_{12}) = n_2 \hat{p}$, and $\hat{E}(n_{22}) = n_2 \hat{q}$. The table of observed and estimated expected cell counts is shown below.

	Treated	Untreated
Improved	$n_1 \hat{p}_1 \ (n_1 \hat{p})$	$n_2 \hat{p}_2 \ (n_2 \hat{p})$
Not Improved	$n_1 \hat{q}_1 \ (n_1 \hat{q})$	$n_2 \hat{q}_2 \ (n_2 \hat{q})$

Then

$$X^2 = \sum_i \sum_j \frac{\left[n_{ij} - \hat{E}(n_{ij})\right]^2}{\hat{E}(n_{ij})}$$

$$= \frac{n_1^2 (\hat{p}_1 - \hat{p})^2}{n_1 \hat{p}} + \frac{n_1^2 (\hat{q}_1 - \hat{q})^2}{n_1 \hat{q}} + \frac{n_2^2 (\hat{p}_2 - \hat{p})^2}{n_2 \hat{p}} + \frac{n_2^2 (\hat{q}_2 - \hat{q})^2}{n_2 \hat{q}}$$

$$= \frac{n_1(\widehat{p}_1 - \widehat{p})^2}{\widehat{p}} + \frac{n_1\left[(1 - \widehat{p}_1) - (1 - \widehat{p})\right]^2}{\widehat{q}} + \frac{n_2(\widehat{p}_2 - \widehat{p})^2}{\widehat{p}}$$

$$+ \frac{n_2\left[(1 - \widehat{p}_2) - (1 - \widehat{p})\right]^2}{\widehat{q}}$$

$$= \frac{(1 - \widehat{p})n_1(\widehat{p}_1 - \widehat{p})^2 + n_1\widehat{p}(\widehat{p}_1 - \widehat{p})^2}{\widehat{p}\,\widehat{q}} + \frac{(1 - \widehat{p})n_2(\widehat{p}_2 - \widehat{p})^2 + n_2\widehat{p}(\widehat{p}_2 - \widehat{p})^2}{\widehat{p}\,\widehat{q}}$$

$$= \frac{n_1(\widehat{p}_1 - \widehat{p})^2}{\widehat{p}\,\widehat{q}} + \frac{n_2(\widehat{p}_2 - \widehat{p})^2}{\widehat{p}\,\widehat{q}}$$

Substituting for $\widehat{p}$, we obtain

$$X^2 = \frac{n_1}{\widehat{p}\,\widehat{q}}\left(\frac{n_1\widehat{p}_1 + n_2\widehat{p}_1 - n_1\widehat{p}_1 - n_2\widehat{p}_2)}{n_1 + n_2}\right)^2 + \frac{n_2}{\widehat{p}\,\widehat{q}}\left(\frac{n_1\widehat{p}_2 + n_2\widehat{p}_2 - n_1\widehat{p}_1 - n_2\widehat{p}_2)}{n_1 + n_2}\right)^2$$

$$= \frac{n_1 n_2(n_1 + n_2)(\widehat{p}_1 - \widehat{p}_2)^2}{\widehat{p}\,\widehat{q}(n_1 + n_2)^2} = \frac{n_1 n_2(\widehat{p}_1 - \widehat{p}_2)^2}{\widehat{p}\,\widehat{q}(n_1 + n_2)}$$

Note that X^2 is identical to Z^2, as defined at the beginning of the exercise.

14.21 The contingency table, including column and row totals and the estimated expected cell counts, follows.

Age Group	More	Less	Same	Total
35–54	90 (65.00)	18 (39.00)	92 (96.00)	200
55+	40 (65.00)	60 (39.00)	100 (96.00)	200
Total	130	78	192	400

The test statistic is

$$X^2 = \frac{(90 - 65)^2}{65} + \frac{(18 - 39)^2}{39} + \ldots + \frac{(100 - 96)^2}{96} = 42.179$$

using computer accuracy. The rejection region with 2 d.f. is $X^2 > 9.21034$ and H_0 is rejected. The investing pattern of the baby-boomer group differs from that of the older group.

14.23 Similar to previous exercises, except that observed cell counts must be obtained from the given information as

$$n_i = \frac{(\text{number of samples})_i \times (\text{percentage})_i}{100}.$$

The approximate observed and estimated expected cell counts are shown in the following table.

	Age 1	2	3	4	5	6	7	Total
Have nodules	23 (54.80)	25 (19.72)	35 (30.14)	18 (11.83)	52 (34.79)	159 (157.77)	11 (13.95)	323
No nodules	366 (334.20)	115 (120.28)	179 (183.86)	66 (72.17)	195 (212.21)	961 (962.23)	88 (85.05)	1970
Total	389	140	214	84	247	1120	99	2293

The test statistic is

$$X^2 = \frac{(23 - 54.80)^2}{54.80} + \ldots + \frac{(88 - 85.05)^2}{85.05} = 38.429$$

and the rejection region with $\alpha = .05$ and $(r-1)(c-1) = 6$ degrees of freedom is

$$X^2 \geq 12.59.$$

The null hypothesis is rejected and we conclude that age and probability of finding nodules are dependent.

14.25 a. We wish to test a hypothesis of equivalence among the proportions of residents with lung disease in the four areas, which implies a null hypothesis of independence between the row and column classifications. The estimated expected and observed cell counts are given below.

	City A	City B	Nonurban Area 1	Nonurban Area 2	Total
Number with	34 (28.75)	42 (28.75)	21 (28.75)	18 (28.75)	115
Number without	366 (371.25)	358 (371.25)	379 (371.25)	382 (371.25)	1485
Total	400	400	400	400	1600

$$X^2 = \frac{(34 - 28.75)^2}{28.75} + \ldots + \frac{(382 - 371.25)^2}{371.25} = 14.19$$

The rejection region with $(r-1)(c-1) = 3$ degrees of freedom and $\alpha = .05$ is $X^2 \geq 7.81$, and the null hypothesis is rejected. There is a dependence between the proportion of people with lung disease and the location (i.e., there is a difference in the proportions with lung disease for the four locations).

b. Cigarette smokers probably should have been excluded from the sample. Cigarette smoking is a major contributor to lung disease, and a greater proportion of smokers in one or more of the locations would affect the proportions with lung disease. The number of people who smoke is an unknown variable in the four locations. It is quite possible that more people who live in urban areas smoke, owing perhaps to increased advertising or tension in the cities. The result may then indicate a dependence between lung disease and location that is not caused by air pollution but by an increased proportion of people in certain locations who smoke.

14.27 The table of estimated cell counts is shown below.

	Rhode Island	Colorado	California	Florida
Participate	63.62	78.63	97.88	97.88
Do Not Participate	131.38	162.37	202.12	202.12

The observed value of the test statistic is calculated to be $X^2 = 21.51$. Since the degrees of freedom associated with X^2 are $(r-1)(c-1) = 1(3) = 3$, the rejection region is $X^2 \geq \chi^2_{3,.01} = 11.3449$. Hence the null hypothesis of independence is rejected.

14.29 This is similar to previous exercises. The estimated expected cell counts are given in the table below.

	Student	Faculty	Administration
Favor	237.44	114.16	50.40
Oppose	153.56	73.84	32.60

The rejection region, based on $(r-1)(c-1) = 1(2) = 2$ degrees of freedom, is $X^2 \geq \chi^2_{2,.05} = 5.99$. The observed value of the test statistic is $X^2 = 6.18$, which falls in the rejection region, and the null hypothesis of independence is rejected. Since $5.99 < 6.18 < 7.38$, $.025 < p\text{-value} < .05$.

14.31 a. The data are analyzed as a 5×3 contingency table with estimated expected cell counts shown in parentheses.

Level	0 to 1	2	3 to 6	Total
1	3 (11.00)	21 (14.67)	20	44
2	2 (5.75)	12 (7.67)	9	23
3	26 (19.25)	20 (25.67)	31	77
4	13 (10.25)	10 (13.67)	18	41
5	7 (4.75)	5 (6.33)	7	19
Total	51	68	84	204

The test statistic is

$$X^2 = \frac{(3 - 11.00)^2}{11.00} + \frac{(21 - 14.67)^2}{14.67} + \ldots + \frac{(7 - 7.92)^2}{7.92} = 20.513$$

using calculator accuracy.

b.   The derived value, $X^2 = 20.513$, lies between $\chi_{.01}^2$ and $\chi_{.005}^2$ with 8 d.f.   Hence $.005 < p\text{-value} < .01$.   The author's value is correct.

c.   Based on the p-value observed in part (b), H_0 is rejected.   The level of angina is dependent on the level of coronary artery obstruction.

14.33   For a chi-square goodness-of-fit test of the given data, it is necessary that the values n_i and $E(n_i)$ be known for each of the five cells.   The n_i (the number of measurements falling in the ith cell) are given.   However, $E(n_i) = np_i$ must be calculated.   Remember that p_i is the probability that a measurement falls in the ith cell.   The hypothesis to be tested will be

H_0: the experiment is binomial   vs. H_a: the experiment is not binomial

Let $Y = $ number of successes and $p = $ probability of success on a single trial.   Then assuming the null hypothesis to be true, we have

$$p_0 = P(Y = 0) = \binom{4}{0} p^0 (1 - p)^4 \qquad\qquad p_3 = P(Y = 3) = \binom{4}{3} p^3 (1 - p)^1$$

$$p_1 = P(Y = 1) = \binom{4}{1} p^1 (1 - p)^3 \qquad\qquad p_4 = P(Y = 4) = \binom{4}{4} p^4 (1 - p)^0$$

$$p_2 = P(Y = 2) = \binom{4}{2} p^2 (1 - p)^2$$

Hence, once an estimate for p is obtained, the expected cell frequencies can be calculated by using the above probabilities.   Note that each of the 100 experiments consists of four trials and hence the complete experiment involves a total of 400 trials.   The maximum likelihood estimator of p is $p = \frac{Y}{n}$ (as in Chapter 10).   Thus,

$$\hat{p} = \frac{y}{n} = \frac{\text{number of successes}}{\text{number of trials}} = \frac{0(11) + 1(17) + 2(42) + 3(21) + 4(9)}{400} = \frac{1}{2}$$

The experiment consisting of four trials was repeated 100 times.   There is a total of 400 trials in which the result "no successes in four trials" was observed 11 times, the result "one success in four trials" was observed 17 times, and so on.   Then

$$p_0 = \binom{4}{0}\left(\frac{1}{2}\right)^0 \left(\frac{1}{2}\right)^4 = \frac{1}{16} \qquad\qquad p_3 = \binom{4}{3}\left(\frac{1}{2}\right)^3 \left(\frac{1}{2}\right)^1 = \frac{4}{16}$$

$$p_1 = \binom{4}{1}\left(\frac{1}{2}\right)^1 \left(\frac{1}{2}\right)^3 = \frac{4}{16} \qquad\qquad p_4 = \binom{4}{4}\left(\frac{1}{2}\right)^4 \left(\frac{1}{2}\right)^0 = \frac{1}{16}$$

$$p_2 = \binom{4}{2}\left(\frac{1}{2}\right)^2\left(\frac{1}{2}\right)^2 = \frac{6}{16}$$

The observed and expected cell frequencies are shown in the table below.

			y		
	0	1	2	3	4
n_i	11	17	42	21	9
$E(n_i)$	6.25	25.00	37.50	25.00	6.25

The test statistic is

$$X^2 = \frac{(11 - 6.25)^2}{6.25} + \ldots + \frac{(9 - 6.25)^2}{6.25} = 8.56$$

In order to set up the rejection region, we must first determine the degrees of freedom associated with the test statistic. Two restrictions are placed upon the cell counts:

(1) $n_1 + n_2 + \ldots + n_5 = 100$.

(2) The binomial parameter p is estimated by using a linear combination of the n_i.

The number of degrees of freedom is equal to the number of cells (k) less 1 degree of freedom for each independent linear restriction placed on the cell frequencies. Therefore, $(k - 2) = 5 - 2 = 3$, and the critical value of $\chi^2 = 7.81$, using $\alpha = .05$. The test statistic falls in the rejection region, and we conclude that the experiment does not fulfill the properties of a binomial experiment. The p-value is between .025 and .05.

14.35 In order to find the maximum likelihood estimator of p_i, the probability of falling in row i, consider row i as a single cell with r_i observations falling in this cell. Then the variables r_1, r_2, ..., r_r follow a multinomial distribution with parameters n, p_1, p_2, ..., p_r. Hence the likelihood function is

$$L = \frac{n!}{r_1! r_2! \cdots r_r!}\, p_1^{r_1} p_2^{r_2} \cdots p_r^{r_r} = K \prod_{j=1}^{r} p_j^{r_j}$$

so that

$$\ln L = \ln K + \sum_{j=1}^{r} r_j \ln p_j \qquad \text{with} \qquad \sum_{j=1}^{r} p_j = 1$$

Notice that because of the above restriction, we may write $p_r = 1 - \sum_{j=1}^{r-1} p_j$ and that p_r is really a function of p_i for $i = 1, 2, \ldots, r - 1$. Also, $r_r = n - \sum_{j=1}^{r-1} r_j$. Hence

$$\ln L = \ln K + \sum_{j=1}^{r-1} r_j \ln p_j + \left(n - \sum_{j=1}^{r-1} r_j \right) \ln \left(1 - \sum_{j=1}^{r-1} p_j \right)$$

Now

$$\frac{d(\ln L)}{dp_i} = \frac{r_i}{p_i} - \frac{n - \sum_{j=1}^{r-1} r_j}{1 - \sum_{j=1}^{r-1} p_j} \qquad \text{for} \qquad i = 1, 2, \ldots, r-1$$

Setting these $(r-1)$ equations equal to zero, we have, for $i = 1, 2, \ldots, r-1$,

$$(^*) \quad r_i \left(1 - \sum_{j=1}^{r-1} \widehat{p}_j \right) = \widehat{p}_i \left(n - \sum_{j=1}^{r-1} r_j \right)$$

In order to solve the $(r-1)$ equations simultaneously, add them together to obtain

$$\sum_{i=1}^{r-1} r_i \left(1 - \sum_{j=1}^{r-1} \widehat{p}_j \right) = \sum_{i=1}^{r-1} \widehat{p}_i \left(n - \sum_{j=1}^{r-1} r_j \right) \qquad \text{or} \qquad n \sum_{j=1}^{r-1} \widehat{p}_j = \sum_{j=1}^{r-1} r_j$$

or

$$\sum_{j=1}^{r-1} \widehat{p}_j = \frac{1}{n} \left(\sum_{j=1}^{r-1} r_j \right)$$

Substituting in (*), we have

$$r_i \left(1 - \frac{1}{n} \sum_{j=1}^{r-1} r_j \right) = \widehat{p}_i \left(n - \sum_{j=1}^{r-1} r_j \right) \Rightarrow \widehat{p}_i = \frac{r_i \left(1 - \frac{1}{n} \sum_{j=1}^{r-1} r_j \right)}{n - \sum_{j=1}^{r-1} r_j} = \frac{r_i}{n}$$

14.37 The problem describes a multinomial experiment with $k = 4$ cells. Under the null hypothesis, the four cell probabilities are

$$p_1 = \frac{p}{2} \qquad\qquad p_2 = \frac{p^2}{2} + pq \qquad\qquad p_2 = \frac{q}{2} \qquad\qquad p_4 = \frac{q^2}{2}$$

where p is unspecified, but $p + q = 1$. In order to use the chi-square goodness-of-fit test for this genetic model, we must first obtain an estimate of p. We will use a maximum likelihood estimate for p, with the likelihood function given by

$$L = \frac{n!}{n_1! \, n_2! \, n_3! \, n_4!} \left(\frac{p}{2} \right)^{n_1} \left(\frac{p^2}{2} + pq \right)^{n_2} \left(\frac{q}{2} \right)^{n_3} \left(\frac{q^2}{2} \right)^{n_4}$$

$$= \ = \frac{n!}{\prod\limits_{i=1}^{4} n_i \, 2^{\Sigma n_i}} \, p^{n_1} \left(p^2 + 2pq \right)^{n_2} (1-p)^{n_3} (1-p)^{2n_4}$$

$$= kp^{n_1} \left[p(2-p) \right]^{n_2} (1-p)^{n_3 + 2n_4} = Kp^{n_1 + n_2} (2-p)^{n_2} (1-p)^{n_3 + 2n_4}$$

so that

$$\ln L = \ln K + (n_1 + n_2) \ln p + n_2 \ln (2-p) + (n_3 + 2n_4) \ln (1-p)$$

and

$$\frac{d(\ln L)}{dp} = \frac{n_1 + n_2}{p} - \frac{n_2}{2-p} - \frac{n_3 + 2n_4}{1-p}$$

Setting the derivative equal to zero and solving for $\hat{p}$, we obtain

$$(n_1 + n_2)(2 - \hat{p})(1 - \hat{p}) - n_2 \hat{p}(1 - \hat{p}) - (n_3 + 2n_4)(\hat{p})(2 - \hat{p}) = 0$$

$$(n_1 + 2n_2 + n_3 + 2n_4)\hat{p}^2 - (3n_1 + 4n_2 + 2n_3 + 4n_4)\hat{p} + 2(n_1 + n_2) = 0$$

That is, $\hat{p}$ is the root (between 0 and 1) of a quadratic equation in the form

$$a\hat{p}^2 + b\hat{p} + c = 0,$$

with

$$a = n_1 + 2n_2 + n_3 + 2n_4 = 3040 \qquad\qquad b = -(3n_1 + 4n_2 + 2n_3 + 4n_4) = -6960$$

$$c = 2(n_1 + n_2) = 3824$$

The solution will be

$$\hat{p} = \frac{-b - \sqrt{b^2 - 4ac}}{2a} = \frac{6960 - \sqrt{1{,}941{,}760}}{6080} = .9155$$

Now the estimated cell probabilities, estimated expected cell counts, and the value of the test statistic can be obtained. (See the table below.)

$\hat{p}_i$	$\hat{E}(n_i)$	n_i
$\dfrac{\hat{p}}{2}$	915.50	880
$\left(\dfrac{\hat{p}}{2} \right) + \hat{p}\hat{q} = .49643$	992.86	1032
$\dfrac{\hat{q}}{2} = .04225$	84.50	80
$\dfrac{\hat{q}^2}{2} = .00357$	7.14	8

Then

$$X^2 = \frac{(880 - 915.5)^2}{915.5} + \ldots + \frac{(8 - 7.14)^2}{7.14} = 3.26$$

The degrees of freedom are $(k - 1 - 1) = 2$, since:

$$\sum_{i=1}^{k} n_i = n$$

and

(2) the observed counts, n_i, were used to calculate $\hat{p}$.

For $\alpha = .05$, the critical value of X^2 is $\chi^2_{2, .05} = 5.99$, and the hypothesized model is not rejected.

14.39 In this exercise there are four binomial experiments performed, one at each of four dosage levels. Let n_i = number of survivors for dose i, $p_i = P(\text{insect survives at dose } i)$, and $q_i = P(\text{insect dies at dose } i)$. The hypothesis of interest is

$$H_0: \ p_1 = 1 + \beta, \ p_2 = 1 + 2\beta, \ p_3 = 1 + 3\beta, \ p_4 = 1 = 4\beta$$

Notice that this automatically implies that $q_1 = -\beta$, $q_2 = -2\beta$, $q_3 = -3\beta$, $q_4 = -4\beta$. It is necessary to obtain an estimate of β, which can be done by using the method of maximum likelihood.

$$L = \prod_{i=1}^{4} \binom{1000}{n_i} (1 + i\beta)^{n_i} (-i\beta)^{1000 - n_i} = K \prod_{i=1}^{4} (1 + i\beta)^{n_i} (\beta)^{1000 - n_i}$$

$$\ln L = \ln K + \sum_{i=1}^{4} n_i \ln (1 + i\beta) + \sum_{i=1}^{4} (1000 - n_i) \ln \beta$$

Then

$$\frac{d(\ln L)}{d\beta} = \sum_{i=1}^{4} \frac{i n_i}{1 + i\beta} + \frac{1}{\beta} \sum_{i=1}^{4} (1000 - n_i)$$

Putting $\dfrac{d(\ln L)}{d\beta} = 0$ and expanding, we obtain a quartic equation in $\hat{\beta}$.

$$\left(4000 - \sum_{i=1}^{4} n_i \right)(1 + \hat{\beta})(1 + 2\hat{\beta})(1 + 3\hat{\beta})(1 + 4\hat{\beta}) + n_1 \hat{\beta}(1 + 2\hat{\beta})(1 + 3\hat{\beta})(1 + 4\hat{\beta})$$

$$+ 2n_2 \hat{\beta}(1 + \hat{\beta})(1 + 3\hat{\beta})(1 + 4\hat{\beta}) + 3n_3 \hat{\beta}(1 + \hat{\beta})(1 + 2\hat{\beta})(1 + 4\hat{\beta})$$

$$+ 4n_4 \hat{\beta}(1 + \hat{\beta})(1 + 2\hat{\beta})(1 + 3\hat{\beta})$$

$$= 0$$

or

$$\left(4000 - \sum_{i=1}^{4} n_i\right) + \left[10\left(4000 - \sum_{i=1}^{4} n_i\right) + n_1 + 2n_2 + 3n_3 + 4n_4\right]\widehat{\beta}$$

$$+ \left[35\left(4000 - \sum_{i=1}^{4} n_i\right) + 9n_1 + 16n_2 + 21n_3 + 24n_4\right]\widehat{\beta}^2$$

$$+ \left[50\left(4000 - \sum_{i=1}^{4} n_i\right) + 26n_1 + 38n_2 + 42n_3 + 44n_4\right]\widehat{\beta}^3$$

$$+ \left[24(4000 - \sum n_i) + 24 \sum_{i=1}^{4} n_i\right]\widehat{\beta}^4$$

$$= 0$$

Substituting $n_1 = 820$, $n_2 = 650$, $n_3 = 310$, and $n_4 = 50$, we obtain

$$217 + 2495\widehat{\beta} + 10{,}144\widehat{\beta}^2 + 16{,}974\widehat{\beta}^3 + 9600\widehat{\beta}^4 = 0$$

This equation can best be solved iteratively, using a "guessed" value for $\widehat{\beta}$ based on a graph of p_i against D_i. Recall that β is the slope of the line $p = 1 + \beta D$, which is a line with intercept 1. Using the four points available to us, a good guess for the slope of the line would be $\beta = -.2$ (see Figure 14.1). We use this as an initial estimate and solve the equation until a value is found such that $f(\widehat{\beta}) = 0$.

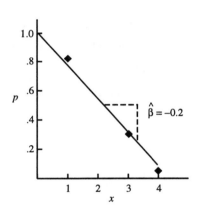

Figure 14.1

$\widehat{\beta}$	$-.2$	$-.22$	$-.225$	$-.226$	$-.228$	$-.230$	$-.232$	$-.233$
$f(\widehat{\beta})$	3.328	.819	.424	.355	.226	.110	.004	$-.040$

Hence, our estimate is

$$\widehat{\beta} = -.232,$$

so that

$$\widehat{p}_1 = 1 + \widehat{\beta} = .768 \qquad\qquad \widehat{p}_3 = 1 + 3\widehat{\beta} = .304$$

$$\widehat{p}_2 = 1 + 2\widehat{\beta} = .536 \qquad\qquad \widehat{p}_4 = 1 + 4\widehat{\beta} = .072$$

and the table of observed and estimated expected cell counts is as shown below.

	Dosage			
	1	2	3	4
Survived	820	650	320	50
	(768)	(536)	(304)	(72)
Died	180	350	690	950
	(232)	(464)	(696)	(928)

The calculated test statistic is

$$X^2 = 74.8$$

and the rejection region based on a chi-square variable with $(8 - 4 - 1) = 3$ degrees of freedom is

$$X^2 \geq \chi^2_{3,\,.05} = 7.81$$

The null hypothesis is rejected and the hypothesis $p = 1 + \beta D$ is contradicted.

Note: There are 8 cells, but 5 restrictions:

(1) $p_i + q_i = 1$ for $i = 1, 2, 3, 4$.

(2) Estimate β using n_i.

CHAPTER 15 NONPARAMETRIC STATISTICS

15.1 It is necessary that α (the probability of rejecting the null hypothesis when it is true) take values between $\alpha = .01$ and $\alpha = .15$. Assuming the null hypothesis to be true, the two populations are identical and consequently $p = P(A$ exceeds B for a given pair of observations) is $\frac{1}{2}$.

The binomial probability distribution was discussed in Chapter 3. In particular, it was noted that the distribution of the random variable Y is symmetrical about the mean np when $p = \frac{1}{2}$. For example, with $n = 25$, $P(Y = 0) = P(Y = 25)$. That is,

$$P(Y = 0) = \binom{25}{0}\left(\frac{1}{2}\right)^0 \left(\frac{1}{2}\right)^{25} = \left(\frac{1}{2}\right)^{25} \quad \text{and} \quad P(Y = 25) = \binom{25}{25}\left(\frac{1}{2}\right)^{25}\left(\frac{1}{2}\right)^0 = \left(\frac{1}{2}\right)^{25}$$

Similarly, $P(Y = 1) = P(Y = 24)$, and so on. Hence is it necessary to determine a rejection region such that $.005 \leq \frac{\alpha}{2} \leq .075$, where $P(Y \leq a) = \frac{\alpha}{2}$ is given in Table 1. Indexing $n = 25$, $p = \frac{1}{2}$ in Table 1, it is found that the critical values of Y that determine the lower tail of the desired rejection region are $Y = 6$, $Y = 7$, and $Y = 8$. The reader may verify that the corresponding values for the upper tail of the rejection region are $Y = 19$, $Y = 18$, and $Y = 17$. The significance levels and corresponding rejection regions are shown in the table below.

$\frac{\alpha}{2}$	α	Rejection Region
.007	.014	$Y \leq 6;\ Y \geq 19$
.022	.044	$Y \leq 7;\ Y \geq 18$
.054	.108	$Y \leq 8;\ Y \geq 17$

15.3 Let $p = P($judge favors mixture $B)$ and $M = $ number of judges favoring mixture B. The hypothesis of interest is

$$H_0:\ p = \frac{1}{2} \qquad \text{vs.} \qquad H_a:\ p \neq \frac{1}{2},$$

with $n = 10$. The observed value of M is m $= 2$. The p-value is

$$2P(M \leq 2) = 2(.055) = .11 > .05.$$

There is no significant difference between the tastes of A and B.

15.5   **a.**   The hypothesis to be tested is

$$H_0: \; p = \tfrac{1}{2} \qquad \text{vs.} \qquad H_a: \; p \neq \tfrac{1}{2}$$

where $p = P(\text{response for stimulus 1 exceeds that for stimulus 2})$. Then the test statistic is M, the number of times the response for stimulus 1 exceeds that doe stimulus 2. Again, denote a positive difference by a plus sign and a negative difference by a minus sign. Then M will be equivalent to the number of plus signs observed. These signs are

Subject	1	2	3	4	5	6	7	8	9
Sign	−	−	+	−	−	−	−	+	−

and $m = 2$. The next step is to select a rejection region such that $\alpha = P\left(\text{reject } H_0 | p = \tfrac{1}{2}\right)$ is close to .05. Using the rejection region $M = 0, 1, 8, 9$, we find that

$$\alpha = P\left(M = 0,\, 1,\, 8,\, 9 \,\big|\, p = \tfrac{1}{2}\right) = \binom{9}{0}\left(\tfrac{1}{2}\right)^0\left(\tfrac{1}{2}\right)^9 + \binom{9}{1}\left(\tfrac{1}{2}\right)^1\left(\tfrac{1}{2}\right)^8 + \binom{9}{8}\left(\tfrac{1}{2}\right)^8\left(\tfrac{1}{2}\right)^1$$

$$+ \binom{9}{9}\left(\tfrac{1}{2}\right)^9\left(\tfrac{1}{2}\right)^0$$

$$= .180$$

Examining the test statistic ($m = 2$), the decision is to not reject the null hypothesis.

b.   The two samples are <u>not</u> random and independent. Rather, the experiment has been conducted in a paired manner, and a paired-difference analysis is used. The differences and the associated t test are given below.

d_i	d_i^2
−.9	.81
−1.1	1.21
1.5	2.25
−2.6	6.76
−1.8	3.24
−2.9	8.41
−2.5	6.25
2.5	6.25
−1.4	1.96

(1) $H_0: \; \mu_1 - \mu_2 = 0$ vs. $H_a: \; \mu_1 - \mu_2 \neq 0$

(2) $\bar{d} = \dfrac{\sum\limits_i d_i}{n} = \dfrac{-9.2}{9} = -1.022$

$$s^2 = \frac{\sum\limits_i d_i^2 - \left[\dfrac{\left(\sum\limits_i d_i\right)^2}{n}\right]}{n-1} = \frac{37.14 - 9.404}{8}$$

$$= \frac{27.736}{8} = 3.467$$

(3) Test statistic: $t = \dfrac{\bar{d} - 0}{\frac{s_d}{\sqrt{n}}} = \dfrac{-1.022}{1.86373} = -1.65$

(4) Rejection region: The critical value of t for a two-tailed test, based on 8 degrees of freedom, will be $t_{.025,\,8} = 2.306$, and the rejection region is $|t| > 2.306$. The

test statistic does not fall in the rejection region. Hence the null hypothesis cannot be rejected.

15.7 **a.** Since two of the pairs are tied, $n = 10$. Let $p = P(\text{before exceeds after})$. Then

$$H_0: \ p = \tfrac{1}{2} \qquad \text{vs.} \qquad H_a: \ p > \tfrac{1}{2}.$$

Large values of M will favor H_a. The observed value of the test statistic is $m = 9$, The p-value is $P(M \geq 9) = 1 - .989 = .011$. H_0 is not rejected at the $\alpha = .01$ level. The campaign was successful. (Note: the p-value is larger than α, although by only a small amount.)

15.9 **a.** Define d_i to be the difference between school A and school B (i.e., $A - B$). The differences, along with their ranks (according to absolute magnitude) are shown in the table at the right. Then the rank sum for positive differences is $T^+ = 49$, and the rank sum for negative differences is $T^- = 6$. Consider the minimum rank sum, $T = 6$. Indexing $n = 10$ in Table 9, $5 < T < 8$ so $.02 < p\text{-value} < .05$. When $\alpha = .05$, the null hypothesis is rejected.

d_i	Rank $|d_i|$
28	13
5	4
-4	3
15	9
12	7
-2	1
7	5
9	6
-3	2
13	8

b. If a one-tailed test of hypothesis is desired, the p-value is between .01 and .025. We therefore conclude that school A is superior to school B. Notice that the rank-sum test leads to rejection of the null hypothesis but the sign test does not. The rank-sum test utilizes more information than the sign test and is more powerful in detecting departures from the null hypothesis.

15.11 The experimenter's design was paired using people as blocks. The Wilcoxon rank-sum test is the appropriate test for this experiment, with the null hypothesis that the populations follow the same distributions. The data are shown below.

Subject	Normal	Stress	Difference $(N - S)$	Rank
1	126	130	-4	5
2	117	118	-1	1
3	115	125	-10	7
4	118	120	-2	2
5	118	121	-3	3.5
6	128	125	$+3$	3.5
7	125	130	-5	6
8	120	120	0	$-$

The rank sum for positive values is $T^+ = 3.5$, and for negative values $= 24.5$. For $n = 7$ (one tie) and for a one-tailed test, we use the rank sum $T^* = 3.5$. Since $2 < T^* < 4$, $.025 < p$-value $< .05$.

15.13 The ranked differences are

Difference untreated-treated	Absolute value of difference	Rank of absolute value
0	0	(eliminated)
2	2	6
2	2	6
1	1	2.5
−1	1	2.5
1	1	2.5
4	4	9
−2	2	6
5	5	10
0	0	(eliminated)
3	3	8
−1	1	2.5

$T^+ =$ sum of positive ranks $= 44$ $T^- =$ sum of negative ranks $= 11$

The Wilcoxon test is

H_0: The probability distributions of the number of cavities are the same for the two populations.

H_a: The distribution of the number of cavities for the untreated teeth is shifted to the right of the distribution of the number of cavities for the treated teeth.

The test statistic is $T^- = 11$. Using Table 9, since $T^- = 11 \geq 11$, we reject H_0 at rhw level $\alpha = .05$.

15.15 Let ξ be the median of a random variable Y with distribution function $F(y)$. By definition, then,

$$P(Y > \xi) = P(Y < \xi) = \frac{1}{2}$$

It is necessary to test

H_0: $\xi = \xi_0$ vs. H_a: $\xi \neq \xi_0$.

Notice that, if H_0 is true,

$$P(Y < \xi_0) = \frac{1}{2}.$$

a. Instead of defining $d_i = X_i - Y_i$, we define $d_i = Y_i - \xi_0$ and let M be the number of negative differences. If H_0 is true, $p = P(\text{negative difference}) = \frac{1}{2}$, and M will have a binomial distribution with n trials and $p = \frac{1}{2}$. If M is too large or too small, H_0 will be rejected. The rejection region for M will be obtained by using the binomial distribution of Chapter 3.

b. We now make use of the magnitude of the differences as well as their sign. Define

$$d_i = Y_i - \xi_0$$

so that if H_0 is true, $P(d_i < 0) = \frac{1}{2}$. Rank the d_i according to absolute magnitude and define T^- to be the su of the negative ranks. T^- will have a distribution identical to the Wilcoxon signed-rank statistic given in Section 15.4. If T^- is too large or too small (using an appropriate rejection region), the null hypothesis will be rejected.

15.17 The hypothesis to be tested is

H_0: the population distributions for plastics 1 and 2 are the same

H_a: the population distributions differ in location

Ranking the 12 observations in order of magnitude and counting the number of observations in sample 1 that precede each observation in sample 2, the test statistic U is obtained. The data, with corresponding ranks, are shown in the table at the right.

Plastic 1	Plastic 2
15.3 (2)	21.2 (9)
18.7 (6)	22.4 (11)
22.3 (10)	18.3 (5)
17.6 (4)	19.3 (8)
19.1 (7)	17.1 (3)
14.8 (1)	27.7 (12)

The two possible values for U are

$$U_A = n_1 n_2 + \frac{n_1(n_1 + 1)}{2} - W_A = 36 + \frac{6(7)}{2} - W_A = 57 - 30 = 27$$

$$U_B = n_1 n_2 + \frac{n_2(n_2 + 1)}{2} - W_B = 36 + \frac{6(7)}{2} - 48 = 9$$

Since we have agreed to use the smaller value of U as a test statistic, a lower-tailed rejection region must be determined so that α is close to .10. Notice that the hypothesis to be tested is actually two-tailed. That is, both large and small values of U will tend to contradict the null hypothesis. Hence, although we will only consider the area below some critical value of U (denoted by U_0) in determining α for the test, there is a similar area in the upper tail of the distribution. Thus the area below the critical U is actually $\frac{\alpha}{2}$ and must be doubled to obtain the value for α.

Referring to Table 8 and indexing $n_1 = n_2 = 6$, a rejection region is determined such that

$$P(U \leq U_0) = \tfrac{\alpha}{2} = .05$$

Hence we will reject if $U \leq 7$, with $\tfrac{\alpha}{2} = .0465$ or $\alpha = .093$. The minimum of U_1 and U_2 is $U = 9$, and H_0 is not rejected. We cannot conclude that the populations differ in location.

15.19 If the alternative hypothesis is true, we would expect the batteries from plant A to fail later than the batteries from plant B. Hence the observations from plant A will be ranked near the end of the sequence and the U statistic (i.e., the number of observations from plant A that precede each observation from plant B) will be small. Hence, small values of U will tend to contradict the null hypothesis, and a lower-tailed rejection region is desired. Remember that the test statistic U has expected value $E(U) = \frac{n_1 n_2}{2}$ and variance $V(U) = \frac{n_1 n_2 (n_2 + n_2 + 1)}{12}$ when the null hypothesis is true. Also, for large values of n_1 and n_2, the quantity

$$Z = \frac{U - E(U)}{\sigma_\mu}$$

will be approximately normal with mean 0 and variance 1. Once this z value has been calculated, the null hypothesis will be rejected if $z < -1.645$ (cf. Chapter 10). The following data are available: $n_1 = n_2 = 15$, $W_A = 276$, and $W_B = 189$. Notice that observations are presented in order of failure, so that

$$W_A = 1 + 5 + 7 + 8 + 13 + 15 + 20 + 21 + 23 + 24 + 25 + 27 + 28 + 29 + 30$$
$$= 276x$$

The value of U is

$$U_A = n_1 n_2 + \frac{n_1(n_1 + 1)}{2} - W_A = 345 - 276 = 69$$

Also,

$$E(U) = \frac{n_1 n_2}{2} = \frac{225}{2} = 112.5$$

and

$$V(U) = \frac{n_1 n_2 (n_1 + n_2 + 1)}{12} = \tfrac{1}{12}(15)(15)(31) = 581.25$$

Thus,

$$z = \frac{U - E(U)}{\sigma_\mu} = \frac{69 - 112.5}{\sqrt{581.25}} = \frac{-43.5}{24.1} = -1.80$$

The null hypothesis is rejected since $z = -1.80$ falls in the rejection region.

15.21 To test for a difference in location, a two-tailed Mann-Whitney U test is used. The data, with corresponding ranks, are shown in the following table.

Sample 1	Sample 2
235 (10)	180 (3.5)
225 (9)	169 (1)
190 (8)	180 (3.5)
188 (7)	185 (6)
	178 (2)
	182 (5)

Calculate

$$U_A = n_1 n_2 + \frac{n_1(n_1 + 1)}{2} - W_A = 4(6) + \frac{4(5)}{2} - 34 = 0$$

$$U_B = n_1 n_2 + \frac{n_2(n_2 + 1)}{2} - W_B = 4(6) + \frac{6(7)}{2} - 21 = 24$$

The test statistic is $U = 0$ and the rejection region, found in Table 8, is $U \leq 3$ with $\alpha = 2(.0333) = .0666$. The null hypothesis is rejected. There is a difference in the distributions of wing stroke frequencies.

The approximate p-value can be found in Table 8. Since the test is two-tailed, the observed level of significance is $2P[U \leq 0] = 2(.0048) = .0096$.

15.23 **a.** The necessary computations follow:

$$T_1 = 209 \qquad\qquad T_2 = 199 \qquad\qquad T_3 = 343$$

$$\sum y = 751 \qquad\qquad \sum y^2 - 51,889$$

$$\text{TSS} = 51,889 - \frac{(751)^2}{15} = 14,288.933$$

$$\text{SST} = \frac{(209)^2}{5} + \frac{(199)^2}{5} + \frac{(343)^2}{5} - \frac{(751)^2}{15} = 2586.1333$$

$$\text{SSE} = \text{TSS} - \text{SST} = 11,702.8$$

We test

$$H_0: \ \mu_A = \mu_B = \mu_C \qquad\qquad \text{vs.} \qquad\qquad H_a: \text{ At least two means differ}$$

The test statistic is

$$F = \frac{\dfrac{2586.1333}{2}}{\dfrac{11,702.8}{12}} - 1.33$$

Since $F < 3.89 = F_{.05}$, we do not reject H_0. There is not sufficient evidence to indicate a significant difference among the means for the three brands. The analysis above is valid if the three population probability distributions are normal and the three population variances are equal. The normality assumption is probably not valid since lifelength data usually follow an exponential distribution (skewed to the right).

b. A composite ranking of the data is

A	B	C
5	8	14
7	4	3
2	12	15
13	1	11
9	10	6
$R_1 = 36$	$R_2 = 35$	$R_3 = 49$

We test

H_0: The population probability distributions of lifelengths are identical for the three brands.

H_a: At least two of the probability distributions differ in location.

The test statistic is

$$H = \frac{12}{15(16)}\left[\frac{(36)^2}{5} + \frac{(35)^2}{5} + \frac{(49)^2}{5}\right] - 3(16) = 1.22$$

Since $H < \chi^2_{2,.05} = 5.99$, we do not reject H_0. There is not enough evidence to conclude that the brands of magnetron tubes tend to differ in length of life under stress.

15.25 A composite ranking of the data is

$38°F$	$42°F$	$46°F$	$50°F$
16	3	2	6.5
18.5	14	22	8.5
4.5	21	14	1
8.5	4.5	10.5	12
10.5	20	18.5	14
	6.5	17	
$R_1 = 58$	$R_2 = 69$	$R_3 = 84$	$R_4 = 42$

We test

H_0: The population probability distributions of weights are identical for the four water temperatures.

H_a: At least two of the probability distributions differ in location.

The test statistic is

$$H = \frac{12}{(22)(23)}\left[\frac{(58)^2}{5} + \frac{(69)^2}{6} + \frac{(84)^2}{6} + \frac{(42)^2}{5}\right] - 3(23) = 2.03$$

Since $H < \chi^2_{3,.1} = 6.25139$, we do not reject H_0. There is insufficient evidence to detect a difference among the weight distributions.

15.27 Expanding H, we have

$$H = \frac{12}{n(n+1)} \sum \left[n_i\left(\overline{R}_i^2 - 2\overline{R}_i \frac{(n+1)}{2} + \frac{(n+1)^2}{4}\right)\right]$$

$$= \frac{12}{n(n+1)} \sum \left[n_i\left(\frac{R_i^2}{n_i^2} - (n+1) \frac{R_i}{n_i} + \frac{(n+1)^2}{4}\right)\right]$$

$$= \frac{12}{n(n+1)} \sum \frac{R_i^2}{n_i} - \frac{12}{n} \sum R_i + \frac{3(n+1)}{n} \sum n_i$$

$$= \frac{12}{n(n+1)} \sum \frac{R_i^2}{n_i} - \frac{12n(n+1)}{2n} + \frac{3(n+1)(n)}{n}$$

$$= \frac{12}{n(n+1)} \sum \frac{R_i^2}{n_i} - 3(n+1)$$

15.29 A summary of the ranked data (ranked within each metal) is

Metal	I	II	III
1	2	1	3
2	3	1	2
3	1.5	3	1.5
4	3	1	2
5	3	1	2
6	2	1	3
7	1.5	1.5	3
8	1	2	3
9	3	1	2
10	3	1	2

Thus, $R_1 = 23$, $R_2 = 13.5$, and $R_3 = 23.5$. We test

H_0: The probability distributions for the amount of corrosion are the same for all three types of sealer.

H_a: At least two of the probability distributions differ in location.

The test statistic is

$$F_r = \frac{12}{10(3)(4)}\left[(23)^2 + (13.5)^2 + (23.5)^2\right] - 3(10)(4) = 6.35$$

Since $F_r > \chi^2_{2,.05} = 5.99$, we reject H_0 with $\alpha = .05$. There is sufficient evidence of a difference in the abilities of the sealers to prevent corrosion.

15.31 a. To carry out the Friedman test, we need the rank sums, R_i, for each model. We can find these by adding the ranks given for each model. Thus for model A, $R_1 = 8(15) = 120$. For model B, $R_2 = 4 + 2(6) + 7 + 8 + 9 + 2(14) = 68$, etc. The R_i values are 120, 68, 37, 61, 31, 87, 100, 34, 32, 62, 85, 75, 30, 71, 67. Thus,

$$\sum R_i^2 = 71,948$$

and

$$F_r = \frac{12}{8(15)(16)}(71,948) - 3(8)(16) = 65.675$$

Since $F_r > \chi^2_{14,.005} = 31.32$, the p-value is less than .005. We reject the hypothesis that the 15 distributions are identical.

b. The highest (best) rank given to model H is lower than the lowest (worst) rank given to model M. Thus, the value of the test statistic is $m = 0$. Since $P(M = 0) = \left(\frac{1}{2}\right)^8 = \frac{1}{256}$, the p-value is $2\left(\frac{1}{256}\right) = \frac{1}{128}$.

c. To use the sign test we must know, for each judge, whether the judge preferred model H or model M. We do not have this information since the rankings "overlap," unlike the situation in part (b).

15.33 If $k = 2$ and $b = n$, then

$$F_r = \frac{2}{n}\left(R_1^2 + R_2^2\right) - 9n$$

If $R_1 = 2n - M$ and $R_2 = n + M$, then

$$F_r = \frac{2}{n}\left[(2n - M)^2 + (n + M)^2\right] - 9n$$

$$= \frac{2}{n}\left[(4n^2 - 4nM + M^2) + (n^2 + 2nM + M^2) - 4.5n^2\right]$$

$$= \frac{2}{n}\left(-.5n^2 - 2nM + 2M^2\right) = \frac{4}{n}\left(M^2 - nM - \tfrac{1}{4}n^2\right) = \frac{4}{n}\left(M - \tfrac{n}{2}\right)^2$$

The Z statistic from Section 15.3 is

$$Z = \frac{M - \frac{n}{2}}{\left(\frac{1}{2}\right)\sqrt{n}} = \frac{2}{\sqrt{n}}\left(M - \frac{n}{2}\right)$$

so

$$Z^2 = \frac{4}{n}\left(M - \frac{n}{2}\right)^2 = F_r$$

15.35 This is similar to Exercise 15.28. As in that exercise, we need only worry about the 3! possible rank pairings. They are listed below, with the R_i values and F_r. When $b = 2$ and $k = 3$,

$$F_r = \frac{\sum R_i^2}{2} - 24.$$

Block

1	2	R_i
1	1	2
2	2	4
3	3	6
	$F_r = 4$	

1	2	R_i
1	2	3
2	1	3
3	3	6
	$F_r = 3$	

Block

1	2	R_i
1	1	2
2	3	5
3	2	5
	$F_r = 3$	

1	2	R_i
1	2	3
2	3	5
3	1	4
	$F_r = 1$	

Block

1	2	R_i
1	3	4
2	1	3
3	2	5
	$F_r = 1$	

Block

1	2	R_i
1	3	4
2	2	4
3	1	4
	$F_r = 0$	

Thus, $P(F_r = 0) = P(F_r = 4) = \frac{1}{6}$ and $P(F_r = 1) = P(F_r = 3) = \frac{1}{3}$

15.37 The runs test is used to test the null hypothesis of randomness. Either a large or a small number of runs indicates nonrandomness, and a two-tailed test is used. The data are shown below:

$$W, W, W, W, B, W, W, W, B, B, W, B, B$$

There are $n_1 = 5$ blacks hired and $n_2 = 8$ whites hired. The number of runs observed is $R = 6$. From Table 10, the p-value is $2P(R \le 6) = 2(.347) = .694$. We do not reject the null hypothesis of randomness. The data do not suggest a nonrandom racial selection in the hiring of the union's members.

15.39 **a.** In this exercise it is necessary to calculate $P(R \leq 11)$, where $n_1 = 11$ and $n_2 = 23$. Since it is known that the quantity

$$z = \frac{R - E(R)}{\sigma_R}$$

is approximately normally distributed for large n_1 and n_2 (say, $n_1 \geq 10$, $n_2 \geq 10$), we may use the normal approximation to calculate $P(R \leq 11)$. The first step is to determine the z value corresponding to an R value of 11. Note that

$$E(R) = \frac{2n_1 n_2}{n_1 + n_2} + 12 = \frac{2(11)(23)}{11 + 23} + 1 = 15.88$$

$$V(R) = \frac{2n_1 n_2 (2n_1 n_2 - n_1 - n_2)}{(n_1 + n_2)^2 (n_1 + n_2 - 1)} = \frac{2(11)(23)[2(11)(23) - 11 - 23]}{(11 + 23)^2 (11 + 23 - 1)} = \frac{238,832}{38,148}$$

$$= 6.2607$$

$$\sigma_R = \sqrt{6.2607} = 2.50$$

Hence the corresponding z value will be

$$z = \frac{R - E(R)}{\sigma_R} = \frac{11 - 15.88}{2.50} = -1.95$$

Thus,

$$P(R \leq 11) = P(Z \leq -1.95) = P(Z \geq 1.95) = .0256$$

b. The hypothesis to be tested is

H_0: randomness of occurrence   vs. H_a: nonrandomness of occurrence

The test statistic is R, the number of runs observed. The reader may verify that the observed value of R is $R = 11$, with $n_1 = 11$ and $n_2 = 23$. Using a large-sample approximation, the standardized test statistic is

$$z = \frac{R - E(R)}{\sigma_R} = -1.95 \quad \text{(calculated above)}$$

Since an unusually small or unusually large number of runs would imply a nonrandomness of defectives, a two-tailed test is employed and the rejection region is $z < -1.96$ or $z > 1.96$. Since the test statistic, $z = -1.95$, does not fall in the rejection region, the null hypothesis is not rejected. Hence, there is not sufficient evidence of a nonrandomness of defectives.

15.41 Let A represent an observation from population A, and let B represent an observation from population B. Now referring to Exercise 15.18, the observations are arranged according to rank, and the population from which they were drawn is noted. Using the ranks obtained in Exercise 15.18 to arrange the observations, the sequence of runs is as follows:

$$\underline{A}\ \underline{B}\ \underline{A}\ \underline{B}\ \underline{A}\ \underline{B\ B\ B}\ \underline{A}\ \underline{B\ B}\ \underline{A\ A}\ \underline{B}\ \underline{A}\ \underline{B}\ \underline{A\ A}$$

Notice that the 9th and 10th and the 13th and 14th letters in the sequence represent the two pairs of tied observations. If the tied observations were reversed in the sequence of runs, we would still obtain $R = 13$. Hence the order of the tied observations is irrelevant.

Consider the alternative situation that asserts that the two distributions are not identical. If the alternative is true, we would expect a small number of runs because most of the measurements for population A will fall below those for population B (or vice versa). Hence small values for R will tend to contradict the null hypothesis. A one-tailed test of hypothesis is employed with a lower-tailed rejection region. Table 10 is then used to find the p-value. For $n_1 = n_2 = 9$ and $R = 13$, the p-value is $P(R \leq 13) = .956$. The null hypothesis is not rejected. This is the same conclusion that was reached when the Mann-Whitney U test was employed.

15.43 The ranks of the two variables are shown below.

Leaf	Rank, x	Rank, y
1	10.5	12
2	5.5	7.5
3	7.5	9
4	7.5	6
5	4	4.5
6	9	10
7	2	3
8	5.5	4.5
9	1	1
10	12	11
11	10.5	7.5
12	3	2

Calculate

$$\sum x_i y_i = 636.25 \qquad \sum x_i^2 = 648.5 \qquad \sum y_i^2 = 649$$

$$n = 12 \qquad \sum x_i = 78 \qquad \sum y_i = 78$$

Then

$$S_{xy} = 636.25 - \frac{(78)^2}{12} = 129.25 \qquad S_{xx} = 648.5 - \frac{(78)^2}{12} = 141.5$$

$$S_{yy} = 649 - \frac{(78)^2}{12} = 142$$

and

$$r_s = \frac{S_{xy}}{\sqrt{S_{xx}S_{yy}}} = \frac{129.25}{\sqrt{141.5(142)}} = .912$$

To test for correlation with $\alpha = .05$, index .025 in Table 11, Appendix III, and the rejection region is $|r_s| \geq .591$. The null hypothesis is rejected and we conclude that there is a correlation between the two variables.

15.45 The objective is to determine whether or not test scores are correlated with interview ratings. Hence a Spearman rank correlation coefficient may be used to test for a relation between two ranked variables. Since the first variable (interview rating) is already in ranked form, we need only rank the second variable (test score). This variable will be ranked from low to high. The ranks (x_i and y_i) are shown in the table below.

Subject	x_i	y_i	Subject	x_i	y_i
1	8	5	6	1	9
2	5	6	7	4	10
3	10	2.5	8	7	4
4	3	7	9	9	1
5	6	2.5	10	2	8

a.
$$r_s = \frac{n \sum_i x_i y_i - \left(\sum_i x_i\right)\left(\sum_i y_i\right)}{\sqrt{n \sum_i x_i^2 - \left(\sum_i x_i\right)^2 \left[n \sum_i y_i^2 - \left(\sum_i y_i\right)^2\right]}}$$

$$= \frac{(10)(233) - (55)(56)}{\sqrt{[10(385) - (55)^2][10(384.5) - (55)^2]}}$$

$$= \frac{-695}{\sqrt{825(820)}} = \frac{-695}{822.5} = -.845$$

b. The hypothesis to be tested is

H_0: no correlation between interview rank and test score

H_a: negative correlation

and the test statistic will be the Spearman rank correlation coefficient, r_s. For a one-tailed test with $\alpha = .05$ and $n = 10$, the critical value for rejection is $-.564$ (see Table 11). The test statistic falls in the rejection region. Hence the null hypothesis is rejected. There is evidence of a significant negative correlation between the two ranked variables. The p-value $= P(r_s \leq -.845)$ $< .005$.

15.47 The ranks for the two variables of interest (x_i and y_i, corresponding to math and art, respectively) are shown in the table below.

Student	x_i	y_i	Student	x_i	y_i
1	1	5	9	10.5	6
2	3	11.5	10	12	15
3	2	1	11	13.5	11.5
4	4	2	12	6	7
5	5	3.5	13	13.5	10
6	7.5	8.5	14	15	14
7	7.5	3.5	15	10.5	8.5
8	9	13			

Then

$$r_s = \frac{15(1148.5) - 120(120)}{\sqrt{\left[15(1238.5) - (120)^2\right]^2}} = .6768$$

Consulting Table 11 for $\alpha = .10$, the critical value of r_s is .441, and the rejection region is $|r_s| > .441$. (Notice that the α for a two-tailed test requires doubling the probabilities tabulated in Table 11.) Since the calculated value of r_s falls in the rejection region, H_0 is rejected, and we conclude that there is a correlation between math and art scores.

15.49 The ranks for the two variables of interest are given below, along with $d_i = x_i - y_i$.

y_i	2	3	1	4	6	8	5	10	7	9
x_i	2	3	1	4	6	8	5	10	7	9
d_i	0	0	0	0	0	0	0	0	0	0

Using the alternative formula given in this section, we have

$$r_s = 1 - \frac{6 \sum d_i^2}{n\left(n^2 - 1\right)} = 1 - 0 = 1$$

Using Table 11, $1 > .794$, so the p-value $< .005$.

15.51 **a.** Since we are interested in a difference in recovery rates, let

$$p = P(\text{recovery rate for } A \text{ exceeds } B \text{ at a given hospital})$$

$$M = \text{number of times } A \text{ exceeds } B$$

The hypothesis to be tested is

$$H_0: \ p = \frac{1}{2} \qquad \text{vs.} \qquad H_a: \ p \neq \frac{1}{2},$$

and the data are shown in the following table.

Hospital	A	B	Sign of $(A - B)$
1	75.0	85.4	−
2	69.8	83.1	−
3	85.7	80.2	+
4	74.0	74.5	−
5	69.0	70.0	−
6	83.3	81.5	+
7	68.9	75.4	−
8	77.8	79.2	−
9	72.2	85.4	−
10	77.4	80.4	−

Various rejection regions are tried in order to find $\alpha = .10$. (Use Table 1, Appendix III).

Rejection Region	α
$M = 0, \ M = 10$	.002
$M \leq 1, \ M \geq 9$	.022
$M \leq 2, \ M \geq 8$	.110

Using the rejection region $M \leq 2$ or $M \geq 8$, the null hypothesis is rejected since the observed value of M is $m = 2$. We conclude that a difference does exist in the recovery rates for the two drugs.

b. In the above analysis we made no assumptions concerning the underlying distributions of the data. To use the t test, we must be able to assume normality of the distributions and equal variances for the two populations. Since the observations given above are percentages, their distributions may be almost mound-shaped, but the variances will not be equal.

15.53 For the Wilcoxon signed-rank test, the differences and the ranks of their absolute values are given below for $n = 17$ differences.

d_i	Rank $|d_i|$	d_i	Rank $|d_i|$
−2	10.5	−3	15
−1	4.5	3	15
3	15	3	10.5
1	4.5	−2	10.5
−1	4.5	−1	4.5
3	15	1	4.5
−1	4.5	−2	10.5
−3	15	1	4.5
1	4.5		

Then $T^+ = 73.5$ and $T^- = 79.5$. With $\alpha = .05$ and $n = 17$, the lower portion of the rejection region is $T \leq 35$ (see Table 9). Since the observed value of T is $T = 73.5$, the null hypothesis is not rejected, as in Exercise 15.52.

15.55 The data along with corresponding ranks are shown in the table at the right. Then

Stimuli 1	Stimuli 2
1 (2.5)	4 (16)
3 (12.5)	2 (7)
2 (7)	3 (12.5)
1 (2.5)	3 (12.5)
2 (7)	1 (2.5)
1 (2.5)	2 (7)
3 (12.5)	3 (12.5)
2 (7)	3 (12.5)
$W_1 = 53.5$	$W_2 = 82.5$

$$U_A = 8^2 + \frac{8(9)}{2} - 53.5 = 46.5$$

$$U_B = 8^2 + \frac{8(9)}{2} - 82.5 = 17.5$$

Consulting Table 8 with $n_1 = n_2 = 8$, the hypothesis of no difference will be rejected if $U \leq 13$ with $\alpha = 2(.0249) = .0498$. We choose the minimum value of U_i, $U = 17.5$, and H_0 is not rejected. There is no evidence of a difference in location for the two distributions. This is the same conclusion as reached in Exercise 13.1.

15.57 Designate the five observations from samples 1 and 2 as A and B, respectively. When the $(n_1 + n_2) = 10$ observations are ordered according to their magnitude, the U statistic is obtained by counting the number of observations in sample A that precede each observation in sample B. Then

$$P(U \leq 2) = P(U = 0) + P(U = 1) + P(U = 2)$$

The only sample point associated with $U = 0$ is

$$B\ B\ B\ B\ B\ A\ A\ A\ A\ A$$

because there are no A's preceding any of the B observations. In order to obtain the probability of observing this sample point, we proceed as follows:

(1) The total number of permutations of the 10 observations is 10!.

(2) The number of different arrangements of the five A observations is 5!. Similarly, the number of different arrangements of the five B observations is 5!.

(3) The total number of distinct arrangements of the five A's and five B's will be $\frac{10!}{5!\ 5!}$. Hence

$$P(U = 0) = \frac{1}{\frac{10!}{5!\ 5!}}$$

The only sample point associated with $U = 1$ is

$$B\ B\ B\ B\ A\ B\ A\ A\ A\ A$$

where only one A observation precedes a B observation. Then

$$P(U = 1) = \frac{1}{\frac{10!}{5!\,5!}}$$

Finally, the event $U = 2$ will occur when one of the following two sample points occurs:

$$B\ B\ BA\ B\ B\ A\ A\ A\ A \qquad \text{or} \qquad B\ B\ B\ B\ A\ A\ B\ A\ A\ A$$

and

$$P(U = 2) = \frac{2}{\frac{10!}{5!\,5!}}.$$

Then

$$P(U \le 2) = P(U = 0) + P(U = 1) + P(U = 2) = \frac{1 + 1 + 2}{\frac{10!}{5!\,5!}} = \frac{4(5!)(5!)}{10!}$$

$$= \frac{1}{63} = .0159$$

15.59 A composite ranking of the data is

Line 1	Line 2	Line 3
19	14	2
16	10	15
12	5	4
20	13	11
3	9	1
18	17	8
21	7	6
$R_1 = 109$	$R_2 = 75$	$R_3 = 47$

We test

H_0: The probability distributions of the production figures are the same for all three production lines.

H_a: At least two of the probability distributions differ in location.

The test statistic is

$$H = \frac{12}{21(22)}\left[\frac{(109)^2}{7} + \frac{(75)^2}{7} + \frac{(47)^2}{7}\right] - 3(22) = 7.154$$

Since $H > \chi^2_{2,\,.05} = 5.99147$, we reject H_0 with $\alpha = .05$. The data provide sufficient evidence to indicate a difference in location for the three sets of production figures.

15.61 The ranked data are best analyzed using the Friedman test with people representing the blocks. The rank sums are

$$R_1 = 19 \qquad R_2 = 21.5 \qquad R_3 = 27.5 \qquad R_4 = 32$$

We test

H_0: The probability distributions are identical for the four items.

H_a: At least two of the probability distributions differ in location.

The test statistic is

$$F_r = \frac{12}{10(4)(5)}\left[(19)^2 + (21.5)^2 + (27.5)^2 + (32)^2\right] - 3(10)(5) = 6.21$$

Since $F_r < \chi^2_{3,.05} = 7.81473$, we do not reject H_0. There is insufficient evidence to indicate that one or more of the items are preferred to the others.

15.63 The necessary probability can be read directly from Table 10, Appendix III, or can be obtained by using the results given in Section 15.9 of the text. We must obtain the probability of observing exactly Y_1 S runs and Y_2 F runs, where $Y_1 + Y_2 = R$. Note that

$$P[y_1, y_2] = \frac{\binom{n_1-1}{y_1-1}\binom{n_2-1}{y_2-1}}{\binom{n_1+n_2}{n_1}} = \frac{\binom{7}{y_1-1}\binom{7}{y_2-1}}{\binom{16}{8}}$$

(1) Consider $P(R = 2)$. The event $R = 2$ will occur when $y_1 = 1$ and $y_2 = 1$, with either the S elements or F elements beginning the sequence. Then

$$P(R = 2) = 2P(Y_1 = 1, Y_2 = 1) = \frac{2\binom{7}{0}\binom{7}{0}}{\binom{16}{8}} = \frac{2}{12,870}$$

(2) The event $R = 3$ will occur when we observe $y_1 = 1$ S run and $y_2 = 2$ F runs, or when we observe $y_1 = 2$ S runs and $y_2 = 1$ F run. Note that if there are two F runs and one S run, there is only one possible ordering because the F's must commence the sequence. Then

$$P(R = 3) = P(Y_1 = 1, Y_2 = 2) + P(Y_1 = 2, Y_2 = 1) = \frac{\binom{7}{1}\binom{7}{0}}{\binom{16}{8}} + \frac{\binom{7}{0}\binom{7}{1}}{\binom{16}{8}}$$

$$= \frac{14}{12,870}$$

(3) Similarly,

$$P(R = 4) = 2P(Y_1 = 2, Y_2 = 2) = \frac{2\binom{7}{1}\binom{7}{1}}{\binom{16}{8}} = \frac{98}{12,870}$$

(4) $P(R = 5) = P(Y_1 = 3, Y_2 = 2) + P(Y_1 = 2, Y_2 = 3) = \dfrac{\binom{7}{2}\binom{7}{1}}{\binom{16}{8}} + \dfrac{\binom{7}{1}\binom{7}{2}}{\binom{16}{8}}$

$$= \frac{294}{12,870}$$

(5) $P(R = 6) = 2P(Y_1 = 3, Y_2 = 3) = \dfrac{2\binom{7}{2}\binom{7}{2}}{\binom{16}{8}} = \dfrac{882}{12,870}$

Then

$$P(R \le 6) = P(R = 2) + P(R = 3) + P(R = 4) + P(R = 5) + P(R = 6)$$

$$= \frac{1290}{12,870} = .100$$

15.65 We have n_1 A observations, $A_1, A_2, \ldots, A_{n_1}$, and n_2 B observations, $B_1, B_2, \ldots, B_{n_2}$. The Mann-Whitney U statistic is defined as

$$U = \sum_{i=1}^{n_2} U_i$$

where u_i is the number of A observations preceding the ith B. Write $B_{(i)}$ to be the ith B in the combined sample when it has been ranked from smallest to largest, and write $R[B_{(i)}]$ to be the rank of the ith ordered B in the total ranking of the A's and B's. Then u_i is the number of A observations preceding $B_{(i)}$. We know that there are $(i - 1)$ B's preceding $B_{(i)}$, and that there are $R[B_{(i)}] - 1$ A's and B's preceding $B_{(i)}$. Hence there are $u_i = R[B_{(i)}] - i$ A's before $B_{(i)}$. Then

$$U = \sum_{i=1}^{n_2} u_i = \sum_{i=1}^{n_2} [R(B_{(i)}) - i] = \sum_{i=1}^{n_2} R[B_{(i)}] - \sum_{i=1}^{n_2} i = W_B - \frac{n_2(n_2 + 1)}{2}$$

since the first term is simply the sum of the B ranks, which is, by definition, W_B. The second term is the sum of the first n_2 integers, whose value is as given. Now if we let $n_1 + n_2 = N$, and define Z_i to be the ith observation in the combined sample of A's and B's, where $i = 1, 2, \ldots, N$, we can write

$$W_A + W_B = \sum_{i=1}^{N} R(Z_i) = \sum_{i=1}^{N} i = \frac{N(N + 1)}{2}$$

so that

$$W_B = \frac{N(N+1)}{2} - W_A$$

and

$$U = \frac{N(N+1)}{2} - \frac{n_2(n_2+1)}{2} - W_A = \frac{N^2 + N - n_2^2 - n_2}{2} - W_A$$

$$= \frac{n_1^2 + 2n_1n_2 + n_2^2 + n_1 + n_2 - n_2^2 - n_2}{2} - W_A$$

$$= n_1n_2 + \frac{n_1(n_1+1)}{2} - W_A$$

15.67 In order to obtain T, the Wilcoxon signed-rank statistic, the differences d_i are calculated and ranked according to absolute magnitude. The rank sum of the positive or negative differences is then calculated. We will work, for the time being, with T^+ and write, as in Exercise 15.66,

$$T^+ = \sum_{i=1}^{n} X_i \qquad \text{where} \qquad X = \begin{cases} R(d_i), & \text{if } d_i \text{ is positive} \\ 0, & \text{if } d_i \text{ is negative} \end{cases}$$

If H_0 is true and the two populations are identical, then the probability of observing a positive difference is $\frac{1}{2}$, and

$$E(X_i) = E(d_i)P[X_i = R(d_i)] = \frac{1}{2}R(d_i)$$

$$E\left(X_i^2\right) = R(d_i)^2 P[X_i = R(d_i)] = \frac{1}{2}R(d_i)^2$$

$$E(X_iX_j) = R(d_i)R(d_j)P[X_i = R(d_i), \ X_j = R(d_j)] = \frac{1}{4}R(d_i)R(d_j)$$

Then

$$V(X_i) = \frac{1}{2}R(d_i)^2 - \frac{1}{4}R(d_i)^2 = \frac{1}{4}R(d_i)^2$$

$$\text{Cov}\,(X_i, \ X_j) = \frac{1}{4}R(d_i)R(d_j) - \frac{1}{4}R(d_i)R(d_j) = 0$$

Finally,

$$E(T^*) = \sum_{i=1}^{n} E(X_i) = \frac{1}{2}\sum_{i=1}^{n} R(d_i) = \left(\frac{1}{2}\right)\frac{n(n+1)}{2} = \frac{n(n+1)}{4}$$

$$V(T^*) = \sum_{i=1}^{n} V(X_i) = \frac{1}{4}\sum_{i=1}^{n} R(d_i)^2 = \left(\frac{1}{4}\right)\frac{n(n+1)(2n+1)}{6} = \frac{n(n+1)(2n+1)}{24}$$

Had we calculated the mean and variance of

$$T^- = \sum_{i=1}^{n} Y_i \qquad \text{where} \qquad Y_i = \begin{cases} R(d_i), & \text{if } d_i \text{ is negative} \\ 0, & \text{if } d_i \text{ is positive} \end{cases}$$

we would find that $E(T^-) = E(T^+)$ and $V(T^-) = V(T^+)$, since the probability of observing a negative difference is also $\frac{1}{2}$ under H_0.